VOLUME FIFTY EIGHT

ADVANCES IN INSECT PHYSIOLOGY

Mechanisms Underlying Microbial Symbiosis

VOLUME FIFTY EIGHT

ADVANCES IN INSECT PHYSIOLOGY

Mechanisms Underlying Microbial Symbiosis

Edited by

KERRY M. OLIVER
Department of Entomology, University of Georgia, Athens, GA, United States

JACOB A. RUSSELL
Department of Biology, Drexel University, Philadelphia, PA, United States

ELSEVIER

ACADEMIC PRESS
An imprint of Elsevier

Academic Press is an imprint of Elsevier
125 London Wall, London, EC2Y 5AS, United Kingdom
The Boulevard, Langford Lane, Kidlington, Oxford OX5 1GB, United Kingdom
525 B Street, Suite 1650, San Diego, CA 92101, United States
50 Hampshire Street, 5th Floor, Cambridge, MA 02139, United States

First edition 2020

ISBN: 978-0-08-102987-9
ISSN: 0065-2806

For information on all Academic Press publications
visit our website at https://www.elsevier.com/books-and-journals

Publisher: Zoe Kruze
Acquisitions Editor: Sam Mahfoudh
Editorial Project Manager: Hilal Johnson
Production Project Manager: Abdulla Sait
Cover Designer: Alan Studholme

Typeset by SPi Global, India

Contents

Contributors

Benoît Béchade
Department of Biology, Drexel University, Philadelphia, PA, United States

Kerri L. Coon
Department of Entomology, The University of Wisconsin-Madison, Madison, WI, United States

Matthew R. Doremus
Graduate Interdisciplinary Program in Entomology & Insect Science, The University of Arizona, Tucson, AZ, United States

Nina E. Fatouros
Biosystematics Group, Wageningen University and Research, Wageningen, The Netherlands

Enric Frago
CIRAD, CBGP; CBGP, CIRAD, INRA, IRD, Montpellier SupAgro, University Montpellier, Montpellier, France

Allison K. Hansen
Department of Entomology, University of California, Riverside, CA, United States

Martha S. Hunter
Department of Entomology, The University of Arizona, Tucson, AZ, United States

Yoshitomo Kikuchi
Bioproduction Research Institute, National Institute of Advanced Industrial Science and Technology (AIST), Hokkaido Center; Graduate School of Agriculture, Hokkaido University, Sapporo, Japan

Peter Mergaert
Université Paris-Saclay, CEA, CNRS, Institute for Integrative Biology of the Cell (I2BC), Gif-sur-Yvette, France

Tsubasa Ohbayashi
Université Paris-Saclay, CEA, CNRS, Institute for Integrative Biology of the Cell (I2BC), Gif-sur-Yvette, France; Institute for Agro-Environmental Sciences, National Agriculture and Food Research Organization (NARO), Tsukuba, Japan

Kerry M. Oliver
Department of Entomology, University of Georgia, Athens, GA, United States

Steve J. Perlman
Department of Biology, University of Victoria, Victoria, BC, Canada

Daniel Pers
Department of Entomology, University of California, Riverside, CA, United States

Jacob A. Russell
Department of Biology, Drexel University, Philadelphia, PA, United States

Kevin J. Vogel
Department of Entomology, The University of Georgia, Athens, GA, United States

John T. Wertz
Department of Biology, Calvin University, Grand Rapids, MI, United States

Alex C.C. Wilson
Department of Biology, University of Miami, Coral Gables, FL, United States

Sharon E. Zytynska
Institute of Integrative Biology, Department of Evolution, Ecology and Behaviour, University of Liverpool, Liverpool, United Kingdom

CHAPTER ONE

Mechanisms underlying microbial symbiosis

Jacob A. Russell[a], Kerry M. Oliver[b]
[a]Department of Biology, Drexel University, Philadelphia, PA, United States
[b]Department of Entomology, University of Georgia, Athens, GA, United States

Contents

Abstract

Like many miniscule phenomena in biology, the importance of intimate, host-associated microbes lay hidden until the timely collaboration between a microscope and a curious mind. Chronicled early in insect groups such as termites, ants, and both sap- and blood-feeders, recent research within the 'omics era has completely transformed our understanding of symbiosis within the hexapods. Bacteria, protists, and fungi shape a range of insect phenotypes, complementing their endogenous abilities for resource acquisition and defence. Other microbes exploit chinks in insect armour, manipulating conserved features of insect cell biology to selfishly spread within host populations. Through the use of ever-improving technology, combined with conceptual and experimental advances, scientists now understand a number of key mechanisms underpinning the function, regulation, and transmission of insect-associated microbes across a range of systems. In this volume of Advances in Insect Physiology we present eight chapters focused on symbiont mechanism, including those involved in resource acquisition among herbivores, detritivores and blood-feeding insects, and those associated with reproductive manipulation and host defence. Within the present introduction, we sandwich summaries of each chapter between a brief historical perspective on insect-microbe symbiosis and a discussion on the implications of growing mechanistic knowledge. We conclude that understanding mechanism brings the potential to manage insects—and their impacts—in a changing world, whether

Advances in Insect Physiology, Volume 58
ISSN 0065-2806
https://doi.org/10.1016/bs.aiip.2020.04.003

(Bost et al., 2018; Burke et al., 2010; Hansen and Moran, 2011; Kešnerová et al., 2017; Poliakov et al., 2011; Zheng et al., 2017). Such efforts have told us much about microbial metabolism and its impacts on host physiology, giving deep insight into the basis of symbiont-conferred traits. In addition, phylogenetics, studies of host gene expression, and increasingly sophisticated microscopy techniques have told us much about the origins and development of insect anatomical structures that sustain essential microbes (Braendle et al., 2003; Currie et al., 2006; Kaltenpoth et al., 2005).

While many of the aforementioned advances stemmed from our abilities to sequence genomes of single organisms (McCutcheon and Moran, 2007; Sabree et al., 2009; Tamas et al., 2002; Wu et al., 2004), advances in symbiosis research continue to grow due to shotgun metagenomics, a valuable tool in characterizing symbiont function without the need for cultivation. Such work has uncovered genes behind digestive and nutritional phenotypes in classical and developing models of symbiosis like termites and honeybees (Engel et al., 2012; Warnecke et al., 2007). Results from metagenomic studies, enabled by computational binning of scaffolds into discrete genomes (e.g. Hu et al., 2018; Meseguer et al., 2017), have also been paired with studies of individual symbiont genomes to reveal fascinating, and complex consequences of symbiosis for the involved microbes. Most prominent here is that obligate endosymbionts, i.e., those required for insect reproduction, have massively reduced genomes compared to their free-living relatives. But always retained are genes for the biosynthetic pathways making essential nutrients that are lacking in host diets (Hansen and Moran, 2014; Manzano-Marín et al., 2015; Moran and Wernegreen, 2000). Additional work continues to add fascinating texture to this biology, as host restriction, a lack of lateral gene transfer, loss of recombination, and small effective population sizes lead to symbiont decay and a range of compensatory mechanisms. Among these are co-obligate symbionts with metabolic capabilities complementary to those of hosts and their impaired symbionts (Bennett and Moran, 2015; Husnik and McCutcheon, 2016; Meseguer et al., 2017).

From the insect perspective, host genome sequencing (e.g. International Aphid Genomics Consortium, 2010) has lagged behind similar efforts focused on symbionts, with their smaller, simpler genomes. But with maturing technologies and computation, insect genome projects are now providing insights into symbiosis that were previously inaccessible. This has included deeper understanding of how insects treat pathogenic vs beneficial microbes (Douglas, 2014; Gerardo et al., 2010; Vigneron et al., 2014). When coupled with other techniques, insect genomes have also shed light on how hosts and symbionts exchange molecules to coordinate the production of essential nutrients (Hansen and Moran, 2011; Feng et al., 2019). Further revealed has been the exciting discovery that microbial genes are often transferred to insect chromosomes (Dunning Hotopp et al., 2007). This has permitted co-option of microbial solutions to normally challenging tasks in some cases (Wybouw et al., 2016), while in others it has seemingly assisted hosts in regulating their complex interactions with endosymbionts (Husnik and McCutcheon, 2016).

4. Chapter summaries

Building off the aforementioned foundation, this volume of *Advances in Insect Physiology* provides a summary and update to our knowledge of the mechanisms underpinning microbial symbiosis in insects. Across a range of insect hosts and biological phenomena (e.g. resource use, defence, transmission, manipulation), fully equipped explorations of mechanism have yielded key findings barely imagined two decades ago. Through this collection, it is our aim to look backward at a large, accumulating body of work, and for our authors to point the way forward in their respective subdisciplines.

We begin our chapter summaries with a synthesis of primary research on host-symbiont specificity, performed by Ohbayashi et al. (2020). In this chapter the authors raise the question of how insects have maintained faithful interactions with microbial lineages across millions of years of symbiosis. One fundamental question from this realm that has puzzled symbiont biologists for years involves mechanisms enabling specificity of extracellular gut microbiomes. To be clear, the gut microbiomes of many, if not most insect species, are largely influenced by the environment. But extracellular gut bacteria within some insect clades have been remarkably specific and stable over time. A range of mechanisms have evolved to improve the transmission fidelity of gut microbes, including coprophagy and egg smearing (Lanan et al., 2016; Salem et al., 2015). But other interactions involving environmentally acquired microbes mirror patterns seen in well-studied symbioses like those between rhizobia and legumes or *Vibrio* and squid. The bean bug/*Burkholderia* interaction reflects such a pattern, and has attracted considerable interest due to pioneering research showing that environmentally acquired *Burkholderia* confer insecticide resistance upon their *Riptortus* bean bug hosts (Kikuchi et al., 2012). After reviewing the major symbiont transmission modes, and providing mechanistic descriptions of model symbioses with environmental acquisition, Ohbayashi et al. zero in on the

Riptortus/Burkholderia symbiosis for an in-depth look at factors promoting its specificity. Their discussion chronicles the phenomena, and underlying mechanisms, of colonization and subsequent maintenance of *Burkholderia* in the insect gut. Bean bugs in nature harbour a limited range of *Burkholderia* species, and experiments in the lab illustrate that only a subset of bacteria from this genus can successfully colonize the gut. Apparent partner choice is enforced through as yet unrecognized mechanisms, although it is associated with bacterial abilities to traverse a narrow section of the bean bug gut, called the constricted region. The ability to cross this region requires *Burkholderia* motility, as ascertained through gene knock-out studies. Their somewhat unique corkscrew flagellar motility is argued to be central, as other motile bacteria with different motility modes cannot navigate the constricted region. After colonizing a specialized portion of the host midgut, the alimentary canal closes off to potentially new bacterial colonizers, maintaining the integrity and exclusivity of the symbiosis. The authors also highlight that symbiotic *Burkholderia* are superior competitors in the gut environment, suggesting yet another mechanism behind their seemingly ancient dominance of host-gut microbiomes. While it represents just one of the many examples of specific host-gut microbe associations, advances within this developing model point to a highly derived system whereby hosts can ensure the acquisition and maintenance of helpful environmental microbes.

In contrast to bean bugs and their environmental symbiont acquisition, several social insects have evolved behaviours enabling fidelity with their extracellular gut bacteria. Chief among these are trophallactic feedings from the abdomens of their siblings. These permit symbiont acquisition from close relatives, homogenizing the microbes within the colony, and thus promoting partner fidelity. In this volume, Wertz and Bechade (2020) review symbioses in social hexapods that assist in the breakdown of dietary carbon (C) sources from plant-based diets. Like ruminant mammals, several social insects derive useful forms of C through metabolic assistance provided by gut symbionts. This may be especially important for insects with diets in which substantial fractions of C are tied up within recalcitrant plant cell wall biomass, including termites (wood), corbiculate bees and some ants (pollen). The metabolic repertoires used by their extracellular gut symbionts include cellulases and xylanases/beta-xylosidases, which aid in C-scavenging from complex lignocellulose molecules. For insects consuming pollen, symbiont-encoded pectinases play key roles, as pectins are major matrix components of the innermost pollen wall. In addition to cataloguing known and candidate digestive mutualisms across chosen eusocial insects, the authors consider whether symbiont-heightened abilities to acquire C are especially important for insects that must support non-foraging individuals within their colonies. While non-foragers perform key tasks, they do not derive external nutritional resources, potentially placing a premium on efficient nutrient acquisition.

Continuing the theme of symbiont roles in the use of plants as a primary source of resource acquisition, Frago et al. (2020) expand the taxonomic breadth of the considered herbivore-microbe symbioses. The authors note that while many of the world's insect herbivores harbour obligate symbionts, many do not, including most thrips, orthopterans, and lepidopterans. And they discuss, in depth, the importance of symbioses in such non-co-dependent groups. One major theme that emerges is the influence of lateral gene transfer, in which insect genomes have acquired key genes from bacteria or fungi. Among a variety of fulfilled functions, laterally-acquired genes can aid insects in their digestion of recalcitrant plant compounds, or in the detoxification of plant secondary metabolites. Beyond these presently host-controlled mechanisms, the authors further describe microbial roles in nutrient provisioning, digestion, and detoxification for herbivorous insects. They also discuss how insect-associated microbes can manipulate plant physiology to tip the balance toward less defended, more nutritive food plants, or how they prevent plants from 'calling for help' with volatiles used to normally recruit herbivore natural enemies. The authors also emphasize symbioses in holometabolous insects, noting that environmentally acquired microbes may be disproportionately important in these orders due to constraints imposed by gut reorganization during metamorphosis (Hammer and Moran, 2019). Hymenopteran social insects have found ways around these limitations, including the aforementioned behavioural transmission chronicled by Wertz and Bechade (2020). Also circumnavigating such constraints are olive flies, whose bacteria spend time in larval bacteriocytes before migrating to the gut of adults (Estes et al., 2009). The authors conclude with a discussion of how symbionts may help and harm insects in a warming world, and how they may influence the transmission of plant pathogenic viruses and otherwise impact human agriculture. Outcomes of host-microbe symbioses are clearly context-dependent (Caragata et al., 2013; Doremus et al., 2018), and with a rapidly changing world, the authors have raised a timely set of questions.

Building on the aforementioned themes of symbiont-enhanced herbivory, Hansen et al. (2020) focus on microbial roles in herbivore and detritivore N economies. Nitrogen is notoriously scarce in most herbivores' diets; while insect body tissues range from ~6% to 12% nitrogen, consumed plant material often contains less than 2% nitrogen (Fagan et al., 2002). A number of innovations have evolved to make up for this N shortfall. Some insects exhibit high volume feeding combined with selective excretion of C, coupled with preferential N retention to maintain stoichiometric balance. Other herbivores or detritivores supplement feeding by eating more N-rich substances such as faeces, exuviae, or even their siblings. Alternatively, or in conjunction with the above, N-provisioning symbionts are a recurring strategy for many insects, in which the products of bacterial or fungal metabolism assist in the acquisition of sufficient amounts or forms of N. The key mechanisms emphasized in the chapter are symbiont fixation of atmospheric N, recycling of waste N, and synthesis of essential amino acids or their precursors. In the latter, hosts complete the terminal steps in the biosynthesis of key essential amino acids. Such integration of host and symbiont metabolism exemplifies just one of several coevolutionary hallmarks seen for such bacteriocyte-based symbioses. Trends from N-fixing and N-recycling symbioses also show signatures of coevolution, although the evidence is typically not as strong as that compiled for amino acid-synthesizing, bacteriocyte-based symbioses. In another area of emphasis, the authors describe the now commonly recognized phenomenon of symbiont replacement. In doing so they focus on sap-feeding hemipteran insects, whose coevolved, amino acid provisioning symbionts have been repeatedly lost and replaced with other bacteria, or N-recycling relatives of pathogenic fungi. Replacement also occurred in the Blattodea, i.e., the cockroaches and termites. The ancestor of this group harboured a bacteriocyte-dwelling, N-recycling bacterium that is still retained in most cockroaches and in one basal termite family; this endosymbiont, however, is absent from all other termites. The functions of this endosymbiont, *Blattabacterium*, were seemingly replaced by activities of gut menageries involving protists and bacteria—with their abilities to recycle and fix N. Combined, these replacements reveal a theme whereby long-standing nutritive requirements can be fulfilled by a rotating set of characters encoding common forms of microbial N-metabolism.

Addressing the mysteries of metabolic collaboration between hosts and symbionts first raised by Hansen et al. (2020), Wilson (2020) discusses the means by which hosts regulate metabolic functions in their intracellular endosymbionts. She places much emphasis on an emerging model for symbiosis involving pea aphids (*Acyrthosiphon pisum*) and their bacteriocyte-dwelling *Buchnera*. Although recent studies suggest mechanisms for post-transcriptional control (Hansen and Degnan, 2014; Thairu et al., 2018), experiments performed over 10 years ago revealed that obligate symbionts like *Buchnera* have limited abilities to mount strong transcriptional responses to changing nutritive conditions. This raised questions on aphid roles in the regulation of symbiont-performed metabolism. Work since the publication of the pea aphid genome has begun to provide answers. As noted above, hosts complete terminal steps of essential amino acid synthesis, identifying one host-controlled lever used to regulate rates of nutrient acquisition. Control of amino acid precursor provisioning to *Buchnera*, including glutamine, provides another lever. Such control takes place, in part, through bacteriocyte localized amino acid transporters. These limit the supply of non-essential amino acids to symbionts upon sensing higher concentrations of essential amino acids (e.g. arginine) in aphid tissues. In doing so, they provide a negative feedback mechanism inhibiting symbiont nutrient biosynthesis when no longer needed. In addition to these mechanisms, aphid genomes also encode miRNAs, including several with known or hypothesized abilities to regulate amino acid transport and biosynthetic capabilities. Before finishing with a lively discussion of cooperation and conflict in obligate nutritional symbioses, and an argument that the strength of conflict is underappreciated, Wilson dives into one further point of symbiont regulation—the mTor pathway. In particular, the mTORC1 pathway integrates amino acid signals to enact a range of cellular and physiological responses across diverse eukaryotes. Prior work has shown that genes specific to this ancient, amino acid sensing pathway were more highly expressed in aphid bacteriocytes than in other body tissues. Wilson notes the intriguing observation that the amino acids sensed by this pathway are some of the most crucial to the nutritional symbiosis. They include amino acids that are produced collaboratively, those serving as raw materials for N-upgrading, and one whose synthesis is now directed by *Buchnera* after ancient loss in function by the host aphid. Future work into whether constraints in mTORC1 sensing dictated the specific forms of N-provisioning for this symbiosis, and for those beyond, will be of great interest to the symbiosis community.

With symbiont roles in ameliorating C- and N-limitation among insect herbivores receiving their due attention, Vogel and Coon (2020) shift to the importance of symbionts for haematophagous insects. One of their primary functions is the provisioning of B-vitamins, which are present in limited quantities in vertebrate blood. Blood-feeding insects are important to

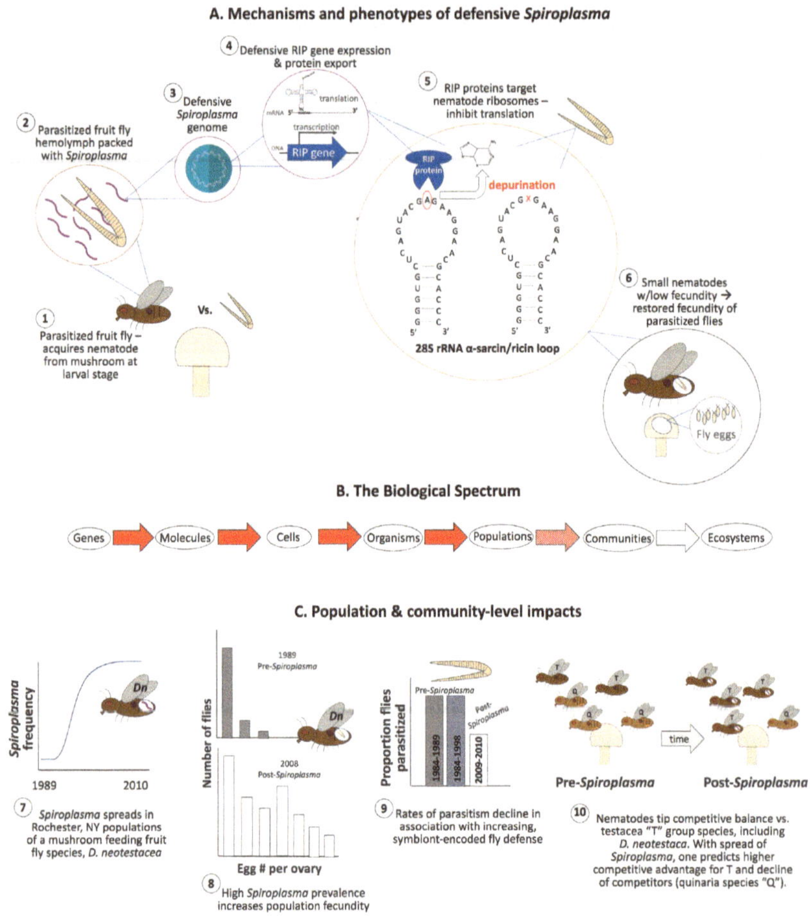

Fig. 1 Our understanding of the defensive *Spiroplasma-Drosophila neotestacea* symbiosis traverses much of the biological spectrum. Note that community-level effects documented thus far (9) stem from a likely impact on population sizes of the parasitic nematode *Howardula aoronymphium*. Additional community-level impacts (10) are predicted. *Information derived from Jaenike and Brekke (2011), Jaenike and Perlman (2002), Jaenike et al. (2010), and Hamilton et al. (2016).*

nematodes, are expected to alter competitive dynamics of mushroom-feeding fly communities in temperate North American forests. Documentation of such effects would enhance the standing of an already powerful system to understand symbiont effects spanning genes through communities.

Beyond this system, interest in community-level impacts has grown in recent years (Hopkins et al., 2017; McLean et al., 2016). And direct demonstration for microbial symbioses with detailed mechanistic underpinnings will represent a major achievement in biology. This will tell us much about the role of microbes in mediating interactions between host genotype, the environment, and the surrounding community (Dean et al., 2014), yielding important insights into the fates and impacts of insects in a rapidly changing world.

Acknowledgements

We thank several anonymous reviewers for help with this volume, including trainees from the Russell and Oliver labs. Our authors were incredibly dedicated, creative, and meticulous in assembling their manuscripts, so we send many thanks to them for working with us this past year. We also owe a debt of gratitude to Becky Neher for contributing the beautiful artwork adoring the cover to this volume. Funding support was provided by collaborative NSF Award #s 1754302 to K.M.O. and 1754597 to J.A.R.

References

Anbutsu, H., Fukatsu, T., 2003. Population dynamics of male-killing and non-male-killing spiroplasmas in *Drosophila melanogaster*. Appl. Environ. Microbiol. 69, 1428–1434.

Bandi, C., Sironi, M., Damiani, G., Magrassi, L., Nalepa, C.A., Laudani, U., Sacchi, L., 1995. The establishment of intracellular symbiosis in an ancestor of cockroaches and termites. Proc. Natl. Acad. Sci. U. S. A. 259, 293–299.

Beckmann, J.F., Ronau, J.A., Hochstrasser, M., 2017. A *Wolbachia* deubiquitylating enzyme induces cytoplasmic incompatibility. Nat. Microbiol. 2, 17007.

Benemann, J.R., 1973. Nitrogen fixation in termites. Science 181, 164–165.

Bennett, G.M., Moran, N.A., 2015. Heritable symbiosis: the advantages and perils of an evolutionary rabbit hole. Proc. Natl. Acad. Sci. U. S. A. 112, 10169–10176.

Bentley, B.L., 1984. Nitrogen fixation in termites: fate of newly fixed nitrogen. J. Insect Physiol. 30, 653–655.

Bost, A., Franzenburg, S., Adair, K.L., Martinson, V.G., Loeb, G., Douglas, A.E., 2018. How gut transcriptional function of *Drosophila melanogaster* varies with the presence and composition of the gut microbiota. Mol. Ecol. 27, 1848–1859.

Bouchon, D., Thierry, R., Pierre, J., 1998. Evidence for widespread *Wolbachia* infection in isopod crustaceans: molecular identification and host feminization. Proc. R. Soc. Lond. Ser. B 265, 1081–1090.

Boyle, L., O'Neill, S.L., Robertson, H.M., Karr, T.L., 1993. Interspecific and intraspecific horizontal transfer of *Wolbachia* in *Drosophila*. Science 260, 1796–1799.

Braendle, C., Miura, T., Bickel, R., Shingleton, A.W., Kambhampati, S., Stern, D.L., 2003. Developmental origin and evolution of bacteriocytes in the aphid-*Buchnera* symbiosis. PLoS Biol. 1, e21.

Brandt, J.W., Chevignon, G., Oliver, K.M., Strand, M.R., 2017. Culture of an aphid heritable symbiont demonstrates its direct role in defense against parasitoids. Proc. R. Soc. Lond. Ser. B 284, 20171925.

Breznak, J.A., Brill, W.J., Mertins, J.W., Coppel, H.C., 1973. Nitrogen fixation in termites. Nature 244, 577–580.

Broderick, N.A., Buchon, N., Lemaitre, B., 2014. Microbiota-induced changes in *Drosophila melanogaster* host gene expression and gut morphology. MBio 5, e01117–14.

Brooks, M.A., Richards, A.G., 1955. Intracellular symbiosis in cockroaches: II. mitotic division of mycetocytes. Science 122, 242.

Buchman, A., Marshall, J.M., Ostrovski, D., Yang, T., Akbari, O.S., 2018. Synthetically engineered Medea gene drive system in the worldwide crop pest *Drosophila suzukii.* Proc. Natl. Acad. Sci. U. S. A. 115, 4725–4730.

Buchner, P., 1965. Endosymbiosis of Animals With Plant Microorganisms, English ed Interscience Publishers, New York.

Burke, G.R., Fiehn, O., Moran, N.A., 2010. Effects of facultative symbionts and heat stress on the metabolome of pea aphids. ISME J. 4, 242–252.

Caragata, E.P., Rancès, E., Hedges, L.M., Gofton, A.W., Johnson, K.N., O'Neill, S.L., McGraw, E.A., 2013. Dietary cholesterol modulates pathogen blocking by *Wolbachia.* PLoS Pathog. 9, e1003459.

Cayetano, L., Rothacher, L., Simon, J.C., Vorburger, C., 2015. Cheaper is not always worse: strongly protective isolates of a defensive symbiont are less costly to the aphid host. Proc. R. Soc. Lond. Ser. B 282, 2014–2333.

Charlat, S., Hurst, G.D.D., Merçot, H., 2003. Evolutionary consequences of *Wolbachia* infections. Trends Genet. 19, 217–223.

Chen, X.A., Li, S., Aksoy, S., 1999. Concordant evolution of a symbiont with its host insect species: molecular phylogeny of genus *Glossina* and its bacteriome-associated endosymbiont, *Wigglesworthia glossinidia.* J. Mol. Evol. 48, 49–58.

Chevignon, G., Boyd, B.M., Brandt, J.W., Oliver, K.M., Strand, M.R., 2018. Culture-facilitated comparative genomics of the facultative symbiont *Hamiltonella defensa.* Genome Biol. Evol. 10, 786–802.

Clark, M.A., Moran, N.A., Baumann, P., Wernegreen, J.J., 2000. Cospeciation between bacterial endosymbionts (*Buchnera*) and a recent radiation of aphids (*Uroleucon*) and pitfalls of testing for phylogenetic congruence. Evolution 54, 517–525.

Colman, D.R., Toolson, E.C., Takacs-Vesbach, C.D., 2012. Do diet and taxonomy influence insect gut bacterial communities? Mol. Ecol. 21, 5124–5137.

Coon, K.L., Valzania, L., McKinney, D.A., Vogel, K.J., Brown, M.R., Strand, M.R., 2017. Bacteria-mediated hypoxia functions as a signal for mosquito development. Proc. Natl. Acad. Sci. U. S. A. 114, E5362–E5369.

Currie, C.R., Poulsen, M., Mendenhall, J., Boomsma, J.J., Billen, J., 2006. Coevolved crypts and exocrine glands support mutualistic bacteria in fungus-growing ants. Science 311, 81–83.

Dale, C., Young, S.A., Haydon, D.T., Welburn, S.C., 2001. The insect endosymbiont *Sodalis glossinidius* utilizes a type III secretion system for cell invasion. Proc. Natl. Acad. Sci. U. S. A. 98, 1883–1888.

de Bary, A., 1879. Die Erscheinung der Symbiose. Verlag von Karl J, Trübner, Strassburg.

De Fine Licht, H.H., Schiøtt, M., Rogowska-Wrzesinska, A., Nygaard, S., Roepstorff, P., Boomsma, J.J., 2013. Laccase detoxification mediates the nutritional alliance between leaf-cutting ants and fungus-garden symbionts. Proc. Natl. Acad. Sci. 110, 583–587.

De Vooght, L., Van Keer, S., van den Abbeele, J., 2018. Towards improving tsetse fly paratransgenesis: stable colonization of *Glossina morsitans morsitans* with genetically modified *Sodalis.* BMC Microbiol. 18, 165.

Dean, S.L., Farrer, E.C., Taylor, D.L., Porras-Alfaro, A., Sunding, K.N., Sinsabaugh, R.L., 2014. Nitrogen deposition alters plant-fungal relationships: linking belowground dynamics to aboveground vegetation change. Mol. Ecol. 23, 1364–1378.

Dennis, A.B., Patel, V., Oliver, K.M., Vorburger, C., 2017. Parasitoid gene expression changes after adaptation to symbiont-protected hosts. Evolution 71, 2599–2617.

Dillon, R.J., Dillon, V.M., 2004. The gut bacteria of insects: nonpathogenic interactions. Annu. Rev. Entomol. 49, 71–92.

Doremus, M.R., Hunter, M.S., 2020. The saboteur's tools: common mechanistic themes across manipulative symbioses. In: Oliver, K.M., Russell, J.A. (Eds.), Mechanisms Underlying Microbial Symbiosis. In: Advances in Insect Physiology, vol. 58. Elsevier, pp. 317–353.

Doremus, M.R., Smith, A.H., Kim, K.L., Holder, A.J., Russell, J.A., Oliver, K.M., 2018. Breakdown of a defensive symbiosis, but not endogenous defenses, at elevated temperatures. Mol. Ecol. 27, 2138–2151.
Douglas, A.E., 1989. Mycetocyte symbiosis in insects. Biol. Rev. Camb. Philos. Soc. 64, 409–434.
Douglas, A.E., 2014. The molecular basis of bacterial-insect symbiosis. J. Mol. Biol. 426, 3830–3837.
Douglas, A.E., Prosser, W.A., 1992. Synthesis of the essential amino acid tryptophan in the pea aphid (*Acyrthosiphon pisum*) symbiosis. J. Insect Physiol. 38, 565–568.
Dunning Hotopp, J.C., Clark, M.E., Oliveira Foster, D.J.M., Fischer, P., Muñoz Torres, M.C., Giebel, J.D., Kumar, N., Ishmael, N., Wang, S., Ingram, J., Nene, R.V., Shepard, J., Tomkins, J., Richards, S., Spiro, D.J., Ghedin, E., Slatko, B.E., Tettelin, H., Werren, J.H., 2007. Widespread lateral gene transfer from intracellular bacteria to multicellular eukaryotes. Science 317, 1753–1756.
Duron, O., Bouchon, D., Boutin, S., Bellamy, L., Zhou, L.Q., Engelstadter, J., Hurst, G.D., 2008. The diversity of reproductive parasites among arthropods: *Wolbachia* do not walk alone. BMC Biol. 6, 27.
Engel, P., Moran, N.A., 2013. The gut microbiota of insects—diversity in structure and function. FEMS Microbiol. Rev. 37, 699–735.
Engel, P., Martinson, V.G., Moran, N.A., 2012. Functional diversity within the simple gut microbiota of the honey bee. Proc. Natl. Acad. Sci. 109, 11002–11007.
Estes, A.M., Hearn, D.J., Bronstein, J.L., Pierson, E.A., 2009. The olive fly endosymbiont, "*Candidatus* Erwinia dacicola" switches from an intracellular existence to an extracellular existence during host insect development. Appl. Environ. Microbiol. 75, 7097–7106.
Ewald, P.W., 1987. Transmission modes and evolution of the parasitism-mutualism continuum. Ann. N. Y. Acad. Sci. 503, 295–306.
Fagan, W.F., Siemann, E., Mitter, C., Denno, R.F., Huberty, A.F., Woods, H.A., Elser, J.J., 2002. Nitrogen in insects: implications for trophic complexity and species diversification. Am. Nat. 160, 784–802.
Feng, H., Edwards, N., Anderson, C.M.H., Althaus, M., Duncan, R.P., Hsu, Y.-C., Luetje, C.W., Price, D.R.G., Wilson, A.C.C., Thwaites, D.T., 2019. Trading amino acids at the aphid–*Buchnera* symbiotic interface. Proc. Natl. Acad. Sci. U. S. A. 32, 16003–16011.
Fieck, A., Hurwitz, I., Kang, A.S., Durvasula, R., 2010. *Trypanosoma cruzi*: synergistic cytotoxicity of multiple amphipathic anti-microbial peptides to *T. cruzi* and potential bacterial hosts. Exp. Parasitol. 125, 342–347.
Flores, H.A., O'Neill, S.L., 2018. Controlling vector-borne diseases by releasing modified mosquitoes. Nat. Rev. Microbiol. 16, 508–518.
Frago, E., Zytynska, S.E., Fatouros, N.E., 2020. Microbial symbionts of herbivorous species across the insect tree. In: Oliver, K.M., Russell, J.A. (Eds.), Mechanisms Underlying Microbial Symbiosis. In: Advances in Insect Physiology, vol. 58. Elsevier, pp. 111–159.
Gerardo, N.M., Altincicek, B., Anselme, C., Atamian, H., Barribeau, S.M., de Vos, M., Duncan, E.J., Evans, J.D., Gabaldón, T., Ghanim, M., Heddi, A., Kaloshian, I., Latorre, A., Moya, A., Nakabachi, A., Parker, B.J., Pérez-Brocal, V., Pignatelli, M., Rahbé, Y., Ramsey, J.S., Spragg, C.J., Tamames, J., Tamarit, D., Tamborindeguy, C., Vincent-Monegat, C., Vilcinskas, A., 2010. Immunity and other defenses in pea aphids, *Acyrthosiphon pisum*. Genome Biol. 11, R21.
Gibson, C.M., Hunter, M.S., 2010. Extraordinarily widespread and fantastically complex: comparative biology of endosymbiotic bacterial and fungal mutualists of insects. Ecol. Lett. 13, 223–234.
Hamilton, P.T., Peng, F.N., Boulanger, M.J., Perlman, S.J., 2016. A ribosome-inactivating protein in a *Drosophila* defensive symbiont. Proc. Natl. Acad. Sci. U. S. A. 113, 350–355.

Hammer, T.J., Moran, N.A., 2019. Links between metamorphosis and symbiosis in holometabolous insects. Philos. Trans. R. Soc. B Biol. Sci. 374, 20190068.
Hammer, T.J., Janzen, D.H., Hallwachs, W., Jaffe, S.P., Fierer, N., 2017. Caterpillars lack a resident gut microbiome. Proc. Natl. Acad. Sci. U. S. A. 114, 9641–9646.
Hansen, A.K., Degnan, P.H., 2014. Widespread expression of conserved small RNAs in small symbiont genomes. ISME J. 8, 2490–2502.
Hansen, A.K., Moran, N.A., 2011. Aphid genome expression reveals host-symbiont cooperation in the production of amino acids. Proc. Natl. Acad. Sci. U. S. A. 108, 2849–2854.
Hansen, A.K., Moran, N.A., 2014. The impact of microbial symbionts on host plant utilization by herbivorous insects. Mol. Ecol. 23, 1473–1496.
Hansen, A.K., Pers, D., Russell, J.A., 2020. Symbiotic solutions to nitrogen limitation and amino acid imbalance in insect diets. In: Oliver, K.M., Russell, J.A. (Eds.), Mechanisms Underlying Microbial Symbiosis. In: Advances in Insect Physiology, vol. 58. Elsevier, pp. 161–205.
Harumoto, T., Lemaitre, B., 2018. Male-killing toxin in a bacterial symbiont of *Drosophila*. Nature 557, 252–255.
Hopkins, S.R., Wojdak, J.M., Belden, L.K., 2017. Defensive symbionts mediate host–parasite interactions at multiple scales. Trends Parasitol. 33, 53–64.
Hornett, E.A., Charlat, S., Duplouy, A.M.R., Davies, N., Roderick, G.K., Wedell, N., Hurst, G.D.D., 2006. Evolution of male-killer suppression in a natural population. PLoS Biol. 4, e283.
Hu, Y., Sanders, J.G., Łukasik, P., D'Amelio, C.L., Millar, J.S., Vann, D.R., Lan, Y., Newton, J.A., Schotanus, M., Kronauer, D.J., Pierce, N.E., Moreau, C.S., Engel, P., Russell, J.A., 2018. Herbivorous turtle ants obtain essential nutrients from a conserved nitrogen-recycling gut microbiome. Nat. Commun. 9, 964.
Hurst, G.D.D., Anbutsu, H., Kutsukake, M., Fukatsu, T., 2003. Hidden from the host: *Spiroplasma* bacteria infecting *Drosophila* do not cause an immune response, but are suppressed by ectopic immune activation. Insect Mol. Biol. 12, 93–97.
Husnik, F., McCutcheon, J.P., 2016. Repeated replacement of an intrabacterial symbiont in the tripartite nested mealybug symbiosis. Proc. Natl. Acad. Sci. U. S. A. 113, E5416–E5424.
Hypša, V., Dale, C., 1997. *In vitro* culture and phylogenetic analysis of "*Candidatus* Arsenophonus triatominarum," an intracellular bacterium from the triatomine bug, *Triatoma infestans*. Int. J. Syst. Bacteriol. 47, 1140–1144.
International Aphid Genomics Consortium, 2010. Genome sequence of the pea aphid *Acyrthosiphon pisum*. PLoS Biol. 8, e1000313.
Ives, A.R., Barton, B.T., Penczykowski, R.M., Harmon, J.P., Kim, K.L., Oliver, K., Radeloff, V.C., 2020. Self-perpetuating ecological–evolutionary dynamics in an agricultural host–parasite system. Nat. Ecol. Evol. https://doi.org/10.1038/s41559-020-1155-0
Jaenike, J., Brekke, T.D., 2011. Defensive endosymbionts: a cryptic trophic level in community ecology. Ecol. Lett. 14, 150–155.
Jaenike, J., Perlman, S.J., 2002. Ecology and evolution of host-parasite associations: mycophagous *Drosophila* and their parasitic nematodes. Am. Nat. 160, S23–S39.
Jaenike, J., Dyer, K.A., Reed, L.K., 2003. Within-population structure of competition and the dynamics of male-killing *Wolbachia*. Evol. Ecol. Res. 5, 1023–1036.
Jaenike, J., Unckless, R., Cockburn, S.N., Boelio, L.M., Perlman, S.J., 2010. Adaptation via symbiosis: recent spread of a *Drosophila* defensive symbiont. Science 329, 212–215.
Jiggins, F.M., Bentley, J.K., Majerus, M.E.N., Hurst, G.D.D., 2002. Recent changes in phenotype and patterns of host specialization in *Wolbachia* bacteria. Mol. Ecol. 11, 1275–1283.
Kaltenpoth, M., Steiger, S., 2014. Unearthing carrion beetles' microbiome: characterization of bacterial and fungal hindgut communities across the Silphidae. Mol. Ecol. 23, 1251–1267.

Kaltenpoth, M., Göttler, W., Herzner, G., Strohm, E., 2005. Symbiotic bacteria protect wasp larvae from fungal infestation. Curr. Biol. 15, 475–479.
Kešnerová, L., Mars, R.A.T., Ellegaard, K.M., Troilo, M., Sauer, U., Engel, P., 2017. Disentangling metabolic functions of bacteria in the honey bee gut. PLOS Biol. 15, e2003467.
Kikuchi, Y., Hayatsu, M., Hosokawa, T., Nagayama, A., Tago, K., Fukatsu, T., 2012. Symbiont-mediated insecticide resistance. Proc. Natl. Acad. Sci. U. S. A. 109, 8618–8622.
King, K.C., Brockhurst, M.A., Vasieva, O., Paterson, S., Betts, A., Ford, S.A., Frost, C.L., Horsburgh, M.J., Haldenby, S., Hurst, G.D.D., 2016. Rapid evolution of microbe-mediated protection against pathogens in a worm host. ISME J. 10, 1915–1924.
Koch, H., Schmid-Hempel, P., 2012. Gut microbiota instead of host genotype drive the specificity in the interaction of a natural host-parasite system. Ecol. Lett. 15, 1095–1103.
Koga, R., Tsuchida, T., Fukatsu, T., 2003. Changing partners in an obligate symbiosis: a facultative endosymbiont can compensate for loss of the essential endosymbiont *Buchnera* in an aphid. Proc. R. Soc. Lond. Ser. B 270, 2543–2550.
Kwong, W.K., Moran, N.A., 2013. Cultivation and characterization of the gut symbionts of honey bees and bumble bees: description of *Snodgrassella alvi* gen. nov., sp. nov., a member of the family *Neisseriaceae* of the *Betaproteobacteria*, and *Gilliamella apicola* gen. nov., sp. nov., a member of *Orbaceae* fam. nov., *Orbales* ord. nov., a sister taxon to the order "*Enterobacteriales*" of the *Gammaproteobacteria*. Int. J. Syst. Evol. Microbiol. 63, 2008–2018.
Lanan, M.C., Rodrigues, P.A.P., Agellon, A., Jansma, P., Wheeler, D.E., 2016. A bacterial filter protects and structures the gut microbiome of an insect. ISME J. 10, 1866–1876.
Leadbetter, J.R., Schmidt, T.M., Graber, J.R., Breznak, J.A., 1999. Acetogenesis from H_2 plus CO_2 by spirochetes from termite guts. Science 283, 686–689.
Leidy, J., 1881. The parasites of the termites. J. Acad. Natl. Sci. Phila. 8, 425–447.
LePage, D.P., Metcalf, J.A., Bordenstein, S.R., On, J., Perlmutter, J.I., Shropshire, J.D., Layton, E.M., Funkhouser-Jones, L.J., Beckmann, J.F., Bordenstein, S.R., 2017. Prophage WO genes recapitulate and enhance *Wolbachia*-induced cytoplasmic incompatibility. Nature 543, 243–247.
Lilburn, T.G., Kim, K.S., Ostrom, N.E., Byzek, K.R., Leadbetter, J.R., Breznak, J.A., 2001. Nitrogen fixation by symbiotic and free-living spirochetes. Science 292, 2495–2498.
Lindsey, A.R.I., Rice, D.W., Bordenstein, S.R., Brooks, A.W., Bordenstein, S.R., Newton, I.L.G., 2018. Evolutionary genetics of cytoplasmic incompatibility genes *cifA* and *cifB* in prophage WO of *Wolbachia*. Genome Biol. Evol. 10, 434–451.
Łukasik, P., van Asch, M., Guo, H.F., Ferrari, J., Godfray, H.C.J., 2013. Unrelated facultative endosymbionts protect aphids against a fungal pathogen. Ecol. Lett. 16, 214–218.
Lynn-Bell, N.L., Strand, M.R., Oliver, K.M., 2019. Bacteriophage acquisition restores protective mutualism. Microbiology 165 (9), 985–989.
Macias, V.M., Ohm, J.R., Rasgon, J.L., 2017. Gene drive for mosquito control: where did it come from and where are we headed? Int. J. Environ. Res. Public Health 14, 1006.
Manzano-Marín, A., Oceguera-Figueroa, A., Latorre, A., Jiménez-García, L.F., Moya, A., 2015. Solving a bloody mess: B-vitamin independent metabolic convergence among gammaproteobacterial obligate endosymbionts from blood-feeding arthropods and the leech *Haementeria officinalis*. Genome Biol. Evol. 7, 2871–2884.
Masson, F., Copete, S.C., Schüpfer, F., Garcia-Arraenz, G., Lemaitre, B., 2018. *In vitro* culture of the insect endosymbiont *Spiroplasma poulsonii* highlights bacterial genes involved in host-symbiont interaction. *mBio 9, e00024-18.*
Mateos, M., Castrezana, S.J., Nankivell, B.J., Estes, A.M., Markow, T.A., Moran, N.A., 2006. Heritable endosymbionts of *Drosophila*. Genetics 174, 363–376.

McCutcheon, J.P., Moran, N.A., 2007. Parallel genomic evolution and metabolic interdependence in an ancient symbiosis. Proc. Natl. Acad. Sci. U. S. A. 104, 19392–19397.

McFall-Ngai, M., Hadfield, M.G., Bosch, T.C., Carey, H.V., Domazet-Loso, T., Douglas, A.E., Dubilier, N., Eberl, G., Fukami, T., Gilbert, S.F., Hentschel, U., King, N., Kjelleberg, S., Knoll, A.H., Kremer, N., Mazmanian, S.K., Metcalf, J.L., Nealson, K., Pierce, N.E., Rawls, J.F., Reid, A., Ruby, E.G., Rumpho, M., Sanders, J.G., Tautz, D., Wernegreen, J.J., 2013. Animals in a bacterial world, a new imperative for the life sciences. Proc. Natl. Acad. Sci. U. S. A. 110, 3229–3236.

McLean, A.H.C., Parker, B.J., Hrček, J., Henry, L.M., Godfray, H.C.J., 2016. Insect symbionts in food webs. Philos. Trans. R. Soc. B 371, 20150325.

Meseguer, A.S., Manzano-Marín, A., Coeur d'Acier, A., Clamens, A.-L., Godefroid, M., Jousselin, E., 2017. *Buchnera* has changed flatmate but the repeated replacement of co-obligate symbionts is not associated with the ecological expansions of their aphid hosts. Mol. Ecol. 26, 2363–2378.

Montllor, C.B., Maxmen, A., Purcell, A.H., 2002. Facultative bacterial endosymbionts benefit pea aphids *Acyrthosiphon pisum* under heat stress. Ecol. Entomol. 27, 189–195.

Moran, N.A., Telang, A., 1998. Bacteriocyte-associated symbionts of insects-a variety of insect groups harbor ancient prokaryotic endosymbionts. Bioscience 48, 295–304.

Moran, N.A., Wernegreen, J.J., 2000. Lifestyle evolution in symbiotic bacteria: insights from genomics. Trends Ecol. Evol. 15, 321–326.

Moran, N.A., Russell, J.A., Koga, R., Fukatsu, T., 2005. Evolutionary relationships of three new species of Enterobacteriaceae living as symbionts of aphids and other insects. Appl. Environ. Mirobiol. 71, 3302–3310.

Mueller, U.G., Gerardo, N.M., Aanen, D.K., Six, D.L., Schultz, T.R., 2005. The evolution of agriculture in insects. Annu. Rev. Ecol. Evol. Syst. 36, 563–595.

Noda, S., Ohkuma, M., Usami, R., Horikoshi, K., Kudo, T., 1999. Culture-independent characterization of a gene responsible for nitrogen fixation in the symbiotic microbial community in the gut of the termite *Neotermes koshunensis*. Appl. Environ. Microbiol. 65, 4935–4942.

Ohbayashi, T., Takeshita, K., Kitagawa, W., Nikoh, N., Koga, R., Meng, X.Y., Tago, K., Hori, T., Hayatsu, M., Asano, K., Kamagata, Y., Lee, B.L., Fukatsu, T., Kikuchi, Y., 2015. Insect's intestinal organ for symbiont sorting. Proc. Natl. Acad. Sci. U. S. A. 112, E5179–E5188.

Ohbayashi, T., Mergaert, P., Kikuchi, Y., 2020. Host-symbiont specificity in insects: underpinning mechanisms and evolution. In: Oliver, K.M., Russell, J.A. (Eds.), Mechanisms Underlying Microbial Symbiosis. In: Advances in Insect Physiology, vol. 58. Elsevier, pp. 27–62.

Oliver, K.M., Martinez, A.J., 2014. How resident microbes modulate ecologically-important traits of insects. Curr. Opin. Insect Sci. 4, 1–7.

Oliver, K.M., Perlman, S.J., 2020. Toxin-mediated protection against natural enemies by insect defensive symbionts. In: Oliver, K.M., Russell, J.A. (Eds.), Mechanisms Underlying Microbial Symbiosis. In: Advances in Insect Physiology, vol. 58. Elsevier, pp. 277–316.

Oliver, K.M., Degnan, P.H., Burke, G.R., Moran, N.A., 2010. Facultative symbionts in aphids and the horizontal transfer of ecologically important traits. Annu. Rev. Entomol. 55, 247–266.

O'Neill, S.L., Giordano, R., Colbert, A.M.E., Karr, T.L., Robertson, H.M., 1992. 16S ribosomal RNA phylogenetic analysis of the bacterial endosymbionts associated with cytoplasmic incompatibility in insects. Proc. Natl. Acad. Sci. U. S. A. 89, 2699–2702.

Parfrey, L.W., Moreau, C.S., Russell, J.A., 2018. Introduction—the host-associated microbiome: pattern, process, and function. Mol. Ecol. 27, 1749–1765.

Poliakov, A., Russell, C.W., Ponnala, L., Hoops, H.J., Sun, Q., Douglas, A.E., van Wijk, K.J., 2011. Large-scale label-free quantitative proteomics of the pea aphid-*Buchnera* symbiosis. Mol. Cell. Proteomics 10, M110.007039.

Powell, J.E., Leonard, S.P., Kwong, W.K., Engel, P., Moran, N.A., 2016. Genome-wide screen identifies host colonization determinants in a bacterial gut symbiont. Proc. Natl. Acad. Sci. U. S. A. 113, 13887–13892.

Prosser, W.A., Douglas, A.E., 1991. The aposymbiotic aphid: an analysis of chloretracycline-treated pea aphid, *Acyrthosiphon pisum*. J. Insect Physiol. 10, 713–719.

Russell, J.A., Latorre, A., Sabater-Muñoz, B., Moya, A., Moran, N.A., 2003. Side-stepping secondary symbionts: widespread horizontal transfer across and beyond the Aphidoidea. Mol. Ecol. 12, 1061–1075.

Russell, J.A., Moreau, C.S., Goldman-Huertas, B., Fujiwara, M., Lohman, D.J., Pierce, N.E., 2009. Bacterial gut symbionts are tightly linked with the evolution of herbivory in ants. Proc. Natl. Acad. Sci. U. S. A. 106, 21236–21241.

Russell, J.A., Funaro, C.F., Milton, Y., Goldman-Huertas, B., Suh, D., Moreau, C.S., Kronauer, D., Pierce, N.E., 2012. A veritable menagerie of heritable bacteria across the ants, lepidopterans, and beyond. PLoS One 7, e51027.

Sabree, Z.L., Kambhampati, S., Moran, N.A., 2009. Nitrogen recycling and nutritional provisioning by *Blattabacterium*, the cockroach endosymbiont. Proc. Natl. Acad. Sci. U. S. A. 106, 19521–19526.

Salem, H., Kreutzer, E., Sudakaran, S., Kaltenpoth, M., 2013. Actinobacteria as essential symbionts in firebugs and cotton stainers (Hemiptera, Pyrrhocoridae). Environ. Microbiol. 15, 1956–1968.

Salem, H., Florez, L., Gerardo, N., Kaltenpoth, M., 2015. An out-of-body experience: the extracellular dimension for the transmission of mutualistic bacteria in insects. Proc. R. Soc. Lond. Ser. B 282, 20142957.

Sanders, J.G., Łukasik, P., Frederickson, M.E., Russell, J.A., Koga, R., Knight, R., Pierce, N.E., 2017. Dramatic differences in gut bacterial densities correlate with diet and habitat in rainforest ants. Integr. Comp. Biol. 57, 705–722.

Sasaki, T., Ishikawa, H., 1993. Nitrogen recycling in the endosymbiotic system of the pea aphid, *Acyrthosiphon pisum*. Zoolog. Sci. 10, 779–785.

Sasaki, T., Aoki, T., Hayashi, H., Isikawa, H., 1990. Amino acid composition of the honeydew of symbiotic and aposymbiotic pea aphids *Acyrthosiphon pisum*. J. Insect Physiol. 36, 35–40.

Sauer, C., Stackebrandt, E., Gadau, J., Hölldobbler, B., Gross, R., 2000. Systematic relationships and cospeciation of bacterial endosymbionts and their carpenter ant host species: proposal of the new taxon *Candidatus* Blochmannia gen. nov. Int. J. Syst. Evol. Microbiol. 50, 1877–1886.

Smith, A.H., Lukasik, P., O'Connor, M.P., Lee, A., Mayo, G., Drott, M.T., Doll, S., Tuttle, R., DiSciullo, R.A., Messina, A., Oliver, K.M., Russell, J.A., 2015. Patterns, causes and consequences of defensive microbiome dynamics across multiple scales. Mol. Ecol. 24, 1135–1149.

Stouthamer, R., Breeuwer, J.A.J., Luck, R.F., Werren, J.H., 1993. Molecular identification of microorganisms associated with parthenogenesis. Nature 361, 66–68.

Sudakaran, S., Kost, C., Kaltenpoth, M., 2017. Symbiont acquisition and replacement as a source of ecological innovation. Trends Microbiol. 25, 375–390.

Tamas, I., Klasson, L., Canback, B., Naslund, A.K., Eriksson, A.S., Wernegreen, J.J., Sandström, J.P., Moran, N.A., Andersson, S.G., 2002. 50 million years of genomic stasis in endosymbiotic bacteria. Science 296, 2376–2379.

Tarpy, D.R., Mattila, H.R., Newton, I.L., 2015. Development of the honey bee gut microbiome throughout the queen-rearing process. Appl. Environ. Microbiol. 81, 3182–3191.

Thairu, M.W., Cheng, S., Hansen, A.K., 2018. A sRNA in a reduced mutualistic symbiont genome regulates its own gene expression. Mol. Ecol. 27, 1766–1776.

Toju, H., Tanabe, A.S., Notsu, Y., Sota, T., Fukatsu, T., 2013. Diversification of endosymbiosis: replacements, co-speciation and promiscuity of bacteriocyte symbionts in weevils. ISME J. 7, 1378–1390.

Turelli, M., Hoffmann, A.A., 1991. Rapid spread of an inherited incompatibility factor in California *Drosophila*. Nature 353, 440–442.

Unckless, R.L., Boelio, L.M., Herren, J.K., Jaenike, J., 2009. *Wolbachia* as populations within individual insects: causes and consequences of density variation in natural populations. Proc. R. Soc. Lond. Ser. B 276, 2805–2811.

Untermann, B.M., Baumann, P., McLean, D.L., 1989. Pea aphid symbiont relationships established by analysis of 16S rRNAs. J. Bacteriol. 171, 2970–2974.

Valovage, W.D., Brooks, M.A., 1979. Uric acid quantities in the fat body of normal and aposymbiotic German cockroaches, *Blattella germanica*. Ann. Entomol. Soc. Am. 72, 687–689.

Vigneron, A., Masson, F., Vallier, A., Balmand, S., Rey, M., Vincent-Monégat, C., Aksoy, E., Aubailly-Giraud, E., Zaidman-Rémy, A., Heddi, A., 2014. Insects recycle endosymbionts when the benefit is over. Curr. Biol. 24, 2267–2273.

Vogel, K.J., Coon, K.L., 2020. Functions and mechanisms of symbionts of insect disease vectors. In: Oliver, K.M., Russell, J.A. (Eds.), Mechanisms Underlying Microbial Symbiosis. In: Advances in Insect Physiology, vol. 58. Elsevier, pp. 233–275.

von Dohlen, C.D., Kohler, S., Alsop, S.T., McManus, W.R., 2001. Mealybug β-proteobacterial endosymbionts contain γ-proteobacterial symbionts. Nature 412, 433–436.

Vorburger, C., 2018. Symbiont-conferred resistance to parasitoids in aphids—challenges for biological control. Biol. Control 116, 17–26.

Warnecke, F., Luginbühl, P., Ivanova, N., Ghassemian, M., Richardson, T.H., Stege, J.T., Cayouette, M., McHardy, A.C., Djordjevic, G., Aboushadi, N., Sorek, R., Tringe, S.G., Podar, M., Martin, H.G., Kunin, V., Dalevi, D., Madejska, J., Kirton, E., Platt, D., Szeto, E., Salamov, A., Barry, K., Mikhailova, N., Kyrpides, N.C., Matson, E.G., Ottesen, E.A., Zhang, X., Hernández, M., Murillo, C., Acosta, L.G., Rigoutsos, I., Tamayo, G., Green, B.D., Chang, C., Rubin, E.M., Mathur, E.J., Robertson, D.E., Hugenholtz, P., Leadbetter, J.R., 2007. Metagenomic and functional analysis of hindgut microbiota of a wood-feeding higher termite. Nature 450, 560–565.

Weeks, A.R., Velten, R., Stouthamer, R., 2003. Incidence of a new sex-ratio-distorting endosymbiotic bacterium among arthropods. Proc. R. Soc. Lond. Ser. B 270, 1857–1865.

Werren, J.H., Windsor, D.M., 2000. *Wolbachia* infection frequencies in insects: evidence of a global equilibrium? Proc. R. Soc. Lond. Ser. B 267, 1277–1285.

Werren, J.H., Baldo, L., Clark, M.E., 2008. *Wolbachia*: master manipulators of invertebrate biology. Nat. Rev. Microbiol. 6, 741–751.

Wertz, J.T., Béchade, B., 2020. Symbiont-mediated degradation of dietary carbon sources in social herbivorous insects. In: Oliver, K.M., Russell, J.A. (Eds.), Mechanisms Underlying Microbial Symbiosis. In: Advances in Insect Physiology, vol. 58. Elsevier, pp. 63–109.

Wilson, A.C.C., 2020. Regulation of an insect symbiosis. In: Oliver, K.M., Russell, J.A. (Eds.), Mechanisms Underlying Microbial Symbiosis. In: Advances in Insect Physiology, vol. 58. Elsevier, pp. 207–232.

Wu, M., Sun, L.V., Vamathevan, J., Riegler, M., Deboy, R., Brownlie, J.C., McGraw, E.A., Martin, W., Esser, C., Ahmadinejad, N., Wiegand, C., Madupu, R., Beanan, M.J., Brinkac, L.M., Daugherty, S.C., Durkin, A.S., Kolonay, J.F., Nelson, W.C.,

Mohamoud, Y., Lee, P., Berry, K., Young, M.B., Utterback, T., Weidman, J., Nierman, W.C., Paulsen, I.T., Nelson, K.E., Tettelin, H., O'Neill, S.L., Eisen, J.A., 2004. Phylogenomics of the reproductive parasite *Wolbachia pipientis* wMel: a streamlined genome overrun by mobile genetic elements. PLoS Biol. 2, e69.

Wybouw, N., Pauchet, Y., Heckel, D.G., Van Leeuwen, T., 2016. Horizontal gene transfer contributes to the evolution of arthropod herbivory. Genome Biol. Evol. 8, 1785–1801.

Zchori-Fein, E., Gottlieb, Y., Kelly, S.E., Brown, J.K., Wilson, J.M., Karr, T.L., Hunter, M.S., 2001. A newly discovered bacterium associated with parthenogenesis and a change in host selection behavior in parasitoid wasps. Proc. Natl. Acad. Sci. U. S. A. 98, 12555–12560.

Zheng, H., Powell, J.E., Steele, M.I., Dietrich, C., Moran, N.A., 2017. Honeybee gut microbiota promotes host weight gain via bacterial metabolism and hormonal signaling. Proc. Natl. Acad. Sci. U. S. A. 114, 4775–4780.

CHAPTER TWO

Host-symbiont specificity in insects: Underpinning mechanisms and evolution

Tsubasa Ohbayashi[a,b], Peter Mergaert[a], Yoshitomo Kikuchi[c,d]
[a]Université Paris-Saclay, CEA, CNRS, Institute for Integrative Biology of the Cell (I2BC), Gif-sur-Yvette, France
[b]Institute for Agro-Environmental Sciences, National Agriculture and Food Research Organization (NARO), Tsukuba, Japan
[c]Bioproduction Research Institute, National Institute of Advanced Industrial Science and Technology (AIST), Hokkaido Center, Sapporo, Japan
[d]Graduate School of Agriculture, Hokkaido University, Sapporo, Japan

Contents

Abstract

How host organisms evolved and maintain specific mutualisms with microorganisms is a fundamental question that is subject to intensive research. In the large majority of insect mutualisms, the host-microbe specificity is maintained by a "partner fidelity" mechanism, mainly through direct symbiont transmission from mother to offspring. Such vertical symbiont transmission is remarkably diverse in insects, including ovarial transmission, milk-gland transmission, coprophagy, egg-smearing, and capsule transmission. In contrast to the insect-microbe symbioses, many animals and plants do

Advances in Insect Physiology, Volume 58
ISSN 0065-2806
https://doi.org/10.1016/bs.aiip.2020.03.002

not vertically transmit their symbionts but acquire symbionts from ambient environments every generation. Sophisticated "partner choice" mechanisms are at play to establish these mutualisms and maintain them over evolutionary timescales. This symbiont transmission mode, called horizontal transmission or environmental acquisition, is rarely found in insects that maintain specific associations, but recent studies have described this type of symbiosis in a few insect groups. The symbiosis between the bean bug *Riptortus pedestris* and its gut symbiont *Burkholderia insecticola* is one of the model systems that is intensively studied to understand how host-symbiont specificity and mutualistic interactions are maintained in insects with horizontal symbiont transmission. Phylogenetic analyses of symbionts in natural insect populations and bacterial inoculation tests in the laboratory revealed a high degree of specificity in this symbiosis while mutant screening of the symbiotic bacterium, genomics and transcriptomics, and histological observations have identified underpinning morphological, genetic and molecular bases. In this chapter, we focus on symbiont transmission modes and mechanisms observed in the amazing diversity of microbial symbioses in insects, highlighting the ways in which they have likely evolved.

1. Introduction

Many, if not most animals and plants are intimately associated with microorganisms. In these symbioses, the microbial partners contribute to host fitness *via* various biological services such as the provisioning of essential nutrients, protection from antagonists, and degradation of toxins (Engel and Moran, 2013; Itoh et al., 2018b; Kikuchi, 2009). Since environments are filled with enormously diverse microorganisms, including not only mutualists but also parasites and pathogens, hosts must winnow out harmful microbes and selectively acquire the desired partners. To favour exclusive colonization by mutualistic symbionts, host organisms have evolved sophisticated mechanisms for symbiont transmission and sorting. In the case of many insect species that harbour specific gut or intracellular symbionts, the specific partnership is maintained by a "partner fidelity" mechanism that is based on "mother-to-offspring" vertical symbiont transmission. The specific vertical transmission mechanisms are remarkably diverse among insect taxonomic groups (Bright and Bulgheresi, 2010; Kikuchi, 2009; Perlmutter and Bordenstein, 2020). In other cases, including in a few insect groups, but more commonly in marine invertebrates and terrestrial plants, symbionts are instead acquired from the ambient environment every generation (called horizontal transmission or environmental acquisition), wherein "partner choice" mechanisms facilitate any host-microbe specificity (Bright and Bulgheresi, 2010; Sachs et al., 2004). In this section, we will review the amazing diversity of symbiont transmission modes in insects and their underlying mechanisms, highlighting how host-symbiont specificity is maintained and has evolved in this most diversified terrestrial animal group.

2. Vertical transmission

Insects that feed exclusively on nutritionally imbalanced diets like plant sap or on indigestible materials like wood typically require specific microbial symbionts that allow their hosts to develop on these limited feeding substrates. These symbionts reside in diverse locales within host body cavities, including the gut and the cytoplasm of specialized cells called mycetocytes or bacteriocytes (Buchner, 1965; Kikuchi, 2009). Symbionts are mostly bacteria, but yeast-like fungi, archaea and protists have also been reported as insect-associated symbionts (Brune and Dietrich, 2015; Buchner, 1965; Engel and Moran, 2013; Kikuchi, 2009; Matsuura et al., 2018; Ohkuma and Brune, 2010; Sudakaran et al., 2017). These symbiotic microorganisms play pivotal metabolic roles in their insect hosts, such as the production of essential metabolites that are scarce in their diets and that the insect cannot synthesize, or the degradation of plant polysaccharides that the insect cannot digest. In such nutritional and digestive symbioses, symbionts are commonly essential for survival, development, and reproduction of the hosts, and therefore, insects show high mortality when symbionts are removed by heat, ethanol or antibiotic treatment (Anbutsu et al., 2017; Bracke et al., 1978; Douglas et al., 2001; Fukatsu and Hosokawa, 2002; Hosokawa et al., 2010; Itoh et al., 2016; Kikuchi et al., 2009; Salem et al., 2017; Sangare et al., 2016; Schlein, 1977; Srivastava and Auclair, 1976; Tada et al., 2011). To ensure acquisition of their partners by the next generation, insects have evolved mechanisms for strict vertical transmission of symbionts. It should be noted here that the vertical manner of symbiont transmission guarantees not only absolute symbiont acquisition by the offspring but also host-symbiont specificity from generation to generation. A dramatic consequence of long-term and strict vertical transmission is that the genomes of the microorganisms are usually strongly eroded, making it therefore difficult or impossible to culture and manipulate these symbionts genetically (Kikuchi, 2009; McCutcheon and Moran, 2012; Moran and Bennett, 2014; Moya et al., 2008). To date, various mechanisms for vertical symbiont transmission have been reported in insects. The transmission mechanisms are fundamentally different depending on the symbiont localization pattern, i.e., intracellular or extracellular.

2.1 Vertical transmission of intracellular symbionts

Intracellular symbionts are broadly known in diverse groups of insects, including the orders Blattaria, Hemiptera, Coleoptera, Hymenoptera and Diptera, in which symbiont harbouring bacteriocytes or mycetocytes form clusters in the insect bodies, called the bacteriome or mycetome. Intracellular symbionts are generally transmitted by ovarial transmission, while in some blood-sucking insects of the Diptera, milk-gland transmission has been reported.

2.1.1 Ovarial transmission

In the ovarial transmission mechanism, obligate symbiotic bacteria directly infect the ovary and/or embryo from maternal bacteriocytes (Baumann, 2005; Szklarzewicz and Michalik, 2017). The detailed infection process has been investigated in the pea aphid *Acyrthosiphon pisum* (Koga et al., 2012). At the interface between a maternal bacteriocyte and embryo, the symbiont *Buchnera aphidicola* is exocytosed from the maternal bacteriocyte, momentarily released into the haemolymph, and then immediately endocytosed by the embryo. Detailed observations have been conducted in other insect groups as well, and similar exocytosis from bacteriocytes and endocytosis into ovaries of symbiotic bacteria play an important role in the studied ovarial transmission mechanisms. Nevertheless, the specific symbiont migration route from the bacteriocytes to the ovaries, as well as the developmental stage of the embryos in the infected ovaries, considerably differ between insects (reviewed in Szklarzewicz and Michalik, 2017).

2.1.2 Milk-gland transmission

In blood-sucking insects of the Diptera, such as tsetse flies and bat flies, symbiotic bacteria, *Wigglesworthia glossinidia* and *Aschnera chinzeii*, respectively, are transmitted to offspring *via* the milk-gland, a specific organ that supplies "milk" to larvae (Attardo et al., 2008; Hosokawa et al., 2012). These blood-sucking insects are unique ovoviviparous species and females grow a single larva in a uterus-like organ, nourishing them from a gland that contains the symbiotic bacteria, and thus enabling transfer to the next generation.

2.2 Vertical transmission of extracellular symbionts

Specific extracellular symbionts, often including gut colonists, are reported in many different insects, including bees, beewolves, cockroaches, termites, stinkbugs, and beetles. These extracellular symbionts can be vertically transmitted either by coprophagy, egg-smearing, or capsule transmission (Onchuru et al., 2018).

2.2.1 Coprophagy

Termites and wood-feeding cockroaches possess specific gut microbiota in the hindgut, consisting of not only bacteria but also specific protists (Brune and Dietrich, 2015; Ohkuma and Brune, 2010). In these social and subsocial insects, aposymbiotic individuals such as newly born hatchlings and newly moulted nymphs acquire the specific microbiota by eating faeces excreted from parents or siblings. Such coprophagy is also reported in the blood-sucking kissing bug *Rhodnius* spp., enabling transmission of the gut symbiont *Rhodococcus rhodonii* (Beard et al., 2002). It was also suggested that honey bees and ants transmit their specific gut microbiota through coprophagy as well as specific food, or gut fluid exchanging behaviours called mouth-to-mouth or anus-to-mouth trophallaxis (Koch and Schmid-Hempel, 2011; Lanan et al., 2016; Onchuru et al., 2018; Powell et al., 2014).

2.2.2 Egg-smearing

Egg-smearing, or egg surface contamination by symbiotic bacteria, has been described in detail in stinkbugs (Bansal et al., 2014; Bistolas et al., 2014; Buchner, 1965; Dasch and Weiss, 1984; Hayashi et al., 2015; Hosokawa et al., 2013, 2016; Itoh et al., 2016; Karamipour et al., 2016; Kashkouli et al., 2019a,b, 2020; Kenyon et al., 2015; Kikuchi et al., 2009, 2012; Miyamoto, 1961; Prado et al., 2006, 2009; Tada et al., 2011; Taylor et al., 2014). Phytophagous species of stinkbugs, particularly members of the infraorder Pentatomomorpha, develop a symbiotic organ composed of rows of crypts, located in the posterior region of the midgut and housing specific extracellular bacterial symbionts (Buchner, 1965; Dasch and Weiss, 1984; Kikuchi et al., 2009; Miyamoto, 1961). These gut symbionts are essential for normal development and reproduction of the host insects, suggesting a pivotal metabolic role of the symbionts. Most species of the superfamily Pentatomoidea of the Pentatomomorpha possess gammaproteobacterial symbionts, which are smeared on the egg surface upon oviposition. In female insects, the last few crypts are bulbous compared with the other crypts and can discharge symbiotic bacteria from the crypt lumen. These crypts are thought to be specialized for housing symbionts for transmission (Hayashi et al., 2015). In the particular case of the stinkbug family Acanthosomatidae however, all crypt entrances are completely sealed in adult insects, making the direct smearing from crypts impossible (Kikuchi et al., 2009). Instead, female insects evolved a novel pair of sac-like "lubricating organs" associated with the female ovipositor, which function to smear symbionts on the egg surface during oviposition

(Kikuchi et al., 2009). It is still enigmatic when and how the symbiotic bacteria migrate from the gut to the lubricating organ and how the new organ specialized for symbiont transmission evolved in the stinkbug family.

Another well-documented case of egg-smearing is found in species of the beetle subfamily Lagriinae. Antibiotic-producing, defensive *Burkholderia gladioli* symbionts protect the vulnerable eggs and larvae of *Lagria* beetles against harmful bacteria and fungi (Kaltenpoth and Flórez, 2019). The extracellular bacterial symbionts are carried by female accessory glands of the reproductive system but are absent from males. During oviposition, females smear the eggs with a *Burkholderia*-containing secretion, derived from the accessory glands. The bacteria then penetrate the egg from the surface and colonize three dorsally located crypts in the cuticle of the embryo that later form symbiont-containing compartments in the larva from which antibiotics are produced (Flórez et al., 2017). Thus contrary to the other above described symbioses where the symbionts provide services to the developing or adult host and the infected eggs serve as a carrier of the symbiont, in the *Lagria*-*Burkholderia* symbiosis, the roles of carrier and beneficiary are inverted. Adult beetles are carriers of symbionts, but require no apparent services from these microbes; but by transmitting symbionts to oviposited eggs, adults provision them with symbionts playing protective roles during early development.

2.2.3 Capsule transmission

In contrast to other Pentatomoidea families, stinkbugs of the family Plataspidae deploy a unique mechanism for vertical transmission, called "capsule transmission" (Buchner, 1965; Fukatsu and Hosokawa, 2002). Together with the egg masses, mothers oviposit brownish capsules containing the symbiotic bacterium *Ishikawaella capsulata*, which hatchlings acquire by sucking up the capsule contents (Fukatsu and Hosokawa, 2002; Hosokawa et al., 2005, 2006, 2007, 2008). Females have a specific organization of the symbiotic region of the midgut that is adapted for capsule production and that is absent in males (Hosokawa et al., 2005). Immediately posterior to the crypt region, harbouring the symbionts, is a so-called "swollen crypt-bearing midgut section" (SCM) where the matrix of the capsules is produced and where the symbionts are embedded in this matrix. The SCM is followed by the "brownish enlarged midgut end section" (BEM) where the filamentous material of the capsule envelope is produced. During oviposition, the contents of the SCM and BEM are expelled, forming the capsules that are deposited next to the eggs (Hosokawa et al., 2005).

A similar mechanism, called "jelly transmission", has been reported in the Urostylididae (also Pentatomoidea), in which mothers oviposit eggs embedded within a symbiont-containing jelly-like matrix (Kaiwa et al., 2014). Hatchlings acquire symbiotic bacteria by consuming the jelly.

In addition to the stinkbug symbioses, another unique type of capsule transmission was discovered in the tortoise leaf beetle *Cassida rubiginosa* (Salem et al., 2017). The leaf beetle harbours a pectin-degrading symbiont, *Candidatus* Stammera capleta, inside a pair of sac-like organs associated with the foregut. Female insects deposit a symbiont-containing capsule or "caplet" on the top of each egg and hatchlings acquire the symbiont by consuming the contents of the caplet.

2.3 Partner choice in insect-microbe symbiosis with vertical transmission

Many insects have developed highly sophisticated mechanisms for vertical symbiont transmission, as shown above. These mechanisms are seemingly both faithful and exclusive, with maintenance of the symbiont lineage over long evolutionary times and with little room for the acquisition of competing, alternative symbionts or of parasites and cheaters. Consequently, many obligate, heritable symbiont phylogenies show strong congruence with phylogenies of their hosts (Moran et al., 2008). The long-term interaction in vertical transmission generates obligate dependency, resulting in the two partners engaged in a fitness feedback loop that strictly aligns the host and symbiont reproductive interests, whereby changes to the fitness of one partner automatically affect the fitness of the other partner in the same direction (Fisher et al., 2017; Sachs et al., 2011).

Despite the benefits of coupled host and symbiont interest provided by vertical transmission, and the capacities of this transmission mode to guarantee symbiont acquisition by offspring, several studies have demonstrated that partner choice can also play a role in maintaining the insect-microbe mutualisms with vertical transmission. For example, the pea aphid is frequently infected with secondary, non-essential intracellular symbionts, such as *Serratia symbiotica*, in addition to the primary symbiont *B. aphidicola* (Moran et al., 2005; Oliver et al., 2010). As described above, *Buchnera* is selectively released by exocytosis from maternal bacteriocytes immediately adjacent to the embryo that takes them up instantly by endocytosis. The facultative symbionts *Serratia* on the other hand do not follow the same path and are non-selectively released into the haemolymph from their specific bacteriocytes. From the haemolymph, *Serratia* likely hijack the pre-existing endocytosis mechanism to gain entry into the blastula (Koga et al., 2012).

Beewolf wasps, that harbour the defensive symbiont *Streptomyces philanthi* in antennal glands, transmit the symbiont by spreading out gland secretions in their nest, where larvae develop through the pupal stage until adult emergence (Goettler et al., 2007; Kaltenpoth et al., 2005, 2010). The bacteria in the nests provide protection against pathogenic fungi and bacteria by producing different antimicrobial compounds (Kroiss et al., 2010). Interestingly, experimental inoculation of a non-symbiotic actinomycete bacterium to aposymbiotic adult females revealed that the related bacterium can stably colonize the antennal glands but cannot be secreted and transmitted from the antennae (Kaltenpoth et al., 2014), suggesting that a partner choice mechanism exists in the vertical transmission process. Both cases highlight that partner choice mechanisms can reinforce partner fidelity to stabilize long-term, strictly specific insect-microbe symbioses.

3. Horizontal symbiont transmission in animals and plants

Apart from the vertically transmitted symbioses common in insects, most animals and plants do not usually transmit their symbionts vertically; instead, they acquire specific partners from the ambient environment every generation. Because of the enormous diversity of microorganisms in the environment, host performance may be improved when specific, service-providing symbionts are selected from the microbial milieu. To efficiently and specifically choose a partner, host organisms have evolved sophisticated mechanisms for symbiont sorting. As mentioned above, vertically transmitted insect symbionts have strongly reduced genomes, making them unlikely to survive outside of their hosts (Kikuchi, 2009; McCutcheon and Moran, 2012; Moran and Bennett, 2014; Moya et al., 2008). Because of their different lifestyles including the free-living state in the environment, the symbionts maintained by horizontal transmission are in contrast generally culturable and genetically manipulatable. These traits have been very useful to clarify the genetic and molecular bases of such symbiotic associations. Partner choice mechanisms have been particularly well investigated and described in two model systems outside of the Insecta: the nitrogen-fixing *Rhizobium* symbiosis in leguminous plants and the bioluminescent *Vibrio* symbiosis in the Hawaiian bobtail squid (Nyholm and McFall-Ngai, 2004; Wang et al., 2018). Because the accumulated knowledge in these two symbioses is a benchmark for comparison as well as a rich source of inspiration spurring research in insect systems, we will first briefly introduce partner choice mechanisms in these two non-insect model symbiotic systems before focusing on an example in the insects, the bean bug-*Burkholderia* symbiosis.

3.1 The legume-rhizobium symbiosis

In response to nitrogen starvation legumes establish a symbiotic association with nitrogen-fixing rhizobia, which are soil and rhizosphere resident bacteria (rhizobium is a generic name for soil bacteria that establish a nitrogen-fixing symbiosis with legumes; they belong to different genera of the Rhizobiaceae or Burkholderiaceae). In the presence of specific compatible rhizobia, legumes form specialized intracellular infection structures, infection threads, and develop symbiotic organs on their roots, called nodules. The infection threads guide the rhizobia inside the developing nodules where they are housed as intracellular symbionts and convert atmospheric nitrogen gas into ammonia, which is used by the plant as a nitrogen source for its growth (Kondorosi et al., 2013). The legume-rhizobium interaction is usually highly specific such that a given rhizobium strain will interact only with a narrow range of legume species and *vice versa*. Host-symbiont compatibility is monitored throughout the symbiotic process, from its initiation in the rhizosphere to the infection of the roots and development of the nodules, followed by nitrogen fixation in mature nodules. Monitoring is achieved by a continuous exchange of signals between the two partners, including plant-derived flavonoids and antimicrobial peptides (AMPs), and bacterial molecules such as lipochitooligosaccharides called Nod factors, extracellular polysaccharides, lipopolysaccharides (LPS), type III and type IV secretion effectors and even small RNAs (D'Haeze and Holsters, 2002; Mergaert, 2018; Poole et al., 2018). These inter-partner signals are matched in the other partner by corresponding response functions and specific receptors. Moreover, in a mature nodule, the host plants monitor the symbiont's nitrogen-fixation activity. If its performance is not optimal, the hosts reduce the oxygen supply to the nodule and limit the survival of the symbiont (Kiers et al., 2003). Such sanction mechanisms are thought to be important to prevent the evolution of cheaters and maintain the mutualistic association with horizontal symbiont transmission (Archetti et al., 2011; Sachs et al., 2004). Thus, the legume-rhizobium mutualism is stabilized by partner choice mechanisms based on signalling cues from host and symbionts, preventing legumes from initiating or completing interactions with incompatible rhizobia, as well as by enforcing sanctions in established interactions, which serve to reduce the costs of maintaining low-quality partners.

3.2 The squid-*Vibrio fischeri* symbiosis

The squid *Euprymna scolopes* forms an association with the bioluminescent bacterium *Vibrio fischeri*, which is housed in the crypts of a specific symbiotic organ, called the light organ. The function of the light organ is to produce an adjustable counter-illumination that masks the animal *via* silhouette reduction when it is foraging at night under moonlight (Jones and Nishiguchi, 2004). Immediately after the squid hatches, *V. fischeri* is selected from among all other bacteria in the seawater and infects the animal. This colonization initiates a developmental programme that results in morphogenesis of the light organ (McFall-Ngai, 2014). Also this squid-*Vibrio* interaction is strictly governed by host-symbiont chemical signalling (Nyholm and McFall-Ngai, 2004; Visick and Ruby, 2006). In the presence of marine bacteria, squid hatchlings start to produce mucus at ciliated epithelia of the light organ, which traps *V. fischeri* from marine water. After aggregating on the mucus and out-competing other bacteria, *V. fischeri* then migrates into the crypts of the light organ. Before reaching the crypts, it uses flagellar motility to pass through a specific infection structure composed of a narrow pore, a ciliated duct and an antechamber that is connected to the crypts *via* a constricted region or bottleneck (Graf et al., 1994; Millikan and Ruby, 2003, 2004; Nyholm et al., 2000). In the duct and crypt of the light organ, a certain concentration of nitric oxide and hypohalous acid is reached and AMPs are produced, further winnowing out contaminating bacteria (Chen et al., 2017; Davidson et al., 2004; Heath-Heckman et al., 2014; Small and McFall-Ngai, 1999; Weis et al., 1996). Tracheal cytotoxin (TCT), a peptidoglycan fragment derived from *V. fischeri* and LPS are bacterial signalling cues that stimulate morphological alterations of the symbiotic organ, leading to its maturation and establishment of the symbiosis (Koropatnick et al., 2004; Nyholm and Graf, 2012). *V. fischeri*-derived outer membrane vesicles (OMVs) are also involved in the morphological alteration and apoptosis of the light organ (Aschtgen et al., 2016a, b). From the host's side, peptidoglycan recognition proteins (PGRPs) and probably also LPS-binding proteins (LBPs) play important roles for recognition of the symbiont-derived signals and triggering of the host developmental responses (Chun et al., 2008; Krasity et al., 2011; Nyholm and Graf, 2012; Troll et al., 2009, 2010).

Despite the obvious major differences between the legume-rhizobium and squid-*Vibrio* symbioses, it is possible to recognize a number of striking parallels between them. Both interactions involve symbiont-induced morphogenesis of symbiotic organs as well as dedicated infection structures, which provide a major contribution to symbiont selection. In both cases, molecular signals between symbionts govern the partners' genetic programmes for symbiosis and the host organisms recognize symbiont-associated molecular patterns *via* immune-related receptors to trigger a developmental response. Both hosts uses antimicrobials, including AMPs to manage the symbiont population and to winnow out contaminants. Competition for the host niche between symbiont strains is another shared parameter that determines the specificity of both symbioses. Strikingly, for many of these symbiosis-governing principles, analogous functions are also observed in the below described horizontally transmitted insect symbiosis of stinkbugs with *Burkholderia*.

4. Horizontal transmission of symbionts in insects

As mentioned above, social insects can transmit gut bacteria through social behaviours such as coprophagy and trophallaxis (Onchuru et al., 2018). This vertical transmission enables the transgenerational maintenance of specialized gut microbiota with essential roles for the host. It even has been argued that the need to faithfully transmit these important mutualists has played a role in the evolution of sociality (Onchuru et al., 2018). In many insects, however, the gut microbiota is dominated by environmental bacteria that colonize the host opportunistically (Engel and Moran, 2013). For example, in *Drosophila* and mosquitos, microbiota are deemed to be inconstant but despite this stochastic variability, there is nevertheless a certain selectivity for retaining some bacterial groups and these symbionts play significant roles in their hosts' nutrition, immunity, and development (Douglas, 2018; Strand, 2018).

In addition, recent studies have discovered a few examples of highly specific, horizontally transmitted associations, which go against the general pattern of vertical transmission for specialized, obligate symbionts in the insects. These groups of insects acquire their symbionts from the environment (Bright and Bulgheresi, 2010; Salem et al., 2015), like in the legume-rhizobium and squid-*Vibrio* symbioses. Among these insect-microbe symbioses with horizontal transmission, the symbiosis between the bean bug *Riptortus pedestris* and *Burkholderia insecticola* has emerged as a powerful model system for investigating the evolution and mechanisms of host-microbe specificity in insects (Kaltenpoth and Flórez, 2019; Takeshita and Kikuchi, 2017).

4.1 The bean bug-*Burkholderia* symbiosis

The bean bug *R. pedestris* (Heteroptera: Pentatomomorpha: Coreoidea: Alydidae) (Fig. 1A) is a serious pest of leguminous crops in India, South-East Asia, China, Korea and Japan (Kikuhara, 2005; Schaefer and Panizzi, 2000). *R. pedestris* is a hemimetabolous insect that develops from egg to adulthood *via* five nymphal instar stages over approximately 20 days (Kikuchi and Fukatsu, 2014). As is typical for stinkbugs, *R. pedestris* possesses numerous crypts at the posterior part of the midgut, in which *Burkholderia* symbionts are harboured (Fig. 1B–G) (Kikuchi et al., 2005). *R. pedestris* eggs, hatchlings and first instars are symbiont free and among the five instar stages, the bean bug acquires the *Burkholderia* symbiont from ambient soil primarily in the 2nd instar stage (Kikuchi et al., 2011b). *Burkholderia insecticola* strain RPE64 is the type strain for the *R. pedestris* symbionts (Takeshita et al., 2018). Infection with *Burkholderia* symbionts is clearly beneficial for the bean bug; symbiotic insects have faster development, a larger body size, and a higher number of eggs than aposymbiotic insects (Kikuchi et al., 2007; Kikuchi and Fukatsu, 2014). Transcriptomic analyses of the *Burkholderia* symbiont suggest that it recycles host's metabolic wastes and provides the host with essential amino acids and vitamins (Ohbayashi et al., 2019a). On the other side, the benefit for the bacteria remains to be more clarified. The uptake of a few tens of bacterial cells by the insect are sufficient to establish a symbiosis and a maximum of a few thousand bacteria can enter the symbiotic organ (Kikuchi and Yumoto, 2013; Kikuchi et al., 2020). This small founder population grows in only two to three days—the time to fully infect the symbiotic organ—to a symbiotic population of several tens of millions of bacteria. This population is kept intact during the whole lifetime of the insect, although fluctuations in its size might be observed. These bacteria remain viable, can be re-cultured and can re-infect a new bean bug host. Together, this suggests that the symbiosis is indeed mutualistic. However, symbionts are not expelled in the droppings of the insects and the escape of symbionts into environmental soil from insect cadavers has not been analyzed yet. Thus, the formal proof of the mutualistic nature of the bean bug-*Burkholderia* symbiosis awaits further experimentation.

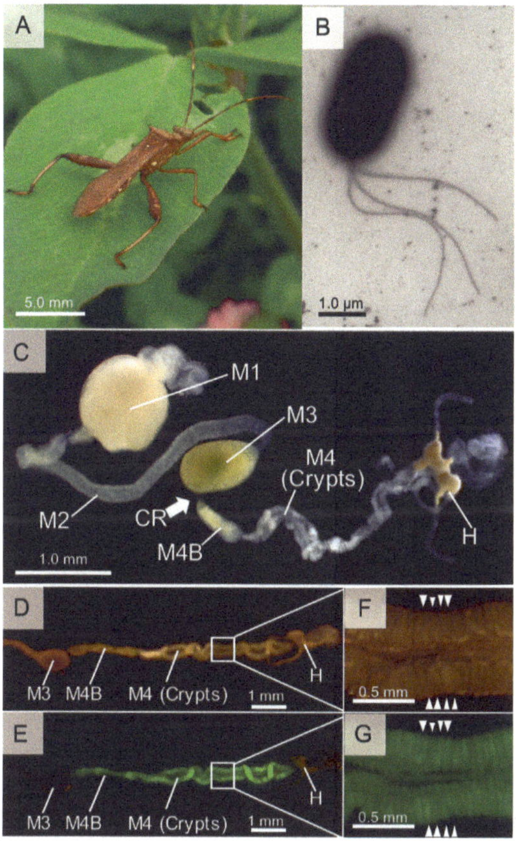

Fig. 1 Gut symbiosis between the bean bug *Riptortus pedestris* and its symbiont *Burkholderia insecticola*. (A) A female adult of the bean bug *R. pedestris*. (B) *B. insecticola*. (C) Whole midgut of *R. pedestris*. (D)–(G) A GFP-labelled strain of *B. insecticola* colonizes the posterior region of the midgut (D)–(E), in which GFP signals are specifically detected in the crypt-bearing symbiotic region (E)–(G). Abbreviations: M1, midgut first section; M2, midgut second section; M3, midgut third section; M4, midgut fourth section with crypts; M4B, M4 bulb; CR, constricted region; H, hindgut. Each crypt is indicated by arrowheads in F and G. *Panels B and C are from Ohbayashi, T., Takeshita, K., Kitagawa, W., Nikoh, N., Koga, R., Meng, X. Y., Tago, K., Hori, T., Hayatsu, M., Asano, K., Kamagata, Y., Lee, B. L., Fukatsu, T. and Kikuchi, Y., 2015. Insect's intestinal organ for symbiont sorting. Proc. Natl. Acad. Sci. U. S. A. 112, E5179–E5188.*

The *Riptortus-Burkholderia* symbiosis is an ideal model to elucidate the molecular bases of insect-microbe symbiosis for several reasons. The *Burkholderia* symbiont is easy to culture in standard bacterial media and to genetically manipulate by standard techniques and tools (Kikuchi and Fukatsu, 2014; Kim et al., 2013; Ohbayashi et al., 2015). The whole genome sequence of the *Burkholderia* symbiont is available (Shibata et al., 2013; Takeshita et al., 2014). In addition, the bean bug host is easily reared in the laboratory and produces large numbers of offspring that require only dried soybean seeds and distilled water containing ascorbic acid for sustenance (Kikuchi et al., 2007). The symbiont inoculation method of adding the desired bacteria to the drinking water, which closely mimics the natural infection process, is also well established (Kikuchi and Fukatsu, 2014). Although the bean bug genome is not yet sequenced, transcriptome data are available, notably for symbiotic conditions and immune responses (Futahashi et al., 2013). RNA interference (RNAi) works very well in *R. pedestris* (Futahashi et al., 2011). Contrary to bacteriocyte-associated symbioses where symbionts are essential, and aposymbiotic insects thus exhibit high mortality, aposymbiotic bean bugs have high survivability, even though growth is delayed (Kikuchi et al., 2007), enabling genetic and physiological comparisons between aposymbiotic and symbiotic insects. These attributes have allowed researchers to investigate the evolution and maintenance of host-symbiont specificity in this insect model with horizontal transmission.

4.2 The bean bug-*Burkholderia* symbiosis is highly specific

The genus *Burkholderia* consists of over 100 species and is an ecologically diverse group (Eberl and Vandamme, 2016). Based on phylogenomic analyses, the genus is grouped into at least three distinct clades. The first one includes human-, animal-, and plant-pathogens, named the "*B. cepacia* complex and *B. pseudomallei*" (BCC&P) clade (Mahenthiralingam et al., 2005). The second clade consists of many plant growth-promoting rhizobacteria and nodule symbionts of leguminous plants, designated as the "plant-associated beneficial and environmental" (PBE) clade (Suárez-Moreno et al., 2012). The third clade mainly consists of environmental species, leaf-nodule symbionts of Rubiaceae plants, and gut symbionts of stinkbugs, and is called the "stinkbug-associated beneficial and environmental" (SBE) clade (Peeters et al., 2016; Takeshita and Kikuchi, 2017). The BCC&P, PBE, and SBE clades were recently designated as different genera called *Burkholderia sensu stricto*, *Paraburkholderia*, and *Caballeronia*, respectively (Beukes et al., 2017). The sister group to this *Burkholderia* trio is the genus *Pandoraea*, mostly consisting of common soil bacteria.

Analyses of the symbionts in natural populations of *R. pedestris* have systematically demonstrated the colonization by *Burkholderia* species

(Jung and Lee, 2019; Kikuchi et al., 2005, 2011a). These *Burkholderia* are genetically diverse, but nearly all of them belong to the SBE clade. Nevertheless, infection with species of the more distant BCC&P clade was also reported in a study of overwintering *R. pedestris* (Jung and Lee, 2019).

As mentioned earlier, the presence of bacterial symbiont-carrying midgut crypts is very common in the heteropteran infraorder Pentatomomorpha (Kikuchi et al., 2011a). In many species of the superfamilies Coreoidea and Lygaeoidea of the Pentatomomorpha, these crypts are also colonized with SBE *Burkholderia* (Boucias et al., 2012; Itoh et al., 2014; Garcia et al., 2014; Kikuchi et al., 2011a, 2005; Kuechler et al., 2016; Ohbayashi et al., 2019b; Olivier-Espejel et al., 2011). Nevertheless, other *Burkholderia*, belonging to the BCC&P and PBE clades, and even *Cupriavidus* and *Pseudomonas* species were occasionally identified in some stinkbug species (Boucias et al., 2012; Garcia et al., 2014; Itoh et al., 2014). However, it should be noted that except for several SBE *Burkholderia*, isolated from *R. pedestris* (Kikuchi et al., 2007), *Coreus marginatus* (Ohbayashi et al., 2019b) and *Blissus insularis* (Xu et al., 2016), none of the other strains (*Burkholderia* or other) have been cultured and tested for their capacity to re-infect and colonize the symbiotic organ of the insect. Finally, the midgut crypts in species of the family Largidae of the superfamily Pyrrhocoroidea also carry *Burkholderia*, but these strains are distinct from the SBE *Burkholderia* and belong strictly to the PBE clade (Gordon et al., 2016; Sudakaran et al., 2015; Takeshita et al., 2015).

Together, these inspections of natural stinkbug species suggest nearly exclusive colonization of the midgut crypts with SBE *Burkholderia* in the Coreoidea and Lygaeoidea superfamilies and with PBE *Burkholderia* in the family Largidae of the Pyrrocoroidea superfamily. Similar to what is firmly established for *R. pedestris* symbionts (Kikuchi et al., 2007), environmental acquisition of *Burkholderia* was postulated or demonstrated in all of the examined stinkbug species, even if occasional vertical transmission was suggested to occur in some (Itoh et al., 2014; Xu et al., 2016). This is in contrast to the aforementioned species of the other stinkbug superfamily, the Pentatomoidea, which possess vertically transmitted gammaproteobacterial symbionts.

Thus, the infections of the midgut crypts in *R. pedestris* and its allied stinkbug species in the superfamilies Coreoidea, Lygaeoidea and Pyrrocoroidea appear to be controlled by partner choice mechanisms that are specific at the broad taxonomic scale of *Burkholderia* groups.

Nevertheless, because of the observed genetic diversity of the symbionts in each insect species, these mechanisms may have a more relaxed specificity at the species or strain levels.

Two types of laboratory experiments in *R. pedestris* have supported this conclusion. First, in laboratory *R. pedestris* reared on soil, infections with PBE *Burkholderia* as well as *Pandoraea* were occasionally identified, but the overwhelming majority of crypt bacteria belonged to the SBE clade (Itoh et al., 2018a). In a second type of experiment, infection tests of aposymbiotic *R. pedestris* were performed with a broad range of bacteria in pure culture, including 34 *Burkholderia* species (13 species of SBE, 12 species of PBE, and 7 species of BCC&P) and an additional 18 taxonomically diverse non-*Burkholderia* bacterial species (Itoh et al., 2019). All tested SBE species efficiently colonized midgut crypts, as did the examined *Pandoraea* species and most PBE, even if the *Pandoraea* and PBE did not fully colonize the midgut crypts. Moreover, the PBE and *Pandoraea* species enhanced the survival and development of the bean bug similar to SBE species despite the incomplete colonization, indicating that these bacteria are not harmful but beneficial to the insect host. In contrast, the tested BCC&P *Burkholderia* species and all other bacterial species, which included other Burkholderiaceae (*Ralstonia*, *Chitinimonas*, *Cupriavidus*) and non-Burkholderiaceae, were unable to establish in the crypts. Thus overall, these laboratory infection experiments using soil or cultured bacteria recapitulated the observed natural specificity of *R. pedestris* for SBE *Burkholderia*, although the specificity seems to be broader in the laboratory than in nature.

What then are the underlying partner choice mechanisms involved in symbiont selection? This seems to be a fundamental issue in light of the enormous diversity of bacteria in soils (Bahram et al., 2018; Delgado-Baquerizo et al., 2018), which suggests that these mechanisms must be particularly powerful, and similar to those in the legume and squid symbioses.

4.3 Partner choice mechanisms in *Riptortus pedestris*

4.3.1 The gut constricted region as a partner choice apparatus

Histological observations and inoculation experiments have revealed that *R. pedestris* develops a pronounced symbiont sorting organ in the midgut (Ohbayashi et al., 2015). The midgut is separated in five morphologically different sections (Fig. 1C): a swollen first section (M1), a tubular second section (M2), an ovoid-shaped third section (M3), the crypt-bearing fourth section (M4) and a bulbous frontal section of the M4 region, called the M4 bulb (M4B). The *Burkholderia* symbionts partially colonize the M4B in addition to the M4 crypts. The M1 to M3 sections constitute the digestive regions of the intestine, while the M4 and M4B are the symbiotic regions.

The junction between M3 and M4B is a remarkably narrow channel, which has a diameter close to the dimensions of *Burkholderia* cells (Fig. 1C; Fig. 2C). This so-called "constricted region" is filled with a mucus-like matrix that strictly prevents food flow from the M3 to the symbiotic sections (Fig. 2A). However, this channel also constitutes the gate through which ingested bacteria pass to enter the symbiotic M4 region (Fig. 2B and C). But the constricted region is not simply an open gate allowing passage to any bacterium; instead, it functions as a selective filter, allowing passage of only those bacteria with the right (still unknown) keys. Indeed, co-inoculation tests of a green fluorescence protein (GFP)-labelled *B. insecticola* and a red fluorescent protein (RFP)-labelled *E. coli* revealed that while *E. coli* can colonize the midgut until the section immediately upstream of the constricted region, only the *Burkholderia* symbiont passes into the symbiotic region (Fig. 2D). The constricted region also prevents bacteria like *Pseudomonas putida*, *Bacillus subtilis*, *Bradyrhizobium japonicum* and other typical soil bacteria from reaching the symbiotic region (Itoh et al., 2019; Ohbayashi et al., 2015). Since these different bacterial species are not strikingly different in cell size, a simple scenario of sieving of bacteria by the narrow channel based on this attribute does not seem to be at play and a more specific mechanism is likely involved.

Interestingly, the constricted region is broadly conserved, not only in the *Burkholderia*-associated lygaeoid, coreoid and largid species, but also in the pentatomoid species that are associated with vertically transmitted gammaproteobacteria (Gordon et al., 2016; Ohbayashi et al., 2015), suggesting evolution of the constricted region in the common ancestor of the stinkbug superfamilies. Thus, even if the symbiotic association in the Pentatomoidea is maintained primarily by partner fidelity mechanisms, i.e., vertical transmission, partner choice contributes as well and may be facilitated by the constricted region, possibly to inhibit environmental contaminants. A striking equivalent function was described in the intestine of the Sonoran Desert turtle ant *Cephalotes rohweri* where the proventriculus filters and blocks bacterial entry into the midgut, which harbours a vertically transmitted, specialized symbiont community (Lanan et al., 2016).

Thus, the constricted region is an infection structure, guiding the symbionts into the symbiotic organ. It has also a major role in the specificity of the symbiosis. These functions of the constricted region are equivalent to

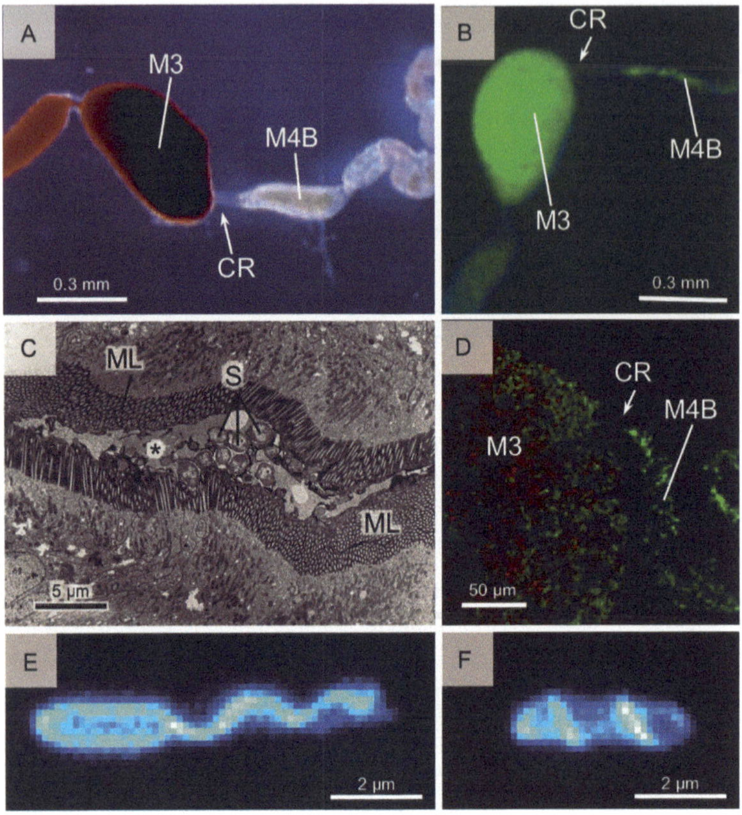

Fig. 2 Partner choice in the gut of *R. pedestris.* (A) A midgut fed with water containing 0.05% food colouring (congo red). Note that the colouring is blocked at the constricted region. (B) A midgut infected with GFP-labelled *B. insecticola* showing that the bacteria can flow from M3 into M4B by passing through the constricted region. (C) Electron microscopy image of the narrow channel in the constricted region. The lumen of the channel is flanked by a layer of microvilli (ML) and contains a few bacterial cells (S) that are passing through the channel. (D) The constricted region in an insect co-inoculated with a GFP-labelled *B. insecticola* and an RFP-labelled *E. coli*. At the constricted region, *E. coli* is sorted out and only the *Burkholderia* symbiont passes through the channel. (E) Normal swimming mode of *B. insecticola* in a liquid environment. (F) Corkscrew motility of *B. insecticola* in a viscous environment. Flagellar filaments are visualized by covalent coupling to the fluorescent dye Cy3-NHS-ester. *Panels (A), (C), and (D) are from Ohbayashi, T., Takeshita, K., Kitagawa, W., Nikoh, N., Koga, R., Meng, X. Y., Tago, K., Hori, T., Hayatsu, M., Asano, K., Kamagata, Y., Lee, B. L., Fukatsu, T. and Kikuchi, Y., 2015. Insect's intestinal organ for symbiont sorting. Proc. Natl. Acad. Sci. U. S. A. 112, E5179–E5188.*

those of the infection structures in legumes, the infection threads, and they are analogous to the conduit formed by the pore, ciliated duct, antechamber and bottleneck in the squid.

4.3.2 Corkscrew flagellar motility

Although the molecular mechanism underpinning bacterial sorting in the constricted region remains unclear, a screening for colonization-defect mutants of *B. insecticola* revealed that flagellar motility is a key factor for the symbiont's ability to colonize the M4 crypts (Ohbayashi et al., 2015). Although non-motile mutants reach the M3 section of the midgut, they do not enter the constricted region and hence never reach the symbiotic organs, clearly demonstrating the importance of flagellar motility for passing through the mucus-filled narrow passage. Interestingly, other tested bacteria with active motility, including *E. coli*, *P. putida*, and *B. subtilis* cannot pass through the constricted region, leaving unanswered why only *Burkholderia* symbionts can do this but demonstrating that motility alone is not sufficient.

To date, two types of flagellar motility have been described in bacteria (Jarrell and McBride, 2008): run-and-tumble motility, described well in bacteria having peritrichous flagella (multiple flagella all over the cell surface), such as *E. coli* and *Pseudomonas* species, but also common in bacteria having polar flagella; and forward-and-reverse motility, found in bacteria having polar flagella, such as *Vibrio* species. *B. insecticola*, which has one to three polar flagella (Fig. 1B), shows normal run-and-tumble motility in a liquid environment. However, in viscous media, a unique, novel type of flagellar motility has been discovered in the *Burkholderia* symbiont, called "corkscrew flagellar motility". In this particular swimming mode, the symbiont wraps the flagellar filaments around the cell body and moves like a drill in the viscous environment (Fig. 2E and F) (Kinosita et al., 2018). Considering that this type of novel motility has been reported in a few polar-flagellated bacteria including the squid symbiont *V. fischeri* and *Shewanella putrefaciens* (Kinosita et al., 2018; Kuhn et al., 2017), it is tempting to speculate that the corkscrew motility enables navigation of the mucus-filled, viscous passage of the constricted region by *Burkholderia* symbionts. To test this hypothesis, it will thus be of interest to analyze the corkscrew flagellar motility in other *Burkholderia* or *Pandorea* strains that can move into the M4 crypts or to identify the genetic determinants of corkscrew motility in *B. insecticola*.

As for the bean bug-*Burkholderia* example, motility is also important in the squid-*Vibrio* symbiosis (Nyholm and McFall-Ngai, 2004), enabling *V. fischeri* to move through the infection structure. The bacteria are guided in their journey by the chemotactic cue chitobiose, which is produced through the degradation of chitin by chitinases present in the light organ (Mandel et al., 2012). Based on the analogy, it is tempting to predict that *B. insecticola* is also guided by chemotaxis and by cues from the crypts to pass through the constricted region or to move through the symbiotic organ to reach the farthest crypts. This hypothesis requires further experimentation. It is worth noting that in the legume-rhizobium symbiosis, motility and chemotaxis are equally important, but only in an earlier stage of the interaction; these allow the rhizobia to migrate from the soil towards to root surface. The passage through the infection structure on the other hand, does not depend on motility but is mediated by bacterial cell divisions that push the bacteria forward through the thin infection threads (Gage et al., 1996).

4.3.3 Midgut closure stimulated by symbiont colonization

In addition to the partner choice achieved by selective passage through the constricted region, a second layer mechanism for maintaining the *Riptortus-Burkholderia* specificity was recently discovered. This mechanism also involves the constricted region and is called the "midgut closure" (Kikuchi et al., 2020). A few hours after the *Burkholderia* symbionts begin entering into the constricted region and colonizing the M4 crypts, the constricted region and the M4B, which are at first permeable for the bacteria, become closed, completely preventing the entrance of additional bacteria. After full occupation of the M4 crypts with symbionts, the M4B region is re-opened and surplus symbiont cells from the crypts flow back to the M4B. Notably, even if the re-opening of M4B occurs, the constricted region is kept firmly closed. This midgut closure prevents additional colonization of the symbiotic region by symbiotic bacteria or any possible contaminants, and it thus contributes to preserve the specific colonization of the crypts. In a strikingly analogous way, the proventriculus of the above-mentioned *Cephalotes* ants allows the passage of bacteria in the young (callow) adult stage; but shortly after enclosed adults acquire bacteria through trophallaxis, the proventriculus develops an impermeable filtering layer sealing off the midgut and preventing further bacterial infections (Lanan et al., 2016).

In the legume-rhizobium interaction, lipochitooligosaccharide Nod factors trigger nodule formation. In the case of the bobtail squid, bacterial peptidoglycan-derived TCT and LPS alters morphogenesis of the light organ. In both cases, these symbiont-associated molecular patterns are recognized by the host using immune-related receptors. The molecular mechanism responsible for the midgut closure remains to be determined but it is likely that also in the *R. pedestris* symbiosis, molecular cues from the *Burkholderia* symbiont are recognized by receptors on the epithelial cells of the constricted region or the M4 to trigger this morphological response in the symbiotic organ.

4.3.4 Competition-based selection in the gut

As mentioned above, the capacity to pass through the constricted region and the infection of the M4 crypts is not limited to the natural symbiont group of SBE *Burkholderia*. For example, *Pandoraea norimbergensis* and *Burkholderia fungorum* (PBE) can also colonize the midgut crypts of *R. pedestris* (Fig. 3A–D) and provide fitness benefits to the host, even though these common soil bacteria are usually not associated with the bean bug in natural conditions (Itoh et al., 2019). An additional partner choice mechanism, called "competition-based selection," was put forward to explain this observed symbiont specificity (Itoh et al., 2019). It was shown that the SBE *Burkholderia* always outcompete PBE *Burkholderia* or *Pandoraea* species in the M4 region when these species are co-infecting the symbiotic organ, achieving the dominance of SBE *Burkholderia* in the bean bug gut (Fig. 3E–G). These experimental data demonstrate that the abilities for colonization and cooperation, usually thought of as specific traits of mutualists, are not unique to the *Burkholderia* symbiont (i.e., SBE *Burkholderia*). On the contrary, competitiveness inside the gut is a trait of the bean bug-symbiont lineage only (Fig. 3H). Thus, competition-based selection of symbionts is a critical determinant of the specificity of the bean bug-*Burkholderia* interaction, and again this mirrors processes in the legume and squid symbioses where competition has a similar impact on the specificity (Boivin et al., 2020; Koehler et al., 2019). However, the timing of the competition-based selection is strikingly different. In *R. pedestris*, the competition happens after passage of the bacteria through the infection structure (Itoh et al., 2019) while in the two other symbioses, it takes place before the infection (Boivin et al., 2020; Koehler et al., 2019).

Although at present, the molecular bases of the *in vivo* competitiveness of the *Burkholderia* symbiont remains unclear, the following four types of mechanisms can be considered.

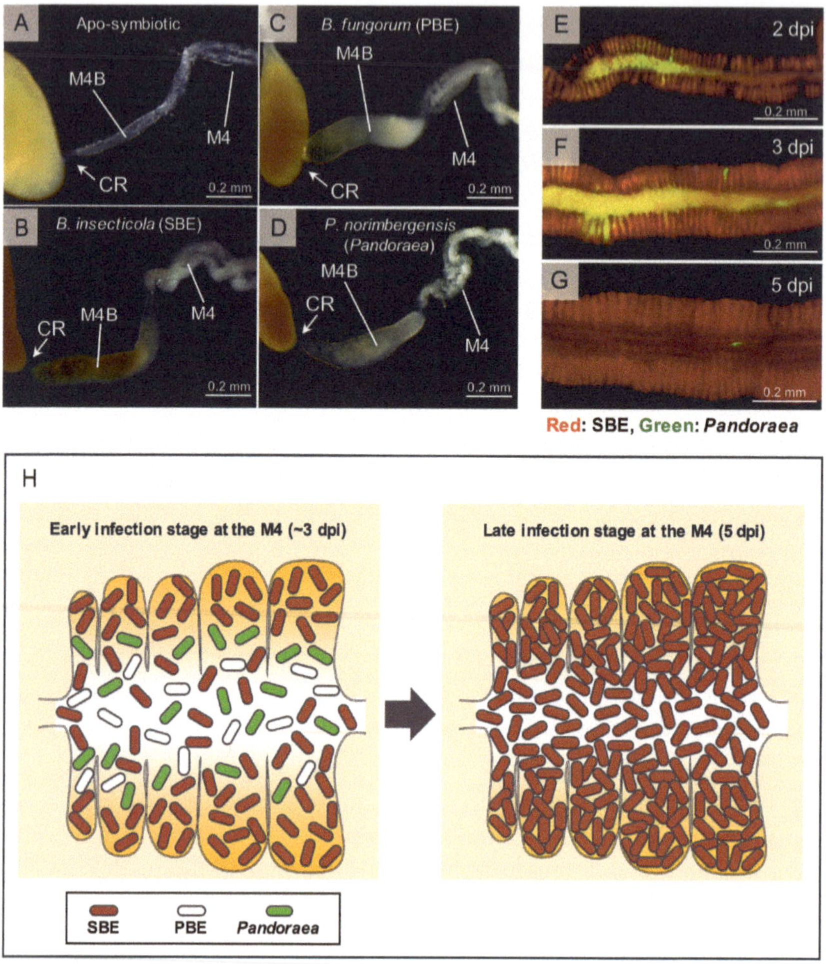

Fig. 3 Microbe-microbe competition in the gut of *R. pedestris*. Midgut M3-M4B-M4 regions of (A) an aposymbiotic insect; (B) an insect infected with a SBE *Burkholderia* (*B. insecticola*); (C) an insect infected with a PBE *Burkholderia* (*B. fungorum*); (D) an insect infected with *Pandoraea* (*P. norimbergensis*). (E)–(G) M4 region of an insect inoculated with an RFP-labelled SBE *Burkholderia* (*B. insecticola*) and an RFP-labelled *Pandoraea* (*P. norimbergensis*) at 2, 3, and 5 days post inoculation (dpi). (H) The schematic drawing of symbiont specificity in the midgut of the bean bug *R. pedestris*. In the early infection stage, SBE *Burkholderia*, PBE *Burkholderia*, and *Pandoraea* co-colonize the M4 (left). In the mature stage, SBE *Burkholderia* outcompetes the other strains and becomes dominant in the M4 (right). *Panels (A)–(G) are from Itoh, H., Jang, S., Takeshita, K., Ohbayashi, T., Ohnishi, N., Meng, X. Y., Mitani, Y. and Kikuchi, Y., 2019. Host-symbiont specificity determined by microbe–microbe competition in an insect gut. Proc. Natl. Acad. Sci. U. S. A. 116, 22673–22682.*

(i) *Nutrient-based mechanisms.* Auxotrophic mutants of the *Burkholderia* symbiont, such as purine biosynthesis mutants, show severe crypt colonization defects (Kim et al., 2014), suggesting that the M4 environment is nutritionally poor or there is a selective repertoire of available nutrients. In fact, transcriptomic analyses suggested that the *Burkholderia* symbiont actively proliferates in M4 by assimilating host metabolic wastes, such as sulphate, allantoin and urea (Ohbayashi et al., 2019a). Such adaptation to the M4 nutritional environment thus conceivably affects the competitiveness and the selective colonization of the *Burkholderia* symbiont.

(ii) *Antimicrobial agent-based mechanisms.* Previous studies demonstrated that AMP resistance of *B. insecticola* plays an important role in stable colonization of the symbiont in M4. For example, a deletion mutant of *uppP* (undecaprenyl pyrophosphate phosphatase; involved in cell wall biosynthesis) becomes susceptible to lysozymes *in vitro* and also shows a severe colonization defect in the M4 crypts (Kim et al., 2013). Similarly, LPS biosynthesis mutants in genes such as *wabO*, *waaC* and *waaF* also show decreased colonization ability in M4 (Kim et al., 2016, 2017). It should be noted that transcriptomic analyses of M4 revealed that a novel type of AMP, called crypt-specific cysteine-rich peptides (CCRs), are highly and specifically expressed in the midgut crypts (Futahashi et al., 2013). Although the principal role of these crypt-specific AMPs remains to be experimentally defined, they may be involved in the symbiont's competitiveness and the *in vivo* selection process. Similar types of peptides operate also in the legume and squid symbioses (Mergaert, 2018).

(iii) *Adhesion-based mechanisms.* Theoretical studies have predicted that bacterial adhesion to gut epithelial cells has a selective advantage for competitive colonization in the gut (McLoughlin et al., 2016). Genomic data indicate that the *Burkholderia* symbiont possesses some Tad pilli and type I fimbriae, which may contribute to the efficient colonization and competitiveness of the *Burkholderia* symbiont.

(iv) *Direct inhibition-based mechanisms.* Characterized bacterial Type VI secretion systems (T6SS) are known to directly inject in the cytosol of target bacteria anti-bacterial effectors that inhibit competitors (Russell et al., 2014). The importance of T6SS in mediating bacterial competition in animal symbioses has been reported in the squid-*Vibrio* system (Speare et al., 2018) and in honey bee gut communities (Steele et al., 2017). Although SBE *Burkholderia* have a single type of T6SS, co-inoculation tests of PBE *Burkholderia* or *Pandoraea* and T6SS deletion mutants of *B. insecticola* demonstrated that T6SS is not involved in the competitiveness of the symbiont in the M4 region (Itoh et al., 2019). Some BCC&P *Burkholderia* utilize another type of microbe-microbe toxin-delivery system, called "contact-dependent growth inhibition (CDI)" (Garcia et al., 2016). However, the SBE *Burkholderia* do not have any CDI genes. Hence, these candidate systems do not seem to be involved in the symbiont's *in vivo* competitiveness, and additional work is needed to identify the responsible factors.

5. Concluding remarks

Although theoretical studies have proposed a variety of mechanisms for the maintenance of host-microbe mutualisms, such as partner fidelity feedback, partner choice and sanctions, competition-based selection, and public goods theory (Archetti et al., 2011), the relative importance of each are unknown, even for relatively well-studied interactions. Furthermore, the genetic and molecular underpinnings responsible for mutualism maintenance are also mostly unknown, although progress has been made in some plant and animal models, such as the legume-rhizobia and squid-*Vibrio* symbioses. The recent body of work on the *Riportus-Burkholderia* symbiosis is now greatly improving our knowledge concerning the molecular bases of insect-microbe symbioses with horizontal symbiont transmission. This useful model system is clarifying how host-symbiont specificity evolved and is maintained in insect gut symbioses. Several, components of this specificity-determining process have been identified, including an infection structure, the constricted region, a motility-based infection of the symbiotic organ, the closure of the constricted region, effectively locking the symbiotic organ once symbionts have established and avoiding super-infection. Competition-based elimination of co-infecting bacteria, within the symbiotic organ, provides a final specificity-supporting mechanism. It is striking that these mechanisms have functional or structural equivalents in other symbioses with environmental acquisition of symbionts like the legume-rhizobia and squid-*Vibrio* interactions. This suggests that nature uses a limited number of options to solve the specificity problem in symbiosis. We expect that the different symbiosis models in plants and animals will continue to cross-fertilize each other. Notably, the molecular bases such as inter-partner signals, regulatory cascades and effectors could also show many similitudes, as suggested for example by the recurrent use of antimicrobial peptides.

In contrast, the genetic and molecular bases of vertical symbiont transmission remain unknown, despite the omnipresence of this transmission mode in insects. At this stage, only a few studies have tackled the question because reverse-genetic approaches like RNAi and genome-editing do not efficiently work in many insect models of endosymbiosis. Stinkbugs and beetles may be promising models because they possess well-developed symbiotic systems and RNAi experiments are feasible in most cases. In fact, using RNAi, recent studies have succeeded in identifying a transcription factor and immune-related genes that play pivotal roles in the maintenance of intracellular symbioses (Login et al., 2011; Maire et al., 2019; Matsuura et al., 2015).

In insects, as summarized in this chapter, there are diverse vertical transmission mechanisms involving highly sophisticated morphological features and behaviours, as well as environmental acquisition mechanisms with well-developed partner choice and competition-based mechanisms. It remains totally unknown how such complex adaptive traits could evolve. But it is our conviction that the symbiotic systems in the Pentatomomorpha opens a window into this fundamental but still enigmatic question. Indeed, the patterns of symbiont phylogeny, transmission mode and type of symbiotic organ of the symbioses in the Pentatomomorpha highlight the existence of several evolutionary transitions in this infraorder. The phylogenetic relationship between the pentatomomorphan superfamilies was determined as (Pentatomoidea + (Pyrrocoroidea + (Coreoidea + Lygaeoidea))) (Fig. 4). Thus, a first transition was the switch between horizontal and vertical transmission that happened at the divergence between the (Pyrrocoroidea + (Coreoidea + Lygaeoidea)) clade and the Pentatomoidea superfamily (Fig. 4 transition 1). Moreover, this transition coincided with a switch in symbionts between *Burkholderia* species and Gammaproteobacteria. Horizontal transmission is believed to be the ancient mode of symbiont transmission while vertical transmission is a derived trait (Kiers and West, 2015; Sachs et al., 2011). It is thus conceivable that in the evolution of the Pentatomomorpha, the horizontally transmitted *Burkholderia* symbiosis was ancestral and was replaced with vertical transmission of Gammaproteobacteria in the Pentatomoidea superfamily. Probably this transition to vertical transmission was made possible by the evolution of new modifications in the posterior regions of the M4 that enabled capsule formation or egg-smearing. A second notable transition, consisting of a switch between PBE and SBE *Burkholderia*, or *vice versa*, occurred at the divergence between the clade (Coreoidea + Lygaeoidea) and the Pyrrocoroidea (Fig. 4 transition 2). This transition is likely the result of a modification in the specificity of the competition-based selection or in the recognition of symbiont-associated molecular patterns. Furthermore, two additional transitions happened,

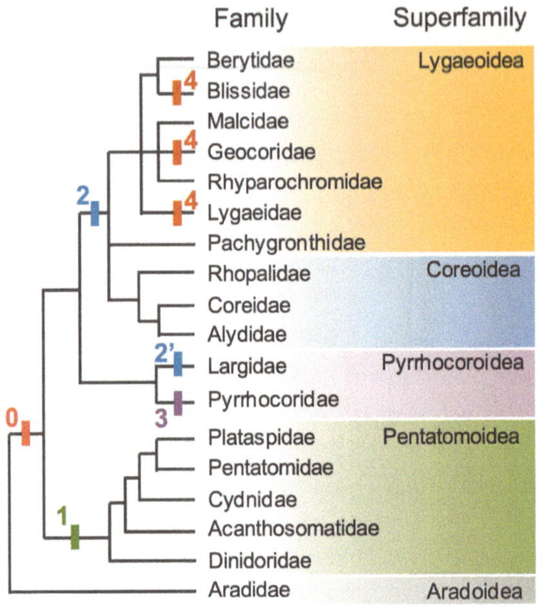

Fig. 4 Major evolutionary transitions in symbiosis in the Pentatomomorpha. The proposed positions of the transitions are numbered: 0, evolution of midgut crypts (possibly with horizontally transmitted *Burkholderia*); 1, switch to Gammaproteobacteria and vertical transmission; 2, switch between PBE and SBE *Burkholderia* (the switch occurred at either 2 or 2′); 3, loss of midgut crypts and acquisition of specific M3 microbiota; 4, loss of midgut crypts and acquisition of bacteriocytes with vertically transmitted Gammaproteobacteria in some species of the branch. See text for more details.

entailing the loss of the midgut crypts altogether. Species of the family Pyrrhocoridae of the superfamily Pyrrhocoroidea have a degenerate M4 crypt region but harbour instead specific microbiota in the M3 midgut region, consisting of Actinobacteria, Firmicutes and Proteobacteria that are transmitted vertically by egg-smearing (Fig. 4 transition 3) (Takeshita and Kikuchi, 2017). And some scattered species from the Lygaeoidea superfamily have instead of the extracellular crypt symbionts, vertically transmitted intracellular Gammaproteobacteria, hosted within bacteriocytes clustered in a single or in paired bacteriomes, similar to the bacteriomes in many insects in the orders Blattaria, Hemiptera, Coleoptera, Hymenoptera and Diptera (Fig. 4 transition 4) (Kuechler et al., 2019). Possibly, these bacteriomes evolved from the crypt symbiosis in successive stages of intracellular infection of M4 crypt epithelial cells, followed by the individualization of these cells in bacteriomes and the subsequent loss of crypts. Thus the amazing diversity of microbial symbioses in the Pentatomomorpha, in conjunction with the established molecular tools and rapidly progressing 'omics techniques, provides us a great opportunity to understand evolutionary transitions in symbiosis at the molecular and genetic level.

Acknowledgements

We thank Yoshiaki Kinoshita for kindly providing an image of the wrapping motility. This study was supported by the JSPS-CNRS Bilateral Open Partnership Joint Research Project to Y.K. and P.M., and the JSPS Research Fellowship for Young Scientists to T.O. (19J01106).

Conflict of interest

The authors declare that they have no competing interests.

References

Anbutsu, H., Moriyama, M., Nikoh, N., Hosokawa, T., Futahashi, R., Tanahashi, M., Meng, X.Y., Kuriwada, T., Mori, N., Oshima, K., Hattori, M., Fujie, M., Satoh, N., Maeda, T., Shigenobu, S., Koga, R., Fukatsu, T., 2017. Small genome symbiont underlies cuticle hardness in beetles. Proc. Natl. Acad. Sci. U. S. A. 114, E8382–E8391.

Archetti, M., Scheuring, I., Hoffman, M., Frederickson, M.E., Pierce, N.E., Yu, D.W., 2011. Economic game theory for mutualism and cooperation. Ecol. Lett. 14, 1300–1312.

Aschtgen, M.S., Lynch, J.B., Koch, E., Schwartzman, J., McFall-Ngai, M., Ruby, E., 2016b. Rotation of *Vibrio fischeri* flagella produces outer membrane vesicles that induce host development. J. Bacteriol. 198, 2156–2165.

Aschtgen, M.S., Wetzel, K., Goldman, W., McFall-Ngai, M., Ruby, E., 2016a. *Vibrio fischeri*-derived outer membrane vesicles trigger host development. Cell. Microbiol. 18, 488–499.

Attardo, G.M., Lohs, C., Heddi, A., Alam, U.H., Yildirim, S., Aksoy, S., 2008. Analysis of milk gland structure and function in *Glossina morsitans*: milk protein production, symbiont populations and fecundity. J. Insect Physiol. 54, 1236–1242.

Bahram, M., Hildebrand, F., Forslund, S.K., Anderson, J.L., Soudzilovskaia, N.A., Bodegom, P.M., Bengtsson-Palme, J., Anslan, S., Coelho, L.P., Harend, H., Huerta-Cepas, J., Medema, M.H., Maltz, M.R., Mundra, S., Olsson, P.A., Pent, M., Polme, S., Sunagawa, S., Ryberg, M., Tedersoo, L., Bork, P., 2018. Structure and function of the global topsoil microbiome. Nature 560, 233–237.

Bansal, R., Michel, A.P., Sabree, Z.L., 2014. The crypt-dwelling primary bacterial symbiont of the polyphagous pentatomid pest *Halyomorpha halys* (hemiptera: Pentatomidae). Environ. Entomol. 43, 617–625.

Baumann, P., 2005. Biology bacteriocyte-associated endosymbionts of plant sap-sucking insects. Annu. Rev. Microbiol. 59, 155–189.

Beard, C.B., Cordon-Rosales, C., Durvasula, R.V., 2002. Bacterial symbionts of the Triatominae and their potential use in control of Chagas disease transmission. Annu. Rev. Entomol. 47, 123–141.

Beukes, C.W., Palmer, M., Manyaka, P., Chan, W.Y., Avontuur, J.R., van Zyl, E., Huntemann, M., Clum, A., Pillay, M., Palaniappan, K., Varghese, N., Mikhailova, N., Stamatis, D., Reddy, T.B.K., Daum, C., Shapiro, N., Markowitz, V., Ivanova, N., Kyrpides, N., Woyke, T., Blom, J., Whitman, W.B., Venter, S.N., Steenkamp, E.T., 2017. Genome data provides high support for generic boundaries in *Burkholderia* sensu lato. Front. Microbiol. 8, 1154.

Bistolas, K.S., Sakamoto, R.I., Fernandes, J.A., Goffredi, S.K., 2014. Symbiont polyphyly, co-evolution, and necessity in pentatomid stinkbugs from Costa Rica. Front. Microbiol. 5, 349.

Boivin, S., Ait Lahmidi, N., Sherlock, D., Bonhomme, M., Dijon, D., Heulin-Gotty, K., Le-Queré, A., Pervent, M., Tauzin, M., Carlsson, G., Jensen, E., Journet, E.P., Lopez-Bellido, R., Seidenglanz, M., Marinkovic, J., Colella, S., Brunel, B., Young, P., Lepetit, M., 2020. Host-specific competitiveness to form nodules in *Rhizobium leguminosarum* symbiovar *viciae*. New Phytol. 226, 555–568. https://doi.org/10.1111/nph.16392.

Boucias, D.G., Garcia-Maruniak, A., Cherry, R., Lu, H., Maruniak, J.E., Lietze, V.U., 2012. Detection and characterization of bacterial symbionts in the heteropteran, *Blissus insularis*. FEMS Microbiol. Ecol. 82, 629–641.

Bracke, J.W., Cruden, D.L., Markovetz, A.J., 1978. Effect of metronidazole on the intestinal microflora of the American cockroach, *Periplaneta americana* L. Antimicrob. Agents Chemother. 13, 115–120.

Bright, M., Bulgheresi, S., 2010. A complex journey: transmission of microbial symbionts. Nat. Rev. Microbiol. 8, 218–230.

Brune, A., Dietrich, C., 2015. The gut microbiota of termites: digesting the diversity in the light of ecology and evolution. Annu. Rev. Microbiol. 69, 145–166.

Buchner, P., 1965. Endosymbiosis of Animals with Plant Microorganims. Interscience, New York.

Chen, F., Krasity, B.C., Peyer, S.M., Koehler, S., Ruby, E.G., Zhang, X., McFall-Ngai, M.J., 2017. Bactericidal permeability-increasing proteins shape host-microbe interactions. mBio 8, e00040-17.

Chun, C.K., Troll, J.V., Koroleva, I., Brown, B., Manzella, L., Snir, E., Almabrazi, H., Scheetz, T.E., de Fatima Bonaldo, M., Casavant, T.L., Soares, M.B., Ruby, E.G., McFall-Ngai, M.J., 2008. Effects of colonization, luminescence, and autoinducer on host transcription during development of the squid-vibrio association. Proc. Natl. Acad. Sci. U. S. A. 105, 11323–11328.

D'Haeze, W., Holsters, M., 2002. Nod factor structures, responses, and perception during initiation of nodule development. Glycobiology 12, 79R–105R.

Dasch, G.E., Weiss, C.K., 1984. Endosymbionts of insects. In: Krieg, N.A. (Ed.), vol. I. Bergey's Manual of Systematic Bacteriology. Williams and Wilkins, Baltimore, pp. 881–883.

Davidson, S.K., Koropatnick, T.A., Kossmehl, R., Sycuro, L., McFall-Ngai, M.J., 2004. No means 'yes' in the squid-vibrio symbiosis: nitric oxide (NO) during the initial stages of a beneficial association. Cell. Microbiol. 6, 1139–1151.

Delgado-Baquerizo, M., Oliverio, A.M., Brewer, T.E., Benavent-González, A., Eldridge, D.J., Bardgett, R.D., Maestre, F.T., Singh, B.K., Fierer, N., 2018. A global atlas of the dominant bacteria found in soil. Science 359, 320–325.

Douglas, A.E., 2018. The *Drosophila* model for microbiome research. Lab. Anim. (NY) 47, 157–164.

Douglas, A.E., Minto, L.B., Wilkinson, T.L., 2001. Quantifying nutrient production by the microbial symbionts in an aphid. J. Exp. Biol. 204, 349–358.

Eberl, L., Vandamme, P., 2016. Members of the genus *Burkholderia*: good and bad guys. F1000Res. 5, 1007.

Engel, P., Moran, N.A., 2013. The gut microbiota of insects-diversity in structure and function. FEMS Microbial. Rev. 37, 699–735.

Fisher, R.M., Henry, L.M., Cornwallis, C.K., Kiers, E.T., West, S.A., 2017. The evolution of host-symbiont dependence. Nat. Commun. 8, 15973.

Flórez, L.V., Scherlach, K., Gaube, P., Ross, C., Sitte, E., Hermes, C., Rodrigues, A., Hertweck, C., Kaltenpoth, M., 2017. Antibiotic-producing symbionts dynamically transition between plant pathogenicity and insect-defensive mutualism. Nat. Commun. 8, 15172.

Fukatsu, T., Hosokawa, T., 2002. Capsule-transmitted gut symbiotic bacterium of the japanese common plataspid stinkbug, *Megacopta punctatissima*. Appl. Environ. Microbiol. 68, 389–396.

Futahashi, R., Tanaka, K., Matsuura, Y., Tanahashi, M., Kikuchi, Y., Fukatsu, T., 2011. Laccase2 is required for cuticular pigmentation in stinkbugs. Insect Biochem. Mol. Biol. 41, 191–196.

Futahashi, R., Tanaka, K., Tanahashi, M., Nikoh, N., Kikuchi, Y., Lee, B.L., Fukatsu, T., 2013. Gene expression in gut symbiotic organ of stinkbug affected by extracellular bacterial symbiont. PLoS One 8, e64557.

Gage, D.J., Bobo, T., Long, S.R., 1996. Use of green fluorescent protein to visualize the early events of symbiosis between *Rhizobium meliloti* and alfalfa (*Medicago sativa*). J. Bacteriol. 178, 7159–7166.

Garcia, J.R., Laughton, A.M., Malik, Z., Parker, B.J., Trincot, C., Chiang, S.S.L., Chung, E., Gerardo, N.M., 2014. Partner associations across sympatric broad-headed bug species and their environmentally acquired bacterial symbionts. Mol. Ecol. 23, 1333–1347.

Garcia, E.C., Perault, A.I., Marlatt, S.A., Cotter, P.A., 2016. Interbacterial signaling via *Burkholderia* contact-dependent growth inhibition system proteins. Proc. Natl. Acad. Sci. U. S. A. 113, 8296–8301.

Goettler, W., Kaltenpoth, M., Herzner, G., Strohm, E., 2007. Morphology and ultrastructure of a bacteria cultivation organ: the antennal glands of female European beewolves, *Philanthus triangulum* (Hymenoptera, Crabronidae). Arthropod Struct. Dev. 36, 1–9.

Gordon, E.R., McFrederick, Q., Weirauch, C., 2016. Phylogenetic evidence for ancient and persistent environmental symbiont reacquisition in Largidae (Hemiptera: Heteroptera). Appl. Environ. Microbiol. 82, 7123–7133.

Graf, J., Dunlap, P.V., Ruby, E.G., 1994. Effect of transposon-induced motility mutations on colonization of the host light organ by *Vibrio fischeri*. J. Bacteriol. 176, 6986–6991.

Hayashi, T., Hosokawa, T., Meng, X.Y., Koga, R., Fukatsu, T., 2015. Female-specific specialization of a posterior end region of the midgut symbiotic organ in *Plautia splendens* and allied stinkbugs. Appl. Environ. Microbiol. 81, 2603–2611.

Heath-Heckman, E.A., Gillette, A.A., Augustin, R., Gillette, M.X., Goldman, W.E., McFall-Ngai, M.J., 2014. Shaping the microenvironment: evidence for the influence of a host galaxin on symbiont acquisition and maintenance in the squid-*Vibrio* symbiosis. Environ. Microbiol. 16, 3669–3682.

Hosokawa, T., Hironaka, M., Inadomi, K., Mukai, H., Nikoh, N., Fukatsu, T., 2013. Diverse strategies for vertical symbiont transmission among subsocial stinkbugs. PLoS One 8, e65081.

Hosokawa, T., Kikuchi, Y., Fukatsu, T., 2007. How many symbionts are provided by mothers, acquired by offspring, and needed for successful vertical transmission in an obligate insect-bacterium mutualism? Mol. Ecol. 16, 5316–5325.

Hosokawa, T., Kikuchi, Y., Meng, X.Y., Fukatsu, T., 2005. The making of symbiont capsule in the plataspid stinkbug *Megacopta punctatissima*. FEMS Microbial. Ecol. 54, 471–477.

Hosokawa, T., Kikuchi, Y., Nikoh, N., Shimada, M., Fukatsu, T., 2006. Strict host-symbiont cospeciation and reductive genome evolution in insect gut bacteria. PLoS Biol. 4, e337.

Hosokawa, T., Kikuchi, Y., Shimada, M., Fukatsu, T., 2008. Symbiont acquisition alters behaviour of stinkbug nymphs. Biol. Lett. 4, 45–48.

Hosokawa, T., Koga, R., Kikuchi, Y., Meng, X.Y., Fukatsu, T., 2010. *Wolbachia* as a bacteriocyte-associated nutritional mutualist. Proc. Natl. Acad. Sci. U. S. A. 107, 769–774.

Hosokawa, T., Matsuura, Y., Kikuchi, Y., Fukatsu, T., 2016. Recurrent evolution of gut symbiotic bacteria in pentatomid stinkbugs. Zool. Lett. 2, 24.

Hosokawa, T., Nikoh, N., Koga, R., Sato, M., Tanahashi, M., Meng, X.Y., Fukatsu, T., 2012. Reductive genome evolution, host-symbiont co-speciation and uterine transmission of endosymbiotic bacteria in bat flies. ISME J. 6, 577–587.

Itoh, H., Aita, M., Nagayama, A., Meng, X.Y., Kamagata, Y., Navarro, R., Hori, T., Ohgiya, S., Kikuchi, Y., 2014. Evidence of environmental and vertical transmission of *Burkholderia* symbionts in the oriental chinch bug, *Cavelerius saccharivorus* (Heteroptera: Blissidae). Appl. Environ. Microbiol. 80, 5974–5983.

Itoh, H., Hori, T., Sato, Y., Nagayama, A., Tago, K., Hayatsu, M., Kikuchi, Y., 2018a. Infection dynamics of insecticide-degrading symbionts from soil to insects in response to insecticide spraying. ISME J. 12, 909–920.

Itoh, H., Jang, S., Takeshita, K., Ohbayashi, T., Ohnishi, N., Meng, X.Y., Mitani, Y., Kikuchi, Y., 2019. Host-symbiont specificity determined by microbe–microbe competition in an insect gut. Proc. Natl. Acad. Sci. U. S. A. 116, 22673–22682.

Itoh, H., Matsuura, Y., Hosokawa, T., Fukatsu, T., Kikuchi, Y., 2016. Obligate gut symbiotic association in the sloe bug *Dolycoris baccarum* (Hemiptera: Pentatomidae). Appl. Entomol. Zool. 52, 51–59.

Itoh, H., Tago, K., Hayatsu, M., Kikuchi, Y., 2018b. Detoxifying symbiosis: microbe-mediated detoxification of phytotoxins and pesticides in insects. Nat. Prod. Rep. 35, 434–454.

Jarrell, K.F., McBride, M.J., 2008. The surprisingly diverse ways that prokaryotes move. Nat. Rev. Microbiol. 6, 466–476.

Jones, B.W., Nishiguchi, M.K., 2004. Counterillumination in the Hawaiian bobtail squid, *Euprymna scolopes* berry (Mollusca: Cephalopoda). Mar. Biol. 144, 1151–1155.

Jung, M., Lee, D.H., 2019. Abundance and diversity of gut-symbiotic bacteria, the genus *Burkholderia* in overwintering *Riptortus pedestris* (Hemiptera: Alydidae) populations and soil in South Korea. PLoS One 14, e0218240.

Kaiwa, N., Hosokawa, T., Nikoh, N., Tanahashi, M., Moriyama, M., Meng, X.Y., Maeda, T., Yamaguchi, K., Shigenobu, S., Ito, M., Fukatsu, T., 2014. Symbiont-supplemented maternal investment underpinning host's ecological adaptation. Curr. Biol. 24, 2465–2470.

Kaltenpoth, M., Flórez, L.V., 2019. Versatile and dynamic symbioses between insects and burkholderia bacteria. Annu. Rev. Entomol. https://doi.org/10.1146/annurev-ento-011019-025025. (Epub ahead of print).

Kaltenpoth, M., Gottler, W., Herzner, G., Strohm, E., 2005. Symbiotic bacteria protect wasp larvae from fungal infestation. Curr. Biol. 15, 475–479.

Kaltenpoth, M., Roeser-Mueller, K., Koehler, S., Peterson, A., Nechitaylo, T.Y., Stubblefield, J.W., Herzner, G., Seger, J., Strohm, E., 2014. Partner choice and fidelity stabilize coevolution in a cretaceous-age defensive symbiosis. Proc. Natl. Acad. Sci. U. S. A. 111, 6359–6364.

Kaltenpoth, M., Schmitt, T., Polidori, C., Koedam, D., Strohm, E., 2010. Symbiotic streptomycetes in antennal glands of the south American digger wasp genus *Trachypus* (Hymenoptera, Crabronidae). Physiol. Entomol. 35, 196–200.

Karamipour, N., Fathipour, Y., Mehrabadi, M., 2016. Gammaproteobacteria as essential primary symbionts in the striped shield bug, *Graphosoma lineatum* (Hemiptera: Pentatomidae). Sci. Rep. 6, 33168.

Kashkouli, M., Fathipour, Y., Mehrabadi, M., 2020. Habitat visualization, acquisition features and necessity of the gammaproteobacterial symbiont of pistachio stink bug, *Acrosternum heegeri* (Hem.: Pentatomidae). Bull. Entomol. Res. 110, 22–33.

Kashkouli, M., Fathipour, Y., Mehrabadi, M., 2019a. Heritable gammaproteobacterial symbiont improves the fitness of *Brachynema germari* Kolenati (Hemiptera: Pentatomidae). Environ. Entomol. 48, 1079–1087.

Kashkouli, M., Fathipour, Y., Mehrabadi, M., 2019b. Potential management tactics for pistachio stink bugs, *Brachynema germari*, *Acrosternum heegeri* and *Acrosternum arabicum* (hemiptera: Pentatomidae): high temperature and chemical surface sterilants leading to symbiont suppression. J. Econ. Entomol. 112, 244–254.
Kenyon, L.J., Meulia, T., Sabree, Z.L., 2015. Habitat visualization and genomic analysis of "*Candidatus Pantoea carbekii*," the primary symbiont of the brown marmorated stink bug. Genome Biol. Evol. 7, 620–635.
Kiers, E.T., Rousseau, R.A., West, S.A., Denison, R.F., 2003. Host sanctions and the legume-rhizobium mutualism. Nature 425, 78–81.
Kiers, E.T., West, S.A., 2015. Evolving new organisms via symbiosis. Science 348, 392–394.
Kikuchi, Y., 2009. Endosymbiotic bacteria in insects: their diversity and culturability. Microbes Environ. 24, 195–204.
Kikuchi, Y., Fukatsu, T., 2014. Live imaging of symbiosis: spatiotemporal infection dynamics of a GFP-labelled *Burkholderia* symbiont in the bean bug *Riptortus pedestris*. Mol. Ecol. 23, 1445–1456.
Kikuchi, Y., Yumoto, I., 2013. Efficient colonization of the bean bug *Riptortus pedestris* by an environmentally transmitted *Burkholderia* symbiont. Appl. Environ. Microbiol. 79, 2088–2091.
Kikuchi, Y., Hosokawa, T., Fukatsu, T., 2007. Insect-microbe mutualism without vertical transmission: a stinkbug acquires a beneficial gut symbiont from the environment every generation. Appl. Environ. Microbiol. 73, 4308–4316.
Kikuchi, Y., Hosokawa, T., Fukatsu, T., 2011a. An ancient but promiscuous host-symbiont association between *Burkholderia* gut symbionts and their heteropteran hosts. ISME J. 5, 446–460.
Kikuchi, Y., Hosokawa, T., Fukatsu, T., 2011b. Specific developmental window for establishment of an insect-microbe gut symbiosis. Appl. Environ. Microbiol. 77, 4075–4081.
Kikuchi, Y., Hosokawa, T., Nikoh, N., Fukatsu, T., 2012. Gut symbiotic bacteria in the cabbage bugs *Eurydema rugosa* and *Eurydema dominulus* (Heteroptera: Pentatomidae). Appl. Entomol. Zool. 47, 1–8.
Kikuchi, Y., Hosokawa, T., Nikoh, N., Meng, X.Y., Kamagata, Y., Fukatsu, T., 2009. Host-symbiont co-speciation and reductive genome evolution in gut symbiotic bacteria of acanthosomatid stinkbugs. BMC Biol. 7, 2.
Kikuchi, Y., Meng, X.Y., Fukatsu, T., 2005. Gut symbiotic bacteria of the genus *Burkholderia* in the broad-headed bugs *Riptortus clavatus* and *Leptocorisa chinensis* (Heteroptera: Alydidae). Appl. Environ. Microbiol. 71, 4035–4043.
Kikuchi, Y., Ohbayashi, T., Jang, S., Mergaert, P., 2020. *Burkholderia insecticola* triggers midgut closure in the bean bug *Riptortus pedestris* to prevent secondary bacterial infections of midgut crypts. ISME J. https://doi.org/10.1038/s41396-020-0633-3.
Kikuhara, Y., 2005. The Japanese species of the genus *Riptortus* (Heteroptera, Alydidae) with description of a new species. Jpn. J. Syst. Entomol. 11, 299–311.
Kim, J.K., Jang, H.A., Kim, M.S., Cho, J.H., Lee, J.B., Lorenzo, F.D., Sturiale, L., Silipo, A., Molinaro, A., Lee, B.L., 2017. The lipopolysaccharide core oligosaccharide of *Burkholderia* plays a critical role in maintaining a proper gut symbiosis with the bean bug *Riptortus pedestris*. J. Biol. Chem. 292, 19226–19237.
Kim, J.K., Jang, H.A., Won, Y.J., Kikuchi, Y., Heum Han, S., Kim, C.H., Nikoh, N., Fukatsu, T., Lee, B.L., 2014. Purine biosynthesis-deficient *Burkholderia* mutants are incapable of symbiotic accommodation in the stinkbug. ISME J. 8, 552–563.
Kim, J.K., Lee, H.J., Kikuchi, Y., Kitagawa, W., Nikoh, N., Fukatsu, T., Lee, B.L., 2013. Bacterial cell wall synthesis gene *uppP* is required for *Burkholderia* colonization of the stinkbug gut. Appl. Environ. Microbiol. 79, 4879–4886.

Kim, J.K., Park, H.Y., Lee, B.L., 2016. The symbiotic role of o-antigen of *Burkholderia* symbiont in association with host *Riptortus pedestris*. Dev. Comp. Immunol. 60, 202–208.

Kinosita, Y., Kikuchi, Y., Mikami, N., Nakane, D., Nishizaka, T., 2018. Unforeseen swimming and gliding mode of an insect gut symbiont, *Burkholderia* sp. Rpe64, with wrapping of the flagella around its cell body. ISME J. 12, 838–848.

Koch, H., Schmid-Hempel, P., 2011. Socially transmitted gut microbiota protect bumble bees against an intestinal parasite. Proc. Natl. Acad. Sci. U. S. A. 108, 19288–19292.

Koehler, S., Gaedeke, R., Thompson, C., Bongrand, C., Visick, K.L., Ruby, E., McFall-Ngai, M., 2019. The model squid-vibrio symbiosis provides a window into the impact of strain- and species-level differences during the initial stages of symbiont engagement. Environ. Microbiol. 21, 3269–3283.

Koga, R., Meng, X.Y., Tsuchida, T., Fukatsu, T., 2012. Cellular mechanism for selective vertical transmission of an obligate insect symbiont at the bacteriocyte-embryo interface. Proc. Natl. Acad. Sci. U. S. A. 109, E1230–E1237.

Kondorosi, E., Mergaert, P., Kereszt, A., 2013. A paradigm for endosymbiotic life: cell differentiation of rhizobium bacteria provoked by host plant factors. Annu. Rev. Microbiol. 67, 611–628.

Koropatnick, T.A., Engle, J.T., Apicella, M.A., Stabb, E.V., Goldman, W.E., McFall-Ngai, M.J., 2004. Microbial factor-mediated development in a host-bacterial mutualism. Science 306, 1186–1188.

Krasity, B.C., Troll, J.V., Weiss, J.P., McFall-Ngai, M.J., 2011. LBP/BPI proteins and their relatives: conservation over evolution and roles in mutualism. Biochem. Soc. Trans. 39, 1039–1044.

Kroiss, J., Kaltenpoth, M., Schneider, B., Schwinger, M.G., Hertweck, C., Maddula, R.K., Strohm, E., Svatos, A., 2010. Symbiotic streptomycetes provide antibiotic combination prophylaxis for wasp offspring. Nat. Chem. Biol. 6, 261–263.

Kuechler, S.M., Fukatsu, T., Matsuura, Y., 2019. Repeated evolution of bacteriocytes in lygaeoid stinkbugs. Environ. Microbiol. 21, 4378–4394.

Kuechler, S.M., Matsuura, Y., Dettner, K., Kikuchi, Y., 2016. Phylogenetically diverse *Burkholderia* associated with midgut crypts of spurge bugs, *Dicranocephalus* spp. (Heteroptera: Stenocephalidae). Microbes Environ. 31, 145–153.

Kuhn, M.J., Schmidt, F.K., Eckhardt, B., Thormann, K.M., 2017. Bacteria exploit a polymorphic instability of the flagellar filament to escape from traps. Proc. Natl. Acad. Sci. U. S. A. 114, 6340–6345.

Lanan, M.C., Rodrigues, P.A., Agellon, A., Jansma, P., Wheeler, D.E., 2016. A bacterial filter protects and structures the gut microbiome of an insect. ISME J. 10, 1866–1876.

Login, F.H., Balmand, S., Vallier, A., Vincent-Monegat, C., Vigneron, A., Weiss-Gayet, M., Rochat, D., Heddi, A., 2011. Antimicrobial peptides keep insect endosymbionts under control. Science 334, 362–365.

Mahenthiralingam, E., Urban, T.A., Goldberg, J.B., 2005. The multifarious, multireplicon *Burkholderia* cepacia complex. Nat. Rev. Microbiol. 3, 144–156.

Maire, J., Vincent-Monegat, C., Balmand, S., Vallier, A., Herve, M., Masson, F., Parisot, N., Vigneron, A., Anselme, C., Perrin, J., Orlans, J., Rahioui, I., Da Silva, P., Fauvarque, M.O., Mengin-Lecreulx, D., Zaidman-Remy, A., Heddi, A., 2019. Weevil *pgrp-lb* prevents endosymbiont TCT dissemination and chronic host systemic immune activation. Proc. Natl. Acad. Sci. U. S. A. 116, 5623–5632.

Mandel, M.J., Schaefer, A.L., Brennan, C.A., Heath-Heckman, E.A., Deloney-Marino, C.R., McFall-Ngai, M.J., Ruby, E.G., 2012. Squid-derived chitin oligosaccharides are a chemotactic signal during colonization by *Vibrio fischeri*. Appl. Environ. Microbiol. 78, 4620–4626.

Matsuura, Y., Kikuchi, Y., Miura, T., Fukatsu, T., 2015. Ultrabithorax is essential for bacteriocyte development. Proc. Natl. Acad. Sci. U. S. A. 112, 9376–9381.
Matsuura, Y., Moriyama, M., Łukasik, P., Vanderpool, D., Tanahashi, M., Meng, X.Y., McCutcheon, J.P., Fukatsu, T., 2018. Recurrent symbiont recruitment from fungal parasites in cicadas. Proc. Natl. Acad. Sci. U. S. A. 115, E5970–E5979.
McCutcheon, J.P., Moran, N.A., 2012. Extreme genome reduction in symbiotic bacteria. Nat. Rev. Microbiol. 10, 13–26.
McFall-Ngai, M.J., 2014. The importance of microbes in animal development: lessons from the squid-vibrio symbiosis. Annu. Rev. Microbiol. 68, 177–194.
McLoughlin, K., Schluter, J., Rakoff-Nahoum, S., Smith, A.L., Foster, K.R., 2016. Host selection of microbiota via differential adhesion. Cell Host Microbe 19, 550–559.
Mergaert, P., 2018. Role of antimicrobial peptides in controlling symbiotic bacterial populations. Nat. Prod. Rep. 35, 336–356.
Millikan, D.S., Ruby, E.G., 2003. FlrA, a 54-dependent transcriptional activator in *Vibrio fischeri*, is required for motility and symbiotic light-organ colonization. J. Bacteriol. 185, 3547–3557.
Millikan, D.S., Ruby, E.G., 2004. *Vibrio fischeri* flagellin A is essential for normal motility and for symbiotic competence during initial squid light organ colonization. J. Bacteriol. 186, 4315–4325.
Miyamoto, S., 1961. Comparative morphology of alimentary organs of Heteroptera, with the phylogenetic consideration. Sieboldia 2, 197–259.
Moran, N.A., Bennett, G.M., 2014. The tiniest tiny genomes. Annu. Rev. Microbiol. 68, 195–215.
Moran, N.A., McCutcheon, J.P., Nakabachi, A., 2008. Genomics and evolution of heritable bacterial symbionts. Annu. Rev. Genet. 42, 165–190.
Moran, N.A., Russell, J.A., Koga, R., Fukatsu, T., 2005. Evolutionary relationships of three new species of Enterobacteriaceae living as symbionts of aphids and other insects. Appl. Environ. Microbiol. 71, 3302–3310.
Moya, A., Pereto, J., Gil, R., Latorre, A., 2008. Learning how to live together: genomic insights into prokaryote-animal symbioses. Nat. Rev. Genet. 9, 218–229.
Nyholm, S.V., Graf, J., 2012. Knowing your friends: invertebrate innate immunity fosters beneficial bacterial symbioses. Nat. Rev. Microbiol. 10, 815–827.
Nyholm, S.V., McFall-Ngai, M., 2004. The winnowing: establishing the squid-*Vibrio* symbiosis. Nat. Rev. Microbiol. 2, 632–642.
Nyholm, S.V., Stabb, E.V., Ruby, E.G., McFall-Ngai, M.J., 2000. Establishment of an animal-bacterial association: recruiting symbiotic vibrios from the environment. Proc. Natl. Acad. Sci. U. S. A. 97, 10231–10235.
Ohbayashi, T., Futahashi, R., Terashima, M., Barriere, Q., Lamouche, F., Takeshita, K., Meng, X.Y., Mitani, Y., Sone, T., Shigenobu, S., Fukatsu, T., Mergaert, P., Kikuchi, Y., 2019a. Comparative cytology, physiology and transcriptomics of *Burkholderia insecticola* in symbiosis with the bean bug *Riptortus pedestris* and in culture. ISME J. 13, 1469–1483.
Ohbayashi, T., Itoh, H., Lachat, J., Kikuchi, Y., Mergaert, P., 2019b. *Burkholderia* gut symbionts associated with European and Japanese populations of the dock bug *Coreus Marginatus* (Coreoidea: Coreidae). Microbes Environ. 34, 219–222.
Ohbayashi, T., Takeshita, K., Kitagawa, W., Nikoh, N., Koga, R., Meng, X.Y., Tago, K., Hori, T., Hayatsu, M., Asano, K., Kamagata, Y., Lee, B.L., Fukatsu, T., Kikuchi, Y., 2015. Insect's intestinal organ for symbiont sorting. Proc. Natl. Acad. Sci. U. S. A. 112, E5179–E5188.
Ohkuma, M., Brune, A., 2010. Diversity, structure, and evolution of the termite gut microbial community. In: Bignell, D.E., Roisin, Y., Lo, N. (Eds.), Biology of Termites: A Modern Synthesis. Springer Netherlands, Dordrecht, pp. 413–438.

Oliver, K.M., Degnan, P.H., Burke, G.R., Moran, N.A., 2010. Facultative symbionts in aphids and the horizontal transfer of ecologically important traits. Annu. Rev. Entomol. 55, 247–266.

Olivier-Espejel, S., Sabree, Z.L., Noge, K., Becerra, J.X., 2011. Gut microbiota in nymph and adults of the giant mesquite bug (*Thasus neocalifornicus*) (Heteroptera: Coreidae) is dominated by *Burkholderia* acquired de novo every generation. Environ. Entomol. 40, 1102–1110.

Onchuru, T.O., Javier Martinez, A., Ingham, C.S., Kaltenpoth, M., 2018. Transmission of mutualistic bacteria in social and gregarious insects. Curr. Opin. Insect Sci. 28, 50–58.

Peeters, C., Meier-Kolthoff, J.P., Verheyde, B., De Brandt, E., Cooper, V.S., Vandamme, P., 2016. Phylogenomic study of *Burkholderia glathei*-like organisms, proposal of 13 novel *Burkholderia* species and emended descriptions of *Burkholderia sordidicola*, *Burkholderia zhejiangensis*, and *Burkholderia grimmiae*. Front. Microbiol. 7, 877.

Perlmutter, J.I., Bordenstein, S.R., 2020. Microorganisms in the reproductive tissues of arthropods. Nat. Rev. Microbiol. 18, 97–111.

Poole, P., Ramachandran, V., Terpolilli, J., 2018. Rhizobia: from saprophytes to endosymbionts. Nat. Rev. Microbiol. 16, 291–303.

Powell, J.E., Martinson, V.G., Urban-Mead, K., Moran, N.A., 2014. Routes of acquisition of the gut microbiota of the honey bee *Apis mellifera*. Appl. Environ. Microbiol. 80, 7378–7387.

Prado, S.S., Golden, M., Follett, P.A., Daugherty, M.P., Almeida, R.P., 2009. Demography of gut symbiotic and aposymbiotic *Nezara viridula* L. (Hemiptera: Pentatomidae). Environ. Entomol. 38, 103–109.

Prado, S.S., Rubinoff, D., Almeida, R.P.P., 2006. Vertical transmission of a pentatomid caeca-associated symbiont. Ann. Entomol. Soc. Am. 99, 577–585.

Russell, A.B., Peterson, S.B., Mougous, J.D., 2014. Type VI secretion system effectors: poisons with a purpose. Nat. Rev. Microbiol. 12, 137–148.

Sachs, J.L., Mueller, U.G., Wilcox, T.P., Bull, J.J., 2004. The quarterly review of biology. Q. Rev. Biol. 79, 135–160.

Sachs, J.L., Skophammer, R.G., Regus, J.U., 2011. Evolutionary transitions in bacterial symbiosis. Proc. Natl. Acad. Sci. U. S. A. 108 (Suppl. 2), 10800–10807.

Salem, H., Bauer, E., Kirsch, R., Berasategui, A., Cripps, M., Weiss, B., Koga, R., Fukumori, K., Vogel, H., Fukatsu, T., Kaltenpoth, M., 2017. Drastic genome reduction in an herbivore's pectinolytic symbiont. Cell 171 (1520–1531), e1513.

Salem, H., Florez, L., Gerardo, N., Kaltenpoth, M., 2015. An out-of-body experience: the extracellular dimension for the transmission of mutualistic bacteria in insects. Proc. Biol. Sci. 282, 20142957.

Sangare, A.K., Rolain, J.M., Gaudart, J., Weber, P., Raoult, D., 2016. Synergistic activity of antibiotics combined with ivermectin to kill body lice. Int. J. Antimicrob. Agents 47, 217–223.

Schaefer, C.W., Panizzi, A.R., 2000. Heteroptera of Economic Importance. CRC Press, Boca Raton.

Schlein, Y., 1977. Lethal effect of tetracycline on tsetse flies following damage to bacterioid symbionts. Experientia 33, 450–451.

Shibata, T.F., Maeda, T., Nikoh, N., Yamaguchi, K., Oshima, K., Hattori, M., Nishiyama, T., Hasebe, M., Fukatsu, T., Kikuchi, Y., Shigenobu, S., 2013. Complete genome sequence of *Burkholderia* sp. strain RPE64, bacterial symbiont of the bean bug *Riptortus pedestris*. Genome Announc. 1, e00441-13.

Small, A.L., McFall-Ngai, M.J., 1999. Halide peroxidase in tissues that interact with bacteria in the host squid *Euprymna scolopes*. J. Cell. Biochem. 72, 445–457.

Speare, L., Cecere, A.G., Guckes, K.R., Smith, S., Wollenberg, M.S., Mandel, M.J., Miyashiro, T., Septer, A.N., 2018. Bacterial symbionts use a type VI secretion system to eliminate competitors in their natural host. Proc. Natl. Acad. Sci. U. S. A. 115, E8528–E8537.

Srivastava, P., Auclair, J., 1976. Effects of antibiotics on feeding and development of the pea aphid, *Acyrthosiphon pisum* (Harris) (Homoptera: Aphididae). Can. J. Zool. 54, 1025–1029.

Steele, M.I., Kwong, W.K., Whiteley, M., Moran, N.A., 2017. Diversification of type VI secretion system toxins reveals ancient antagonism among bee gut microbes. mBio 8, e01630-17.

Strand, M.R., 2018. Composition and functional roles of the gut microbiota in mosquitoes. Curr. Opin. Insect. Sci. 28, 59–65.

Suárez-Moreno, Z.R., Caballero-Mellado, J., Coutinho, B.G., Mendonca-Previato, L., James, E.K., Venturi, V., 2012. Common features of environmental and potentially beneficial plant-associated *Burkholderia*. Microb. Ecol. 63, 249–266.

Sudakaran, S., Kost, C., Kaltenpoth, M., 2017. Symbiont acquisition and replacement as a source of ecological innovation. Trends Microbiol. 25, 375–390.

Sudakaran, S., Retz, F., Kikuchi, Y., Kost, C., Kaltenpoth, M., 2015. Evolutionary transition in symbiotic syndromes enabled diversification of phytophagous insects on an imbalanced diet. ISME J. 9, 2587–2604.

Szklarzewicz, T., Michalik, A., 2017. Transovarial transmission of symbionts in insects. In: Kloc, M. (Ed.), vol. 63. Oocytes. Results and problems in cell differentiation. Springer International Publishing, Cham, Switzerland, pp. 43–67.

Tada, A., Kikuchi, Y., Hosokawa, T., Musolin, D.L., Fujisaki, K., Fukatsu, T., 2011. Obligate association with gut bacterial symbiont in Japanese populations of the southern green stinkbug *Nezara viridula* (Heteroptera: Pentatomidae). Appl. Entomol. Zool. 46, 483–488.

Takeshita, K., Kikuchi, Y., 2017. *Riptortus pedestris* and *Burkholderia* symbiont: an ideal model system for insect-microbe symbiotic associations. Res. Microbiol. 168, 175–187.

Takeshita, K., Matsuura, Y., Itoh, H., Navarro, R., Hori, T., Sone, T., Kamagata, Y., Mergaert, P., Kikuchi, Y., 2015. *Burkholderia* of plant-beneficial group are symbiotically associated with bordered plant bugs (Heteroptera: Pyrrhocoroidea: Largidae). Microbes Environ. 30, 321–329.

Takeshita, K., Shibata, T.F., Nikoh, N., Nishiyama, T., Hasebe, M., Fukatsu, T., Shigenobu, S., Kikuchi, Y., 2014. Whole-genome sequence of *Burkholderia* sp. strain RPE67, a bacterial gut symbiont of the bean bug *Riptortus pedestris*. Genome Announc. 2, e00556-14.

Takeshita, K., Tamaki, H., Ohbayashi, T., Meng, X.-Y., Sone, T., Mitani, Y., Peeters, C., Kikuchi, Y., Vandamme, P., 2018. *Burkholderia insecticola* sp. nov., a gut symbiotic bacterium of the bean bug *Riptortus pedestris*. Int. J. Syst. Evol. Microbiol. 68, 2370–2374.

Taylor, C.M., Coffey, P.L., DeLay, B.D., Dively, G.P., 2014. The importance of gut symbionts in the development of the brown marmorated stink bug, *Halyomorpha halys* (Stal). PLoS One 9, e90312.

Troll, J.V., Adin, D.M., Wier, A.M., Paquette, N., Silverman, N., Goldman, W.E., Stadermann, F.J., Stabb, E.V., McFall-Ngai, M.J., 2009. Peptidoglycan induces loss of a nuclear peptidoglycan recognition protein during host tissue development in a beneficial animal-bacterial symbiosis. Cell. Microbiol. 11, 1114–1127.

Troll, J.V., Bent, E.H., Pacquette, N., Wier, A.M., Goldman, W.E., Silverman, N., McFall-Ngai, M.J., 2010. Taming the symbiont for coexistence: a host PGRP neutralizes a bacterial symbiont toxin. Environ. Microbiol. 12, 2190–2203.

Visick, K.L., Ruby, E.G., 2006. *Vibrio fischeri* and its host: it takes two to tango. Curr. Opin. Microbiol. 9, 632–638.

Wang, Q., Liu, J., Zhu, H., 2018. Genetic and molecular mechanisms underlying symbiotic specificity in legume-rhizobium interactions. Front. Plant Sci. 9, 313.

Weis, V.M., Small, A.L., McFall-Ngai, M.J., 1996. A peroxidase related to the mammalian antimicrobial protein myeloperoxidase in the *Euprymna-Vibrio* mutualism. Proc. Natl. Acad. Sci. U. S. A. 93, 13683–13688.

Xu, Y., Buss, E.A., Boucias, D.G., 2016. Environmental transmission of the gut symbiont *Burkholderia* to phloem-feeding *Blissus insularis*. PLoS One 11, e0161699.

CHAPTER THREE

Symbiont-mediated degradation of dietary carbon sources in social herbivorous insects

John T. Wertz[a,*], Benoît Béchade[b]
[a]Department of Biology, Calvin University, Grand Rapids, MI, United States
[b]Department of Biology, Drexel University, Philadelphia, PA, United States
*Corresponding author: e-mail address: jwertz@calvin.edu

Contents

Abstract

In this chapter we present the current state of research on the contribution of the gut microbiome to the degradation of dietary sources of carbon in three social insects: termites, honeybees and herbivorous ants. Insects from these groups can face similar dietary challenges, and an emerging body of research illustrates that their gut microbes

Advances in Insect Physiology, Volume 58
ISSN 0065-2806
https://doi.org/10.1016/bs.aiip.2020.04.001

have provided symbiotic solutions for the degradation of a variety of carbon sources. Termites, particularly 'lower' termites have a diet enriched in perhaps the most recalcitrant molecule found in the diet of social insects—lignocellulose. Lower termites have a diverse (>400 species/gut) community of protists and bacteria that work together with endogenous host cellulases to provision acetate to the host. Honeybees, feeding on pollen, pollen-rich beebread and nectar, have a relatively simple gut microbiome (8–10 bacterial species). Recently it has been demonstrated that some of these bacteria possess pectinases and an assortment of other sugar- and flavonoid-degrading enzymes and may provision the host with several molecules, including butyrate. In leaf-cutter ants, feeding on their fungal cultivar, bacteria may be involved in the depolymerization of chitin and fermentation of citrate to acetate. In *Cephalotes* turtle ants, metagenomic studies have shown that some gut bacteria encode genes for the degradation of recalcitrant polysaccharides found in pollen and, possibly, other foods from their generally cryptic diets. Across these social insect hosts, faithful symbioses with specialized microbes were enabled by modifications to social behaviour that facilitated faithful spread of microbes within colonies and across generations.

1. Introduction

Social insects are faced with a paramount challenge: how to efficiently feed non-foraging counterparts while fuelling the activity of foragers. The energy requirements of foragers can be met by assimilating adequate amounts of carbohydrates (Paoli et al., 2014; Saran and Rust, 2005; Shik et al., 2018). At the colony level however, the brood and non-foraging caste members also require a substantial amount of dietary carbon (Dussutour and Simpson, 2008; Poissonnier et al., 2020; Shik et al., 2018; Wright et al., 2018). Therefore, in social insects, high levels of carbohydrate intake support entire colony fitness (Dussutour and Simpson, 2009; Paoli et al., 2014; Wills et al., 2015; Zhou et al., 2008). Yet, providing carbohydrates to entire colonies can be hampered by plant-based diets. In such diets, carbon is often found within complex and difficult to degrade molecules such as lignocellulose and pectins, while freely available forms of carbon may be in insufficient supply to meet the energy needs of the colony.

Many herbivores employ symbiotic microbes to outsource the digestion of plant carbohydrates (Douglas, 2013; Söllinger et al., 2018). These organisms typically use a set of depolymerizing enzymes that can be categorized into three groups: (i) enzymes that debranch side chains from the backbone of a polymer (such as hemicelluloses, resistant starch or complex pectins) producing two separate polymeric chains, (ii) enzymes cutting internal bonds in the backbone of a polymer using endoglycosidases and producing oligosaccharides (2–10 monomers), and (iii) enzymes cleaving polymers at their extremities, through use of exoglycosidases, enabling the release of monosaccharides (i.e. simple carbohydrates). The simple carbohydrates produced in these reactions can either be directly assimilated by the animal host to be stored or converted to energy (e.g. González-Peña et al., 2002) or used by gut symbionts, and released as waste products (e.g. organic acids) of potential energy value to hosts (e.g. Odelson and Breznak, 1983). In termites, ruminants, and humans, fermentation can occur in anoxic gut regions and it has been shown that host cells have adapted to using the products of microbial fermentation, mostly short-chain fatty acids (SCFAs), as a primary energy source (den Besten et al., 2013; Donohoe et al., 2011; Odelson and Breznak, 1983). We similarly propose that microbially-produced SCFAs are used as important carbon and energy sources in eusocial corbiculate bees and a subset of herbivorous ants.

Given the substantial role that microbial symbionts can play in plant-carbohydrate digestion, animal hosts that retain them through generations (vertical transmission), and those sharing them among relatives within colonies (social transmission), have likely been favoured by natural selection (Onchuru et al., 2018; Sinotte et al., 2020; Wills et al., 2015). And while C-metabolism may not be the sole means of symbiont benefits, social insects with useful C-metabolizing microbes have evolved stereotyped social transmission mechanisms. These usually differ from mechanisms used by non-social insects (reviewed in Salem et al., 2015), and regularly include anal-oral (also termed abdominal or proctodeal) trophallaxis, a behaviour in which microbes are passed from siblings to their younger nestmates (Kwong et al., 2017; Nalepa, 2015; Onchuru et al., 2018; Salem et al., 2015). The combined effects of social and vertical transmission, which have evolved independently in termites, honeybees and some ant clades, promote host-symbiont fidelity, aligning host and microbe interests around colony survival and reproduction. Such high reliability in the transmission of gut microbes enables the use of niches (e.g. difficult to degrade and/or nitrogen poor food sources) that are inhospitable to other insects (Kwong and Moran, 2016a).

In this chapter, we capitalize on a glut of recent research to review such symbioses. While fascinating examples of C-metabolic-centred symbioses can be found in other social insects (e.g. ambrosia beetles), we focus on plant-carbohydrate digestion and carbon provisioning by microbes of three charismatic social insects, each with high ecological impacts across diverse terrestrial ecosystems: (1) the lower termites, (2) honeybees and (3) herbivorous ants.

2. Termites

For over 130 years, termites and their symbiotic gut microbes have been the focus of scientific study. Undoubtedly, this started with two observations—their uncommon ability to thrive on a diet rich in woody material, and the ubiquitous presence of 'parasites' in their gut (Leidy, 1881). While the contributions by the gut microbes to termite vitality are less mysterious today than in the 19th century, they are no less astonishing.

Termites can be categorized into two groups. Lower termites comprise the first. This polyphyletic grouping includes all but one termite family. Members feed on sound or partially sound wood, and harbour bacterial, archaeal, and protist symbionts within their hindguts (Ohkuma and Brune, 2011). The second group, the higher termites (Termitidae), comprise an ecologically diverse clade exhibiting wood, soil, litter, or fungus diets. Their guts differ from those of lower termites in being highly compartmentalized and in lacking protists (Ohkuma and Brune, 2011). Here we focus on the lower termites, as they are the most well-studied.

2.1 Diet

Together with their gut-associated microbial symbionts, termites have evolved to thrive on the most abundant form of biomass on the planet, 'lignocellulose' (lignins, cellulose, and hemicelluloses)—the primary components of wood (Fig. 1) and other terrestrial plant material. Lignocellulose is approximately 20–50% cellulose, 15–35% hemicelluloses, and 18–35% lignins (Ni and Tokuda, 2013). Cellulolytic enzymes that can degrade the former polysaccharide typically fall into two groups: *endo*-β-1,4-glucanase (i.e. endoglycosidase) and exoglucanases like cellodextrinases and/or cellobiohydrolases (i.e. exoglycosidases). The former will degrade cellulose in a random fashion, the latter will attack the chains from the ends. The resulting carbohydrates—typically the di-saccharide cellobiose—will then be cleaved into two molecules of glucose by β-glucosidase.

Hemicelluloses are more amorphous than cellulose (Fig. 1), consisting of long chains of xylose, or a combination of xylose, mannose and glucose linked by β-1,4-glycosidic bonds. Hemicelluloses often contain short branches of arabinose or galactose, but there is a high degree of variability in their structures. Hence an organism wishing to degrade these polysaccharides must possess an arsenal of enzymes, at the very least xylanases (i.e. endoglycosidases) and β-xylosidases (i.e. exoglycosidases) (Juturu and Wu, 2012; Ni and Tokuda, 2013; Saha, 2003).

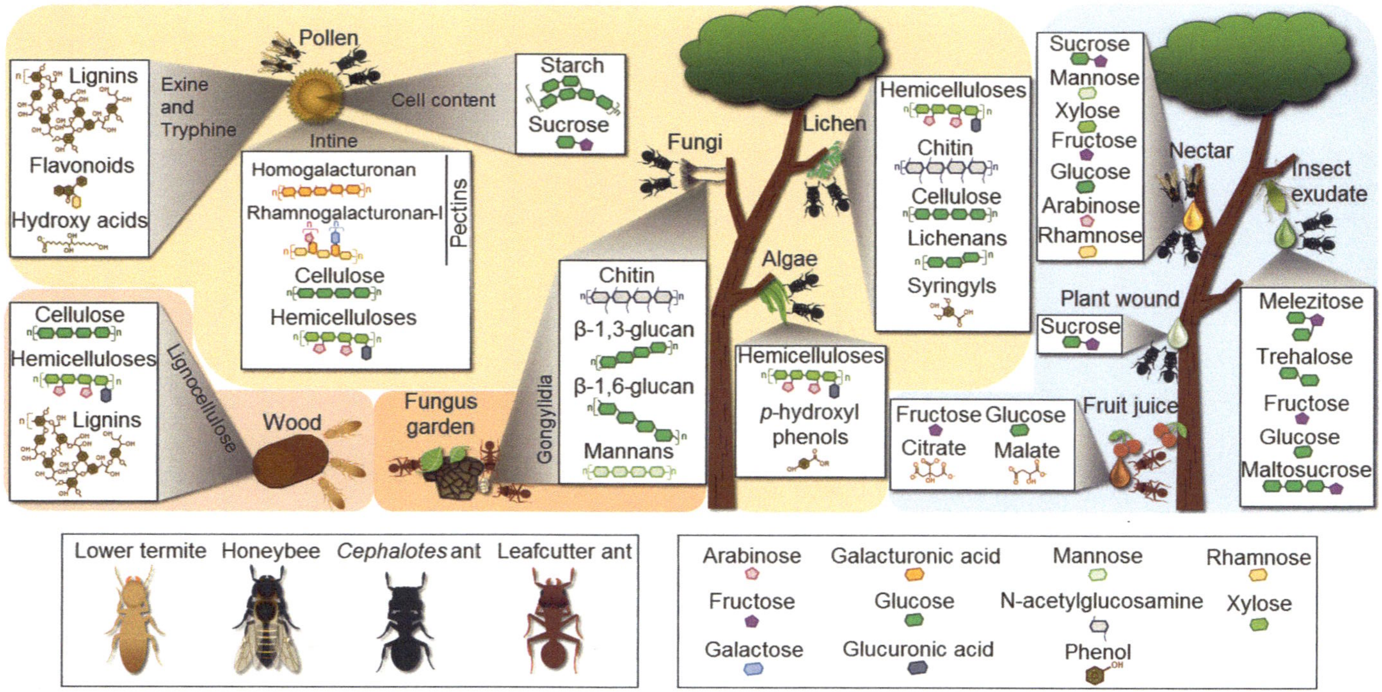

Fig. 1 Carbohydrate and phenol contents of food sources consumed by lower termites, honeybees, *Cephalotes* ants and leaf-cutter ants. The pastel orange background colour corresponds to wood mainly made with lignocellulose, a highly recalcitrant composite polymer, consumed by lower termites. The coral-orange background colour corresponds to fungivory with a farmed fungus being a digestive symbiont of leaf-cutter ants. The cream-yellow background colour corresponds to food sources collected by plant-surface foragers (e.g. pollen, free-living fungi, lichen and non-vascular plants including algae, mosses and liverworts). The pastel blue background colour corresponds to food sources in an exudate diet (e.g. floral and extrafloral nectar, sap-sucking insect exudate, plant wound secretions and fruit juice). Insect cartoons next to a food item indicate which social insect has been recorded to collect it. The list of carbohydrate and phenol contents for each food source is not exhaustive. For polymorphic compounds (e.g. lignins, flavonoids, hydroxy acids, hemicelluloses, pectins, lichenans, monophenols), only one or two common types are shown here. A list of the monomers composing the aforementioned oligomers and polymers can be found at the bottom.

Lignins are incredibly complex phenolic polymers (Fig. 1). They lack a defined primary structure, but their phenylpropane-based monomers are all hydroxylated or methylated (methoxyphenylpropane or hydroxyphenylpropane). Lignins are notoriously resistant to enzymatic digestion by any domain of life except fungi and some bacteria encoding peroxidases and/or laccases that can both depolymerize lignins (Pollegioni et al., 2015).

As difficult as these complex molecules are to degrade on their own, the intertwined, three-dimensional nature of lignocellulose makes decomposition a seemingly herculean task (Breznak and Brune, 1994; Janusz et al., 2017). The hydrophobic cellulose microfibrils are only accessible to exoglucanases at the ends of the fibres because hemicelluloses almost entirely encase cellulose fibres, preventing endoglucanase action (Brune, 2014). Lignins then surround and infiltrate the cellulose/hemicellulose formations in a 'chicken wire' matrix of polyphenols. Despite this, termites are the most rapid and efficient lignocellulose mineralizers known, capable of removing 74–99% of the cellulose and 65–87% of the hemicellulose components (Brune, 2014). This remarkably high efficiency of mineralization requires both host- and microbiota-derived processes.

2.2 Diversity of lower termite gut microbiota

2.2.1 Protists

A defining characteristic of lower termites is the presence of eukaryotic protists in the hindgut. The vast majority are represented by highly flagellated parabasalids (Metamonada: Trichozoa) (Ohkuma and Brune, 2011). These amitochondriate eukaryotes can range from approximately 10 μm in length with six or fewer flagella to >100 μm in length covered in ~50,000 flagella (James et al., 2013; Radek et al., 2014). Many lower termites also harbour oxymonads (Metamonada: Anaeromonadea) some of which are attached to the gut wall (Tamschick and Radek, 2013). Most likely this is to prevent washout but is notable considering that the protists are killed upon exposure to oxygen (Cleveland, 1925a) and microelectrode measurements have detected oxygen in the gut periphery 150–200 μm into the hindgut paunch (see discussion of the importance of oxygen below) (Brune, 1998; Brune et al., 1995a). The parabasalid and oxymonad protists are difficult to maintain in culture, making this method of determining the exact number of species of each group very difficult (Odelson and Breznak, 1985). Molecular methods such as 18S ribosomal RNA (rRNA) gene sequencing and metagenomics suggest that the species pool may be much greater than originally recognized (Brune and Deitrich, 2015; Noda et al., 2012).

2.2.2 Bacteria and archaea

Most of the volume and surface area found within the termite gut is comprised by cytoplasm and outer cell membranes of the symbiotic protists. Hence, the majority of bacteria live as ecto-, *endo-* and intranuclear symbionts of protists (Breznak and Prankratz, 1977; Brune and Deitrich, 2015; Sato et al., 2014). The free-living luminal community, as well as the epithelial-associated biofilm community, is large and diverse (Breznak and Prankratz, 1977; Ohkuma and Brune, 2011; Wertz and Breznak, 2007a). Only recently through high-throughput sequencing has the full scope of prokaryotic diversity in the termite gut been revealed. Analysis of 14,327 16S rRNA sequences from *Reticulitermes flaviceps* led to the conclusion that these termites may harbour 1254 bacterial phylotypes (using the criterion of 3% 16S rRNA sequence difference to separate phylotypes) (Su et al., 2016). Consistently dominant in the lower termites are members of the bacterial genus *Treponema* within the Spirochaetes. Also consistently found are Bacteroidetes, including *Dysgonomonas* and *Candidatus* Azobacteroides; Proteobacteria, such as *Desulfovibrio, Citrobacter, Enterobacter* and Endomicrobia (*Elusimicrobia*); and members of the minor candidate phyla TG3 and TG5 (Brune and Deitrich, 2015; Mikaelyan et al., 2017; Su et al., 2016). Archaea in lower termites appear to be mostly gut wall-associated methanogenic *Methanobrevibacter* (Leadbetter and Breznak, 1996; Shi et al., 2015b). Although progress has been made, most termite gut bacteria have yet to be cultivated.

2.3 Digestion of cellulose and hemicelluloses

2.3.1 Protists

Early on it was recognized that the 'parasites' described by Leidy (1881) could have an important role in the digestion of lignocellulose. Though not the first to propose the idea, Cleveland was the first to systematically study it (Cleveland, 1923, 1925b). By incubating termites at 36 °C, Cleveland could effectively remove hindgut protists without causing immediate death to the host. These 'defaunated' termites could then be used for feeding experiments (Cleveland, 1923, 1925a). He noted that when defaunated termites were fed wood, they died within 3–4 weeks. However, if fed fungus-digested cellulose or humus, they displayed a normal phenotype for at least 4 months. In lab-constructed colonies consisting of equal parts defaunated and non-defaunated termites fed a diet of pure cellulose, the termites showed no noticeable difference in behaviour or colony longevity vs. colonies with 100% non-defaunated insects. Microscopic observation of the guts from the previously defaunated termites revealed the presence of protists—which Cleveland correctly surmised were transmitted by the social behaviour of the insects (anal-oral trophallaxis). Hence, Cleveland provided strong evidence for the large role of protists in the digestion of lignocellulose by termites, as well as direct evidence of social transfer of gut microbes within the colony (Cleveland, 1923, 1925b, 1926). Because defaunated termites died when fed cellulose, and no bacteria could be recovered when non-defaunated termite gut contents were plated on media containing cellulose as the sole carbon source, Cleveland concluded that neither the insect itself nor its gut bacteria were involved in cellulose depolymerization.

2.3.2 The discovery of endogenous cellulases

The question of whether termites synthesized endogenous cellulases to complement those encoded by protists lingered until it was demonstrated that removal of the gut fauna from the lower termite *Coptotermes lacteus* led to loss of cellulase activity in the hindgut, but not the foregut or midgut (O'Brien et al., 1979). In fact, it was calculated that 30–50% of the cellulolytic activity remained in termites from which all gut bacteria and protists had been removed. This suggested that lower termites may be able to produce cellulases in their fore- and midgut tissues to supplement the activity of cellulases in the protists. The use of gene cloning, tissue localization, metagenomic and metatranscriptomic techniques, among others, has confirmed O'Brien's hypothesis that lower termites express endogenous cellulases that work in tandem with those from protists to efficiently degrade cellulose (Feng et al., 2015; Nakashima et al., 2002; Ni and Tokuda, 2013; Tokuda et al., 1999; Watanabe et al., 1998; Wu et al., 2019). Endogenous cellulose degradation is facilitated by a reduction in particle size to less than 20 μm via mastication immediately upon ingestion (Fujita et al., 2010), with some of it degraded into an amorphous state. The mechanical disruption of the cellulose fibres enhances the access of *endo*-β-1,4-glucanases produced in the salivary glands. Also produced in the salivary glands are β-glucosidases. There is little absorption of the degradation products in the foregut due to its thick cuticle (Noirot and Noirot-Thimothee, 1969), but once in the midgut, degradation products such as glucose are readily assimilated and metabolized through the glycolytic and tricarboxylic acid pathways (Fig. 2) (Tokuda et al., 2014).

The importance of endogenous cellulase activity upon the health of the host was demonstrated when researchers used a high-dose double-stranded (ds) RNA feeding approach to silence several termite genes (Zhou et al., 2008). One set of (ds)RNA was designed to complement an endogenous

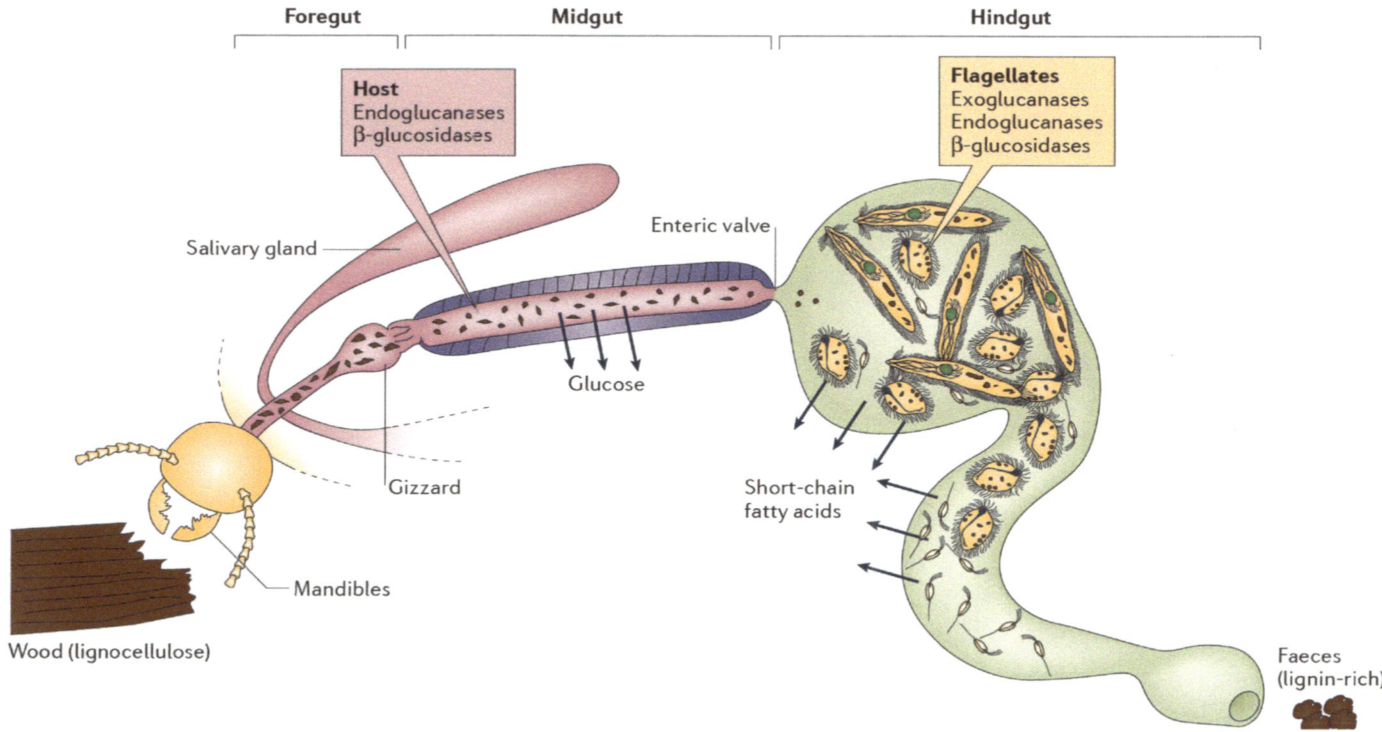

Fig. 2 Digestion of lignocellulose in termites requires both host- and microbe-derived functions. The host will grind the woody material into small pieces, which are enacted upon by host-derived endoglucanases and β-galactosidases. Resulting glucose molecules are free to pass through the foregut epithelium. The material then passes into the hindgut paunch, where it is endocytosed by gut flagellates. The flagellates and perhaps protist-associated bacteria express exoglucanases, endoglucanases and β-glucosidases which free glucose from the cellulose microfibers. The sugar is fermented by the flagellates and some bacteria. The short-chain fatty acid products (mostly acetate) are absorbed by the host and used for growth and reproduction. Most of the lignins are evacuated in faeces. *Figure from Brune, A., 2014. Symbiotic digestion of lignocellulose in termite guts. Nat. Rev.® Microbiol. 12, 168–180. Used with permission.*

digestive cellulase *Cell*-1. The (ds)RNA was then fed to termites via paper disks containing nile blue to track the food as it progressed through the digestive tract. Every 8 days the weight and mortality of termites was assessed. Termite colonies fed the *Cell*-1 (ds)RNA experienced a significantly (25–50%) higher mortality than controls after 24 days (Zhou et al., 2008).

2.3.3 Degradation of cellulose/hemicelluloses by protists

In the anterior hindgut (paunch), a significant increase in cellulose and hemicellulose hydrolysis occurs. Here, these particles are phagocytosed by the parabasalid protists (Fig. 2) (Kitade et al., 2012; Noda et al., 2012; Ohkuma and Brune, 2011). Numerous and detailed studies, enhanced by the ability to grow a few protists as axenic cultures (i.e. isolated microbial cultures), showed that the protists could indeed break down cellulose and hemicelluloses, with the primary end products being acetate, carbon dioxide, and hydrogen (Gutierrez, 1956; Hungate, 1938, 1939, 1943; Inoue et al., 1997; Odelson and Breznak, 1985; Trager, 1934; Yamin, 1978, 1980, 1981). Further, studies using ^{14}C-labelled cellulose and quantitative measurement of cellulase/cellobiase activity demonstrated that they are the primary cellulose degraders and provisioners of acetate in the termite hindgut (Mauldin et al., 1972; McEwen et al., 1980; O'Brien et al., 1979; Yamin, 1978, 1980). It has now been demonstrated by multiple laboratories that the protists express endoglucanases, exoglucanases like cellobiohydrolase, β-glucosidases, and xylanases (Fig. 2) (Ni and Tokuda, 2013). Thus, the hindgut protists fulfil two essential roles in lower termites: (i) their phagocytic activity increases lignocellulose retention in the hindgut, allowing a majority of cellulose and hemicelluloses to be metabolized and (ii) their production of acetate provides a utilizable source of carbon and energy for the termite (Fig. 3).

2.3.4 Degradation of cellulose/hemicellulose by bacteria

The degree of bacterial contribution to cellulose degradation in lower termites has long been debated. The historical hypothesis was that bacteria are primarily present to be consumed by the protists or for nitrogen provisioning (Mikaelyan et al., 2017). However, the large diversity of bacteria in the gut, the differences in their axial and radial distributions, and evidence of coevolution with protists together suggest a direct or indirect role in cellulose degradation (Breznak and Prankratz, 1977; Noda et al., 2018; Schmitt-Wagner et al., 2003a,b). Indeed, many cellulolytic bacteria have been isolated from termites. However, the majority are from higher termites

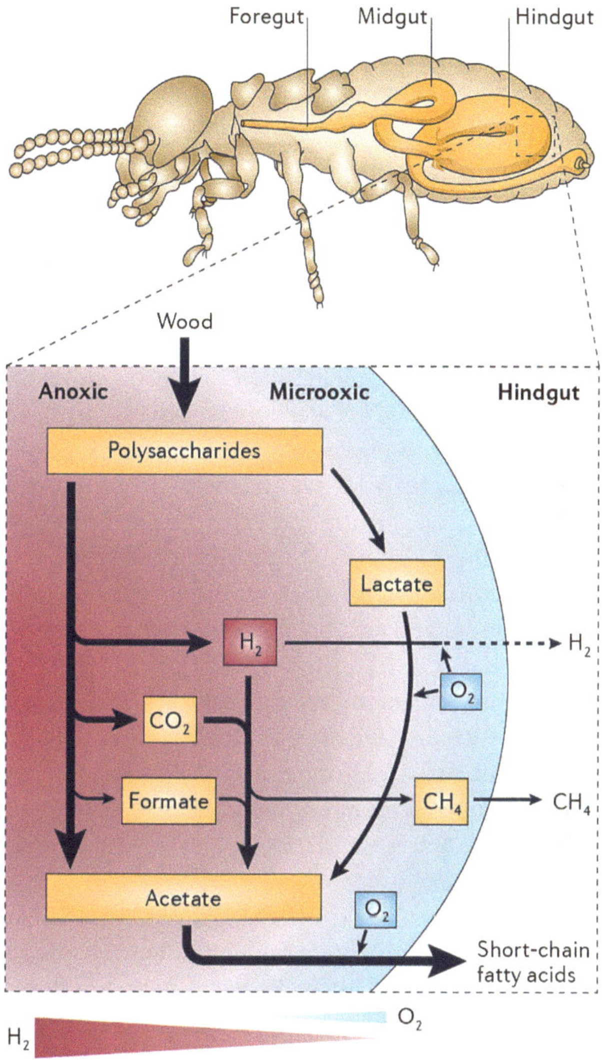

Fig. 3 Microbial carbon provisioning in the hindguts of lower termites. Once small polysaccharides are released from cellulose and hemicellulose particles, they are fermented into short-chain fatty acids such as acetate, lactate and formate. In the process high amounts of CO_2 and H_2 are produced. Acetate is produced from lactate in an oxygen-dependent manner by homoacetogens, and H_2–CO_2 reductive acetogenesis also occurs. The primary short-chain fatty acid found in the haemolymph is acetate. These processes would not be possible without the reduction of inwardly diffusing oxygen by wall-associated acetate, lactate, and hydrogen oxidizers. *Figure from Brune, A., 2014. Symbiotic digestion of lignocellulose in termite guts. Nat. Rev. Microbiol. 12, 168–180. Used with permission.*

(König et al., 2013; Pasti and Belli, 1985; Paul et al., 1993), including the recent exciting evidence that many of them are treponemes, spirochete parasites of humans and other mammals (Tokuda et al., 2018). Abundance estimates of the cellulolytic bacteria isolated from lower termites are scarce. Hence, whether they persist in a stable symbiosis, rather than as transient inhabitants present on wood particles, cannot be determined (Ali et al., 2019; Kuhnigk et al., 1995; Thayer, 1976). One exception is an estimate of approximately 10^5 cellulolytic *Bacillus* per gut in *Zootermopsis angusticollis* (Wenzel et al., 2002). Further, recent studies using metatranscriptomics in *Reticulitermes flavipes* have revealed that gut bacteria express transcripts for cellulases, hemicellulases and other enzymes processing downstream degradation products (Do et al., 2014; Peterson and Scharf, 2018). Ideally more isolates of the numerically dominant bacterial groups need to be collected and cultured so that (hemi)cellulolytic capability can be directly measured. This has been difficult, due to the fastidious nature of most termite gut microbes, especially those associated with protists. Of the very few lower termite gut spirochetes that have been cultured, one, *Treponema primitia* ZAS-2, contains cellulase and cellobiose phosphorylase genes and can grow on cellobiose (Ballor et al., 2012; Graber and Breznak, 2004). In the near term, cultivation-independent methods, such as stable isotope probing, would likely provide deeper insights into this question.

That gut bacteria have a role in hemicellulose digestion is more clear. Several studies have demonstrated hemicellulase activity in bacteria (Juturu and Wu, 2012; Schäfer et al., 1996; Slaytor, 2000). Further, genomic sequencing has revealed that bacterial ectosymbionts of two different gut protists harbour genes coding for endoglucanases and hemicellulases and may be capable of fully decomposing hemicelluloses to acetate (Utami et al., 2019; Yuki et al., 2015). Together with data from axenic cultures of protists (described above), this highlights the obstacle of separating the contribution to cellulose/hemicellulose decomposition by the protists from their bacterial ecto- and endosymbionts. The use of antibiotics intended to remove only bacteria can also cause the loss of protists due to food scarcity or the disruption of their ecto- and endosymbionts—bacteria that may be essential for protist survival. A recent study, recognizing this obstacle, used multiple types of antibiotic treatments, resulting in variable losses of protists and bacteria (Peterson et al., 2015). This study revealed a more detailed picture of the bacterial contribution to hemicellulose digestion, concluding that bacteria do in fact play a large role in degradation of hemicelluloses.

2.4 Lignin degradation

There is no evidence that gut protists have the genetic tools to degrade lignins (Xie et al., 2012), but research investigating the ability of gut bacteria to degrade lignins into utilizable substrates has provided conflicting conclusions (Bugg et al., 2011; Ni and Tokuda, 2013). Two studies, separated by nearly 20 years, demonstrated that the *Coptotermes formosanus* microbiota had little ability to degrade natural lignins, although wood polysaccharides were readily decomposed (Hyodo et al., 2005; Tarmadi et al., 2018). However, feeding trials with lignin-derived aromatics like ^{14}C-labelled benzoic acid and cinnamic acid, and tracking of lignins throughout the digestive process demonstrated core-ring cleavage of lignin-derived molecules was possible in lower termites (Brune et al., 1995b; Ke et al., 2010, 2011a,b). The degradation/modification began as early as during foregut passage, likely in order to 'open up' lignins such that endoglucanases and xylanases can better access wood polysaccharides (Breznak and Brune, 1994; Ni and Tokuda, 2013). Significantly, the ability to do so was oxygen dependent (Brune et al., 1995b). Hence, variable outcomes in these studies likely arise from the fact that pure lignins are difficult to feed to termites, such that lignin precursors or lignin-derived aromatics are often substituted. Further, an assumption that lignin degradation would occur under anoxic conditions likely led to false-negative results. The hypothesis that portions of the gut may be oxic opened a completely new avenue of termite gut research, but also presented a problem with the traditional cellulose fermentation model (see discussion on oxygen below) (Brune et al., 1995a).

2.5 Bacterial acetogenesis

Though the contribution by bacteria to cellulolytic processes in situ has not been quantified, it is known that termite gut bacteria contribute to the total acetate pool via CO_2-reductive acetogenesis (Fig. 3). Hydrogen and CO_2 are produced in abundance by gut flagellates during cellulolysis and it has been demonstrated that bacteria such as spirochetes and others are capable of reducing CO_2 to acetate via the Wood-Ljungdahl pathway (Breznak and Brune, 1994; Breznak and Switzer, 1986; Leadbetter et al., 1999).

Original experiments done by Odelson and Breznak (1983) using six different genera of lower termites suggested that the termite hindgut microbiota is acetogenic. They concluded that acetate, produced fermentatively by protists and through reductive CO_2 acetogenesis by bacteria, dominates the hindgut fatty acid pool at concentrations as high as 80 mM (Fig. 3). The rate of acetate production in *R. flavipes* was as high as 43 nmol/termite/h, enough to support up to 100% of the daily respiratory requirements of the termite (Odelson and Breznak, 1983). Further, acetate was found in mM concentrations in the termite haemolymph but not in the faeces, and microbe-free termite tissue accumulated $^{14}CO_2$ when the termites were fed on ^{14}C-labelled acetate. Hence, in *R. flavipes* (and by assumption, most lower termites due to the fact that most produce acetate as the major end product of cellulose degradation), acetate is absorbed through the hindgut epithelium, oxidized in termite tissues, and serves as the major energy source for the insect (Fig. 3) (Hogan et al., 1985; Odelson and Breznak, 1983).

Later, in a seminal study, Tholen and Brune questioned prior calculations of reductive acetogenesis, asserting the calculations were made with old suppositions about the termite gut environment. To address this, radio-labeled substrates were microinjected into intact hindguts and the products measured at different time intervals (Tholen and Brune, 2000). From this, the authors demonstrated that up to one-third of the carbon flux in termites proceeded via a previously unrecognized pool of lactate (Fig. 3). The absence of measurable lactate within the hindgut implied a very rapid turn-over to acetate, possibly by lactate-fermenting acetogens or lactate-oxidizing aerobes (Bauer et al., 2000; Schultz and Breznak, 1979; Tholen et al., 1997; Tholen and Brune, 2000). Therefore, approximately 90% of the acetate is synthesized by bacteria/protists in the lower termite by saccharide fermentation and oxidation of lactate. Approximately 10% of acetate production occurs via reductive acetogenesis (Tholen and Brune, 2000).

2.6 The importance of O_2 consumption in termite guts

The major contributors to acetogenesis in the termite gut die in the presence of oxygen (Breznak and Brune, 1994; Brune, 2004; Cleveland, 1925a). Therefore, it is no surprise that for approximately 110 years the termite gut was considered a completely anoxic system.

Other data, however, suggested at least some portion of the hindgut was oxic (Bignell, 1984; Brune, 1998; Eutick et al., 1978; Schultz and Breznak, 1978; Tholen et al., 1997; Veivers et al., 1982). The anoxic gut paradigm was not challenged, however, until Veivers et al. (1982), employing redox indicator dyes, noted that termites fed antibiotics were no longer able to maintain the low redox potential (Eh = −150 to −250 mV) typical of normal guts. The observation that in situ degradation of lignin-derived compounds was O_2 dependent (Brune et al., 1995b; Kuhnigk et al., 1994) prompted Brune and coworkers (Brune et al., 1995a; Ebert and Brune, 1997) to use microelectrodes to provide a micrometre scale resolution of O_2 gradients within agarose-embedded, intact hindguts of *R. flavipes* and *Nasutitermes lujae*. The results clearly revealed the presence of a peripheral, hypoxic zone where O_2 partial pressures of 30 mbar (30% air saturation) occur at the epithelial surface then rapidly decrease to anoxia 150–200 μm inward (Fig. 2). The presence of a steep O_2 gradient at the epithelial wall supported the previous conclusion by Veivers et al. (1982) that some members of the microbiota on or near the hindgut wall were a vital 'oxygen sink' without which lignocellulose degradation and acetogenesis would not occur.

Several novel members of the Betaproteobacteria and Verrucomicrobia have been isolated from the hindgut periphery and their contribution to oxygen consumption measured (Wertz et al., 2011, 2018; Wertz and Breznak, 2007a,b). Though the amount of oxygen reduced by each individual species was low, it was collectively significant. Interestingly, there is an apparently important contribution to O_2 consumption by either oxymonad protists (most likely those attached to the lumen wall) or their ecto/endosymbionts that remains unresolved (Wertz and Breznak, 2007a).

2.7 Summary

Termites are the most well-studied group of social insects with regard to the processing and fate of dietary carbon. A complex mix of host and gut microbe-derived enzymes systematically digest cellulose and hemicellulose into short oligosaccharides or monosaccharides that are then used as energy sources for the diverse community of anaerobic protists and bacteria. The primary fermentation end product is lactate, which is rapidly oxidized to acetate. This acetate pool is supplemented by microbes capable of reductive acetogenesis. The termite cells can import the acetate and use it to support up to 100% of their energy requirements. This level of understanding has made termites a model system to which researchers of other insects, like honeybees and ants, turn to gain clues about the processing of dietary carbon in those systems.

3. Honeybees

In contrast to the remarkably diverse community of microbes in the termite gut, the hindgut (ileum + rectum) of eusocial corbiculate bees (Apini, Bombini and Meliponini tribes) is populated by similar core bacteria, with seven to eight species especially well conserved among *Apis* spp. and

Bombus spp. (Kwong et al., 2014; Martinson et al., 2011). Further, all of these microbes are cultivable, and gnotobiotic honeybees (*Apis mellifera*) can be raised in the laboratory. These features make the honeybees an extremely tractable system for studying microbial contributions to the carbon requirements of the host.

3.1 Diet

Like termites, honeybees are strictly herbivorous. Foraging workers collect nectar as the primary source of carbohydrates for the colony, which is comprised of differing amounts of sucrose, glucose and fructose (Fig. 1) (Barker and Lehner, 1974). Also, natural nectar may include sugars like mannose, xylose, arabinose and rhamnose that can be toxic (Sols et al., 1960). Pollen also provides a source of diverse carbohydrates (Alaux et al., 2011; Ricigliano et al., 2017). The composition of pollen varies widely between plant source, geography, climactic conditions and soil type (Mullin et al., 2010). Pollen walls are made of three main layers (reviewed in Shi et al., 2015a): the tryphine, exine, and intine (Fig. 1). The two outermost ones are the tryphine and the exine, with the tryphine filling the interspaces of the exine. Main components of the pollen tryphine include lipids (wax) and phenolic compounds such as lignins and flavonoids. The exine mainly contains sporopollenin, a tough molecule composed of hydroxylated aliphatic compounds and phenolic compounds.

The innermost pollen wall layer, surrounding the plasma membrane, is the intine. The major matrix components of the pollen intine are pectins, cellulose and hemicelluloses (Fig. 1) (Bárány et al., 2010). Pectins can be highly complex molecules, but the most common pectin structure is homogalacturonan, which is a homogenous chain of galacturonates with α-1,4 bonds (Bonnin et al., 2014). Another major pectin structure is rhamnogalacturonan-I formed with a backbone of repeated galacturonate and rhamnose dimers with alternation of α-1,2 and α-1,4 linkages. Rhamnogalacturonan-I possesses side chains composed most frequently with arabinose and galactose residues in proportions variable among plant species. In total, the pollen walls contain molecules with sugar side chains and phenolic glycosides that incorporate arabinose, glucose, galactose, mannose and xylose residues (Ricigliano et al., 2017). Also importantly, the internal tissues of pollen grains are rich in sucrose, starch and often amino acids, proteins and lipids (Roulston and Cane, 2000; Speranza et al., 1997).

3.2 Diversity of the gut microbiota

The study of the honeybee microbiota started with a focus on disease-causing microorganisms (e.g. Katznelson, 1950). Then, in the 1970s, researchers began to study the non-infectious components of the microbiota on the health and growth of bee colonies (Dillon and Dillon, 2004; Gilliam, 1979, 1997; Gilliam et al., 1988). The earliest studies were focused on the bacteria now recognized as *Bifidobacterium*, and the effects of pollen, season, and pathogens on microbial abundance in the gut. Later studies focused on the antimicrobial effects of the culturable microbiome and its protective functions both inside the gut and within the colony itself. It was also suggested that gut bacteria may be able to degrade toxic compounds found in the honeybee diet (Barker, 1977).

With the broader availability and lower cost of culture-independent molecular methods, the first surveys of the gut microbiomes in bees became possible. These studies revealed low overall diversity of gut bacteria, but detected previously unknown Alpha-, Beta-, and Gammaproteobacteria, Bacteroidetes, and Firmicutes bacteria (Jeyaprakash et al., 2003; Mohr and Tebbe, 2006). This broader understanding of the bee gut microbiome was first used to study the impacts of insecticides, insecticidal proteins, and colony collapse disorder on the gut microbiome (Babendreier et al., 2005, 2007; Cox-Foster et al., 2007).

More recently, deep-sequencing of bacterial 16S rRNA genes in individual bee guts revealed the consistent presence of core groups of bacteria in bees collected in widely varying locations (Moran et al., 2012; Sabree et al., 2012). Bacteria consistent within each bee gut were members of the Gammaproteobacteria, Betaproteobacteria, and Firmicutes phyla. Most bees also carried members of the *Bifidobacterium* genus, and of the Alphaproteobacteria class. Deep sequence analyses demonstrated that individual bees harboured numerous distinct strains of these phylotypes, raising the possibility of large metabolic diversity (Ellegaard and Engel, 2019). This, along with findings that the gut bacteria are socially transferred and divergent from free-living relatives, suggested important functional roles (Engel et al., 2012; Koch and Schmid-Hempel, 2011).

Attempts to obtain pure cultures of honeybee gut bacteria have been very successful. At least one representative from each core group has been cultured: *Snodgrassella alvi* (Betaproteobacteria) (Kwong and Moran, 2013), *Gilliamella apicola* and *Frischella perrara* (Gammaproteobacteria) (Engel et al., 2013; Kwong and Moran, 2013), *Bartonella apis* (Alphaproteobacteria)

(Kešnerová et al., 2016), *Bifidobacterium asteroides* (Actinobacteria), *Apibacter adventoris* (Bacteroidetes) (Kwong and Moran, 2016b), and several novel *Lactobacillus* species (Firmicutes) (Olofsson and Vasquez, 2008). In situ localization experiments revealed the bacteria reside within the hindgut (ileum + rectum) (Engel et al., 2013; Kwong and Moran, 2013, 2016b; Zheng et al., 2017). *S. alvi* is strongly associated with the wall of the ileum whereas *G. apicola*, *F. perrara*, and *B. apis* are found in the ileal lumen. The *Lactobacillus* and *B. asteroides* species dominate the rectum. The crop and midgut are mostly devoid of bacteria. With these cultures in hand, in vitro studies have started to hint at in situ functions, such as pectin degradation, sugar fermentation and oxidation of fatty acids (Engel et al., 2012; Kwong et al., 2014; Kwong and Moran, 2013; Martinson et al., 2011).

3.3 Digestion of complex carbohydrates

Comparing the transcriptomes of bees fed with pollen + sugar with bees restricted to a sugar diet, researchers found that pollen activates nutrient-sensing and metabolic pathways of the bees (Alaux et al., 2011). Hypopharyngeal glands present in the head secrete glycoside hydrolases necessary for hydrolysis of glycosidic bonds in complex carbohydrates (Costa and da Cruz-Landim, 2005). It does not appear that any of these can hydrolyse cellulose, pectins or other recalcitrant carbohydrates (Costa and da Cruz-Landim, 2005). A prominent enzyme is α-glucosidase which catalyses the α(1–4) hydrolysis of sugar substrates such as sucrose, maltose, and α-D-glucose (Ohashi et al., 1996). This shows that the honeybee itself contributes to degradation of dietary sugars, but only relatively simple ones. Glucose produced from oligosaccharide hydrolysis is absorbed in the anterior, but not posterior midgut, suggesting host-mediated sugar digestion stops there (Crailsheim, 1988). What happens to the significant amount of pollen that accumulates in the hindgut and the effects of the gut microbiota found there were initially investigated using metagenomics.

3.3.1 Inference of carbohydrate metabolism via metagenomics

The first metagenomic study of honeybee gut microflora was undertaken by Engel et al. (2012). Among the metabolic pathways enriched in the honeybee metagenome, compared to the gut communities of four other insects and five mammals, were for carbohydrate metabolism and transport. Fructosidase enzymes, which hydrolyse sucrose into glucose and fructose, were especially enriched within the predicted proteome, as were gammaproteobacterial pectinases. Additionally, pathways for multidrug efflux pumps and mannose transporters were significantly enriched. These findings are in line with the fact that mannose is toxic to young workers (Barker and Lehner, 1974).

Despite having many bacteria in culture, and despite present knowledge of metabolic capabilities, it was not known whether the bacterial products from the breakdown of the pollen cell wall were used for reasons other than sustaining bacterial growth. However, starting in 2017, Zheng et al. and Kešnerová et al. performed an impressive series of experiments that demonstrated that the gut microbiota is widely involved in food digestion and carbon provisioning to the host, and that degradation tasks are divided among symbionts (Kešnerová et al., 2017; Zheng et al., 2017, 2019).

3.3.2 Inference of microbial metabolism and host nutrient acquisition via metabolomics

Zheng et al. (2017) utilized 'germ-free' (GF) honeybees by separating pupae from the nest and isolating them in a sterile environment with sterile food. As a comparison group, 'conventionally raised' (CV) bees were obtained in the same manner as GF, except the food given to the CV bees was inoculated with the hindgut contents of normally raised nurse bees. As a result, the CV bees had an established microbiome nearly identical to that seen in bees raised in a natural environment, whereas the GF bees possessed a microbial community acquired from the surroundings that was not representative of a 'normal' bee microbiome (Zheng et al., 2017). The first notable difference between GF and CV bees was that the CV bees were 82% heavier than GF bees. A possible reason for this includes direct carbon provisioning of molecules like simple sugars and/or SCFAs (Anderson et al., 2014; Ricigliano et al., 2017).

The hypothesis that gut microbial activity reclaims energy from indigestible or toxic carbohydrates in the form of SCFAs was tested using targeted metabolomics (Zheng et al., 2017). High-performance liquid chromatography was used to identify and quantify the SCFAs present in the ileum, rectum, and haemolymph of GF (i.e. bacterial substrates accumulating in the absence of bacteria) and CV (i.e. bacterial products accumulating in the presence of bacteria) bees. In the CV ileum and rectum, acetate was the most abundant SCFA (114–137 mM). Alternatively, malate was more abundant in GF bee guts, which also possessed large amounts of glucose and fructose. In the CV haemolymph, a significant amount of butyrate was detected (22.8 mM), whereas none was found in the GF haemolymph. This, along with a lack of detection of butyrate in other CV gut compartments, led to the conclusion that members of the normal bee gut microbiota utilize glucose and fructose, fermenting it primarily to acetate and butyrate, with butyrate being rapidly assimilated by the host, thus only detected in its haemolymph (Zheng et al., 2017) (Fig. 4). The hypothesis of a rapid assimilation of butyrate by the host needs to be reconciled, however, with the data of Kešnerová et al. (2017), who documented an accumulation of butyrate in CV honeybee guts.

Building upon their aforementioned results, Zheng et al. (2017) also used untargeted metabolomics via gas chromatography–mass spectrometry to identify any other molecules that varied between the two treatments (GF vs CV). The results showed enrichment of galacturonate, the major

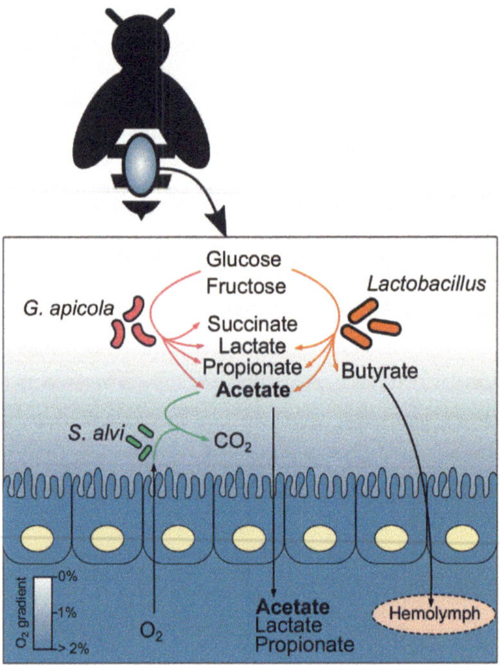

Fig. 4 Dietary carbon source digestion in the hindguts of honeybees. Glucose and fructose, as well as pollen-derived products of flavonoid, phenolamine, and ω-hydroxy acid degradation are fermented by ileum-associated bacteria into succinate, lactate, propionate and acetate. Rectum-associated *Lactobacillus* bacteria also produce lactate, propionate and acetate, in addition to butyrate which crosses the gut epithelium into the haemolymph. These fermentation processes are supported by efficient removal of inwardly diffusing oxygen by wall-associate acetate oxidizing *Snodgrassella alvi*. *Figure modified from Zheng, H., Powell, J.E., Steele, M.I., Dietrich, C., Moran, N.A., 2017. Honeybee gut microbiota promotes host weight gain via bacterial metabolism and hormonal signaling. Proc. Natl. Acad. Sci. U. S. A. 114, 4775–4780. Used with permission.*

product from the breakdown of pectins, in the CV ileum (Zheng et al., 2017). This supports earlier cultivation-based work documenting that honeybee Gammaproteobacteria can degrade homogalacturonan, the most common component of pectin (Engel et al., 2012). The discovery of galacturonate in the CV ileum, but not in the rectum, implies that this compound is either absorbed across the host gut or degraded before reaching the rectum. The latter is more likely, since *G. apicola* localized in the ileum has the genomic capacities to catabolize galacturonate (Zheng et al., 2016).

A very recent study shows evidence of symbiotic niche specialization in the degradation of pollen intine polymers, including pectins (Zheng et al., 2019). The authors of this study used genomic analysis of bee gut isolates, followed by metabolomics after manipulation of the gut microbiota. Broadly, *Gilliamella* sp. was confirmed to be the main degrader of pectins and rectum-inhabiting *Bifidobacterium* sp. could depolymerize hemicelluloses and potentially the side chains of complex pectin constituents, such as rhamnogalacturonans.

Metabolomic studies by Kešnerová et al. (2017) gave additional insights into honeybee gut carbon metabolism. The authors used untargeted metabolomics via flow injection analysis–quadrupole time-of-flight (FIA-qTOF, a type of liquid chromatography–mass spectrometry) to compare GF and CV honeybees produced similarly as in Zheng et al. (2017) and Kešnerová et al. (2017). The most significantly enriched compounds in homogenates of GF midgut, ileum and rectum were complex molecules known to compose the pollen tryphine and exine, such as flavonoids, purines, pyrimidines, ω-hydroxy acids and phenolamides (Kešnerová et al., 2017). This was the first study directly demonstrating that compounds from the outer pollen wall, rather than only from the pollen intine, can be processed by the honeybee gut microbiota. Among the compounds enriched in CV bees, some are known to be products of fermentation of the flavonoid aromatic core (known as the aglycone) such as hydroxy- and di-hydroxyphenylpropionate. Other compounds, such as maleylacetate and hydroxy-3-oxoadipate are products of downstream aglycone degradation (Kešnerová et al., 2017).

In the same study, some sterile honeybee pupae were fed a diet containing a single species of honeybee symbiont (Kešnerová et al., 2017). This enabled the authors to track particular carbon utilization/production by specific bacteria, allowing for the potential metabolic role of each gut member to be individually studied. Of the compounds derived from the pollen tryphine and exine, flavonoids were degraded in bees colonized with *Lactobacillus* sp. or *B. asteroides*. Interestingly, the *Lactobacillus* sp. and

B. asteroides utilized different flavonoids, suggesting, again, niche specialization among the bee gut symbionts. In addition, two compounds found in sporopollenin (i.e. main constituent of the pollen exine) were independently utilized by gut symbionts: phenolamides were exploited by a subset of *Lactobacillus* sp. and ω-hydroxy acids by *B. asteroides* and a different subset of *Lactobacillus* sp. The breakdown products succinate and pimelate can be attributed to a subset of *Lactobacillus* sp., and valerate (i.e. a SCFA) production was exclusive to *B. asteroides* (Kešnerová et al., 2017).

In vitro studies of bacterial isolates grown in pollen-enriched media yielded more data on the ileum-residing bacteria (Kešnerová et al., 2017). The deglycosylation of flavonoids (i.e. separation of the aglycone from the glycosyl group) was observed for *B. apis*, *F. perrara*, *G. apicola* and *S. alvi*. These bacteria likely work in tandem to deglycosylate specific flavonoids and release a monosaccharide and an aglycone. The monosaccharide can then be fermented into organic acids and the aglycone broken down into products like hydroxyphenylpropionate. Future analyses of the bee haemolymph could reveal if any of these breakdown products cross the gut epithelium.

3.4 The importance of oxygen consumption in honeybee guts

Similar to termites, microelectrode studies found a radial oxygen gradient in the ileum of CV honeybees (Zheng et al., 2017). Oxygen was approximately 2% at the gut wall, decreasing to 0%, roughly 150–200 μM inward. In GF bees, this gradient did not exist; instead the entire gut was approximately 2% oxygen (Zheng et al., 2017). In situ, *S. alvi* is closely associated with the wall of the ileum, and grows on acetate in vitro (Kwong and Moran, 2013). Hence, *S. alvi* may be using acetate to reduce inwardly diffusing oxygen, much like *Stenoxybacter acetivorans* and other symbionts do in termites (Wertz and Breznak, 2007a,b). When in acetate buffer, *S. alvi* reduces oxygen at rates greater than *S. acetivorans* (Wertz and Breznak, 2007b; Zheng et al., 2017). The fermentation of glucose by *G. apicola* and other bacteria to acetate is possible only when anoxic conditions exist (Zheng et al., 2017). When in the presence of oxygen, *G. apicola* ferments glucose to malate. Similarly, *Lactobacillus* sp. isolates from the honeybee rectum shift from lactate production to acetate production with decreasing oxygen levels (Zheng et al., 2017). Therefore, removal of inwardly diffusing oxygen by *S. alvi* in the honeybee gut is critical to maintain digestive processes conducted by other symbionts.

3.5 Summary

Just as in termites, honeybees contain enzymes that convert dietary polysaccharides to glucose, galactose, and fructose among others. However, the bacteria found in the hindgut have been shown to have the metabolic capabilities to decompose the components abundant in the cell walls of pollen. Moreover, significant amounts of symbiont fermentation products were shown to accumulate in the bee hindgut (acetate) and possibly the haemolymph (butyrate). Each bee caste consumes different amounts of pollen, beebread, and nectar (Kwong and Moran, 2016a). Therefore, the relevance of the gut microbiota to the energy requirements of the host may be caste-dependent, or seasonal. In this case the bacteria may be augmenting the host's ability to use simple dietary carbohydrates. We look forward to future studies looking at the fate of hindgut SCFAs and how much they can support the energy requirements of the honeybee.

4. Ants

Ants are often seen as the best representatives of successful eusocial insects given their abundance and diversity in almost all terrestrial ecosystems (Hölldobler and Wilson, 1990). Unlike lower termites and honeybees, the diet of some ant species is fairly cryptic. Ants were thought to be unique among eusocial insects in exhibiting predominantly predaceous diets (Floren et al., 2002; Hölldobler and Wilson, 1990). However, although ants have diversified from a predatory ancestor (Nelsen et al., 2018), stable isotope analyses of entire ant communities living in tropical tree canopies demonstrate that besides some specialist predators, the majority of these arboreal ants are omnivorous or herbivorous (Blüthgen et al., 2003; Davidson et al., 2003; Fiedler, 2007; Ottonetti et al., 2008).

Many herbivorous ants are not traditional herbivores—the ants never directly consume vascular plant stems or leaves (Hölldobler and Wilson, 1990). Even the leaf-cutter fungus-growing ants (i.e. *Atta* spp. and *Acromyrmex* spp.) never feed on the leaf parenchymatic tissue, but they do drink sap from leaves when they cut or transport them to their fungal gardens (Littledyke and Cherrett, 1976). Other non-animal derived foods that ants do eat include exudates of sap-feeding insects, plant wound secretions, fruit juice, food bodies, elaiosomes, seeds, pollen, fungi and potentially lichen (Bass and Cherrett, 1995; Blatrix et al., 2012; Cembrowski et al., 2015;

Czechowski et al., 2011; Gordon, 2012; Hölldobler and Wilson, 1990; Lanan, 2015; Powell, 2008). While several of these food sources are liquid, with directly digestible nutrients, some contain recalcitrant polymers that are difficult to digest. For herbivorous ants that regularly or seasonally consume such foods, gut bacterial symbionts could play a role in the digestion of recalcitrant polymers. Insect gut symbionts can also ferment simple carbohydrates from a sugar-based liquid diet, participating in the carbon budgets of their hosts and improving their survival (e.g. Crotti et al., 2010). While these ideas are intriguing, studies on ant symbiont C-catabolism and provisioning are nascent, compared to those in termites and honeybees, as elaborated further below.

4.1 Diet

Like honeybees and termites, many ants have specialized diets, including acacia-associated *Pseudomyrmex* ants, relying entirely on the Beltian bodies and extrafloral nectar (EFN) of their host-plant (Heil et al., 2004). Some army ants like *Eciton dulcium* and *E. mexicanum* are also specialized, feeding only on specific poneroid ant species (Powell and Franks, 2006). But beyond these ants, many ants are contrastingly opportunistic, consuming a wide variety of food resources. This is true for several ant herbivores.

A majority of herbivorous ant diets can be separated into five categories based on the types of substrate ingested: (i) myrmecophytic organ feeding, (ii) granivory, (iii) plant or insect exudate feeding including nectar, wound secretion, fruit juice and insect exudate consumption, (iv) fungivory and (v) plant-surface foraging, including consumption of cryptic particles such as fungal spores, bryophytes, algae and pollen grains. Here we focus on the latter three, as specialized granivores and myrmecophytic organ feeders may be unlikely to require symbiotic bacteria to process their diet (e.g. Martins and Moreau, 2020; Pringle, 2019; Rubin et al., 2019).

4.1.1 Non-recalcitrant diets—Plant and insect exudate feeding

Whether it comes from plants or sap-sucking insects, exudates (i.e. any non-solid substance secreted by plants or insects) constitute abundant liquid food for ants, especially for ants foraging in trees. For example, the most abundant arboreal ants in tropical forests (*Camponotus* sp., *Oecophylla* sp., *Dolichoderus* sp., *Azteca* sp., *Crematogaster* sp.) are exudate feeders (Davidson, 1997; Davidson et al., 2003). Even known predators foraging on trees such as the bullet ant *Paraponera clavata*, sometimes called the 'omnivorous giant tropical ant' (Larson et al., 2014), occasionally rely on liquid food such as EFN (Hölldobler and Wilson, 1990).

Plant wound secretions, and both floral and extrafloral nectar, contain a high abundance of sucrose (Fig. 1) (Baker et al., 1998; Lortzing et al., 2016). Ants can consume fruits and fruit juices (Evison and Ratnieks, 2007; Lanan, 2015; Passos and Oliveira, 2004) known to contain fructose, glucose, organic acids such as citrate and malate and many vitamins, phenolic compounds and terpenoids (Gardner et al., 2000; Romero Rodriguez et al., 1992; Viljakainen et al., 2002). Many ants consume exudates from sap-feeding hemipterans and some lepidopteran larvae (Buckley, 1987; Hölldobler and Wilson, 1990). Aphid honeydew is the most studied insect exudate. Similar to nectar, it contains common simple carbohydrates in addition to trehalose, maltosucrose and melezitose (Fischer et al., 2005; Hölldobler and Wilson, 1990).

Exudate foods appear beneficial for ants, as EFN has been shown to increase ant survival and colony growth (Byk and Del-Claro, 2010; Lach et al., 2009), while access to hemipteran honeydew increases ant activity (Brightwell and Silverman, 2009). The diverse sources of accessible carbohydrates in these liquids may enable ant success, and it is possible that they assimilate monosaccharides from catabolized oligosaccharides without assistance from microbes. But ants could benefit from gut bacteria if they receive energetic compounds, such as SCFAs, produced by bacteria through their own catabolism of exudate sugars.

Given that exudates are highly attractive to many animals and constitute a valuable limited resource, ants not monopolizing myrmecophytic plants or sap sucking trophobionts could suffer from high competition (Blüthgen and Fiedler, 2004; Dáttilo et al., 2014; Fagundes et al., 2016; Lach, 2005) or seasonal variation in exudate availability (Belchior et al., 2016). Thus, when exudates become unavailable, food sources containing recalcitrant carbohydrates might be preferred by some ants. The reliance on gut bacteria to degrade these recalcitrant molecules and release byproducts in the gut lumen would then be valued by the ant hosts. We next review several diets enriched for recalcitrant molecules.

4.1.2 Recalcitrant diets

Several ants consume fungal tissues as part, or most, of their diet, and such fungivory is not restricted the fungus-growing attine ants (e.g. *Petalomyrmex phylax*, *Tetraponera aethiops*, *Pseudomyrmex penetrator*, *Euprenolepis* sp.,

Megalomyrmex sp.) (Blatrix et al., 2012; Lapolla, 2009; Wheeler, 1925; Witte and Maschwitz, 2008). Fungal cell walls are composed of up to 20% chitin (Fig. 1) (Bowman and Free, 2006), a particularly recalcitrant polymer, formed with chains of *N*-acetylglucosamine (i.e. acetylated glucose molecule with an amine group) (Gow et al., 2017). Additionally, fungal cell walls always contain β-1,3-glucan and can be supplemented with α-1,3-glucan, β-1,6-glucan, mannans and proteins.

Distinct from fungivory is a tendency for ants to feed opportunistically from plant surfaces, a feeding habit sometimes known as 'leaf foraging', but here referred to as 'plant-surface foraging'. Diets of plant-surface foraging ants remain somewhat enigmatic, but are likely more diverse than those of fungivores or plant and insect exudate feeders. Indeed, leaf surfaces are complete ecosystems comprised of microorganisms such as protists and bacteria, plant materials such as pollen grains, mosses, algae and liverworts, wastes such as invertebrate and bird faeces and mammalian urine, heterotrophs such as small arthropods, yeasts and filamentous fungi, and composite organisms like lichens (Andrews et al., 1980; Andrews and Harris, 2000; Coley et al., 1993; Davidson et al., 2004; Rodríguez-Zaragoza, 1994).

Although the diet of plant-surface foraging ants resembles an omnivorous diet, encompassing scavenging, detritivory and true herbivory, it is interesting to note that tissues of such ants exhibit low δ15N values, indicating a low trophic position (Davidson et al., 2003). Organic particles found on the surface of plants may constitute carbohydrate-rich food sources if their cell walls or cuticles can be degraded or circumvented to access more valuable internal nutrients. Beyond pollen and fungi, with their aforementioned components, non-vascular plants found on the surface of vascular plants have a partially overlapping set of recalcitrant and carbon-enriched molecules (Fig. 1). For instance, all such organisms, including mosses, liverworts and algae, are rich in diverse hemicelluloses (e.g. xyloglucans, mannans, xylans, arabinoxylans, β-(1,3-1,4)-glucans) (Scheller and Ulvskov, 2010). They further contain small amounts of lignin phenol derivatives, mainly *p*-hydroxyl phenols (Zavarzina et al., 2015). Lichens represent another possible source of food. As symbiotic organisms representing the merger of fungi and algae (or cyanobacteria), their cell walls possess characteristics of both plants and fungi. Found at particular abundance are hemicelluloses, while chitin, cellulose, lichenans and lignin phenol derivatives (syringyls) are present in lower concentrations (Fig. 1) (Kaya et al., 2015; Podterob, 2008; Zavarzina et al., 2015).

4.2 Fungus-growing ants

4.2.1 Fungus-growing ants and their fungal garden

Although different from the previously described extracellular gut symbioses, the mutualism between fungus-growing ants (former tribe Attini, referred to as 'attines') and their fungus gardens can be seen as a digestive symbiosis. In these relationships, ants deposit substrates (e.g. organic wastes or leaf fragments) onto the fungi, which break down recalcitrant plant molecules, assimilating nutrients that are obtained by ants upon fungal consumption (Weber, 1966, 1972). This symbiosis has been documented since the 19th century and is well-studied from a perspective of function (Belt, 1874). Fungus-growing ants display intermediate δ15N values characteristic of an omnivorous-like diet (Davidson et al., 2003). This could be explained by the non-exclusive fungal diet of adult fungus-growing ants. Indeed, while larvae feed exclusively on the fungal cultivar, only 9–50% of the nutritional requirements of adult worker fungus-growing ants come from fungal digestion. The rest originates from plant sap and fruit juices (Bass and Cherrett, 1995; Silva et al., 2003).

In higher attines (e.g. *Acromyrmex* spp., *Atta* spp., *Sericomyrmex* spp., *Trachymyrmex* spp.), the association with the fungal partner appears obligate and the digestive processes are more efficient than in lower attines (De Fine Licht et al., 2010). The hyphal swellings of these fungi, named gongylidia, are organized in clusters (i.e. 'staphylae') and constitute the food items for the ants (Quinlan and Cherrett, 1979; Schultz and Brady, 2008). Among higher attines, the leaf-cutters, namely *Atta* spp. and *Acromyrmex* spp., deposit leaf fragments to the top layer of their fungal gardens. The degradation of these substrates by the fungus is enhanced by fungal enzymes secreted in gongylidia, ingested by the ants and excreted on the leaf fragments (Boyd and Martin, 1975). Thus, the ants act as transporters of fungal enzymes, revealing a complex, and coevolved, digestive mutualism.

Fungal gardens have been extensively investigated for nutritional benefits provided to the ants. Especially in leaf-cutters, metagenomic, metatranscriptomic, proteomic and metaproteomic studies have shown that the Agaricaceae fungal partner synthesizes a plethora of enzymes to degrade plant recalcitrant polymers, including cellulose, hemicelluloses, pectins, lignins, starch and proteins (Aylward et al., 2015; De Fine Licht et al., 2013; De Fine Licht and Boomsma, 2014; Schiott et al., 2010). In addition, fungal gardens host a high abundance of bacteria, dominated by members of the Enterobacteriaceae and Pseudomonadaceae families, which can synthesize cellulose-, hemicellulose-, and pectin-degrading enzymes (Aylward et al., 2012; Moreira-Soto et al., 2017).

In a recent study, Shik et al. (2018) experimentally assessed the timing and caste- and stage-specificity of nutrient assimilation in *Atta colombica* by feeding the ants with diets enriched in carbon isotopes, then quantifying the accumulation of ^{13}C in the fungal garden, brood and adult tissues. They found that adult foragers continuously ingest and assimilate carbon from harvested glucose, bypassing the fungal garden usually believed to be fundamental in fungus-growing ants' digestion (Shik et al., 2018). Although roles of gut symbionts in facilitating glucose digestion have not been tested, this study demonstrates that elements in dietary carbohydrates can be labelled and traced from ant diet to ant assimilates.

4.2.2 Leaf-cutter ants and their entomoplasmatales gut symbionts

Bacterial symbionts in ants were first detected as early as at the end of the 19th century, when Blochmann (1882) described the endosymbiotic bacterium inhabiting bacteriocytes in the midgut of Camponotini ants. Recently, the gut bacterial symbionts of leaf-cutter fungus-growing ants have been studied for their potential in nutrition using next-generation sequencing analyses (Sapountzis et al., 2015, 2018). Entomoplasmatales bacteria associated with leaf-cutter ants and *Trachymyrmex cornetzi* (Sapountzis et al., 2019) are of particular interest in carbohydrate processing, though their abundance varies among colonies of *Acromyrmex* sp. (Sapountzis et al., 2015). In a recent study, two Entomoplasmatales strains were isolated from the fat body tissues and gut lumen of *Ac. echinatior* and subject to genome sequencing. One of these two strains, from the ant gut lumen, appears to be an obligate anaerobe, related to the Entomoplasmatales bacteria associated with neotropical army ants (Sapountzis et al., 2018). This strain, EntAcro1, possesses genes coding for the import and degradation of *N*-acetylglucosamine, the chitin monomer. While it has already been demonstrated that, in *Acromyrmex* ants, an endogenous chitinase is secreted in the infrabuccal pocket to depolymerize chitin (Febvay et al., 1984). Sapountzis et al. (2018) showed that the bacterium EntAcro1 expresses the genes coding for the import of *N*-acetylglucosamine in both midgut and hindgut regions of *Ac. echinatior* ants. Though evidence suggests that EntAcro1 utilizes *N*-acetylglucosamine, it is not yet known whether the host ants benefit from this fungal cell wall catabolism.

EntAcro1 also possesses genes encoding a citrate degradation pathway potentially acquired via horizontal gene transfer with other bacteria (Sapountzis et al., 2018). Citrate is abundant in fruit juices that can supplement leaf-cutter ant workers' diet. EntAcro1 can ferment citrate to produce acetate that could then be assimilated by the *Ac. echinatior* host. Indeed, Sapountzis et al. (2018) have experimentally shown that the expression of an ant-encoded acetate importer gene is higher when the ants are both fed with a diet enriched in citrate and possess EntAcro1 in their gut. Given its abilities to digest citrate, EntAcro1 is well suited to accommodate the partially frugivorous diet of its ant host. This citrate fermentation could provide supplements of energy, fuelling the high level of foraging activity by these ecologically dominant insects.

4.3 *Cephalotes* ants

4.3.1 Diet of Cephalotes *ants*

Cephalotes turtle ants do not display typical dominant behaviours towards other species. Instead, other ants are behaviourally dominant in neotropical arboreal habitats (Cole, 1983). One possible explanation for the persistence of *Cephalotes* ants in this highly competitive habitat is their adoption of a generalist herbivorous and opportunistic diet (Gordon, 2012) to escape interference competition (Davidson, 1998; van Velzen and Etienne, 2013).

Cephalotes ants are known to be plant and insect exudate feeders and plant-surface foragers (Fig. 1). Especially, two main elements of the *Cephalotes* ant diet are pollen (Baroni Urbani and De Andrade, 1997; Cole, 1980) and EFN (Byk and Del-Claro, 2011). *Cephalotes* ants are also known to lick plant wounds and to collect hemipteran honeydew, fungi and lichen (Gordon, 2012; Powell, 2008; Wilson, 1976).

4.3.2 The diverse gut bacterial community of Cephalotes *ants*

Relatively recent has been the characterization of ant gut symbionts in *Cephalotes* ants. First glimpses into a unique symbiosis were gleaned through microscopy, enabling detection of huge masses of extracellular bacteria in the midgut and ileum of several *Cephalotes* species (Caetano and da Cruz-Landim, 1985; Roche and Wheeler, 1997). These bacteria were, at first, proposed to have a role in ant nutrition (Caetano and da Cruz-Landim, 1985; Roche and Wheeler, 1997). Hölldobler and Wilson (1990) even suggested that the abundant bacterial communities observed in *Cephalotes* guts could be involved in the digestion of pollen compounds.

Similar bacterial symbionts are now believed to be associated with most, if not all, species of *Cephalotes* (Sanders et al., 2014). Recently, Hu et al. (2018) studied gut microbiomes through metagenomics on adult workers from 17 *Cephalotes* species. The gut bacterial community was shown to be conserved among *Cephalotes* species, and a subset of these bacteria were involved in the recycling of nitrogen wastes into amino acids. The *Cephalotes* core gut community is composed of members of the Opitutales (*Cephaloticoccus* sp.), Xanthomonadales, Burkholderiales, Rhizobiales, and Pseudomonadales (e.g. *Ventosimonas gracilis*) orders (Anderson et al., 2012; Hu et al., 2018; Russell et al., 2009). Members of the Sphingobacteriales, Flavobacteriales and Campylobacterales can also be found in these ants, but their presence varies among colonies, and possibly species (Hu et al., 2018; Sanders et al., 2014).

A sophisticated anatomical feature could have facilitated the retention of gut symbionts and favoured intimate relationships in these hosts. As in other frequent exudate feeders (ants in the subfamily Formicinae and small ants within the Dolichoderinae), *Cephalotes* ants possess a derived proventriculus, preventing large polymers from passing the crop and reaching the midgut, but enabling the ants to swallow and process large quantities of liquid (Cook and Davidson, 2006; Davidson et al., 2004; Eisner, 1957). The proventriculus of *Cephalotes* ants is additionally covered with a thick, non-cellular layer of a mucus-like substance, resembling the consistency of the peritrophic membrane (Bution and Caetano, 2008; Lanan et al., 2016). This layer acts as a barrier to ingested bacteria that may otherwise disturb the already established gut bacterial community that is acquired just before its formation (Caetano and da Cruz-Landim, 1985; Lanan et al., 2016).

Thanks to this selective proventriculus filter in *Cephalotes* ants, at the intraspecific level, the gut bacterial community is stable. But the relative abundance of some symbionts varies, at least slightly, with diet in lab-reared colonies (Hu et al., 2014). In particular, *C. varians* colonies fed a diet enriched in bee pollen showed increased abundance of some Rhizobiales symbionts, and concomitant decreases in the proportions of some Burkholderiales. The reasons for these responses remain unknown, but Hu et al. (2014) proposed that Rhizobiales either directly benefits from nutrients in pollen or is indirectly affected by host physiological adjustments to different diets.

4.3.3 *Culture of isolated bacteria from* Cephalotes *guts*

Despite difficulties in cultivating ant gut bacterial symbionts (e.g. Brown and Wernegreen, 2019; Eilmus and Heil, 2009), many *Cephalotes* gut bacteria have now been cultured, which has enabled their genome sequencing (Hu et al., 2018). Some of these *Cephalotes* core gut bacterial isolates have been tested for various metabolic activities while acquiring a species name (Lin et al., 2016a,b, 2017).

For example, gut symbionts from the Opitutales, known to dominate *Cephalotes* midguts (Lanan et al., 2016), were isolated and assigned to a novel genus, termed *Cephaloticoccus* (Lin et al., 2016b). *Cephaloticoccus* sp. bacteria utilize organic acids such as lactate and succinate, as well as monosaccharides and disaccharides including sucrose and cellobiose. In addition, the Burkholderiales-derived *Saccharedens versatilis* grow on monosaccharides, lactate and acetate, and were demonstrated to produce butyrate (Lin et al., 2017). Thus, *Cephalotes* midgut and hindgut bacterial symbionts could play a role in the processing of disaccharides, monosaccharides and organic acids originating from an exudate diet, producing valuable metabolites (e.g. butyrate) for either bacteria, the ant host or both.

4.3.4 Metagenomics of Cephalotes *guts*

We inferred the presence of genes encoding enzymes to catabolize recalcitrant polymers potentially originating from the *Cephalotes* ant diet, by inspecting publicly available adult gut metagenomes from three *Cephalotes* species (Hu et al., 2018) in the online tool IMG/M-ER (Chen et al., 2019). In two out of three *Cephalotes* species examined, similar gut bacterial symbionts possess genes coding for enzymes to initiate the degradation of cellulose, resistant starch and chitin (Fig. 5). In each *Cephalotes* host examined, laccase domain-containing proteins were encoded by a variety of *Cephalotes* gut symbionts. Laccases are versatile enzymes that can oxidize aromatic compounds found in polyphenols such as lignins (Janusz et al., 2017) and plant secondary metabolites including tannic acid and flavonols (Botta et al., 2017; Carbajo et al., 2002). For the three types of pectinase-encoding genes assessed, *C. atratus* ants possess unidentified gut bacteria that can remove methyl esters from homogalacturonan (pectin methylesterase), while *C. eduarduli* ants possess unclassified Verrucomicrobia (likely related to *Cephaloticoccus* sp.) with genes coding for the depolymerization of rhamnogalacturonan-I.

These results denote prominent gut bacterial capacities to degrade and process recalcitrant polymers from the diet of a plant-surface forager, especially pollen intine and fungal compounds. However, because of their derived narrow proventriculus (Cook and Davidson, 2006; Davidson et al., 2004; Eisner, 1957) it is unlikely that recalcitrant polymers reach midgut and hindgut chambers of adult *Cephalotes* ants, where the screened bacterial symbionts are known to settle. The recalcitrant polymers from food sources collected on plant surfaces could then be digested in two mutually non-exclusive ways (Fig. 5). (i) Since ant larvae are considered as the

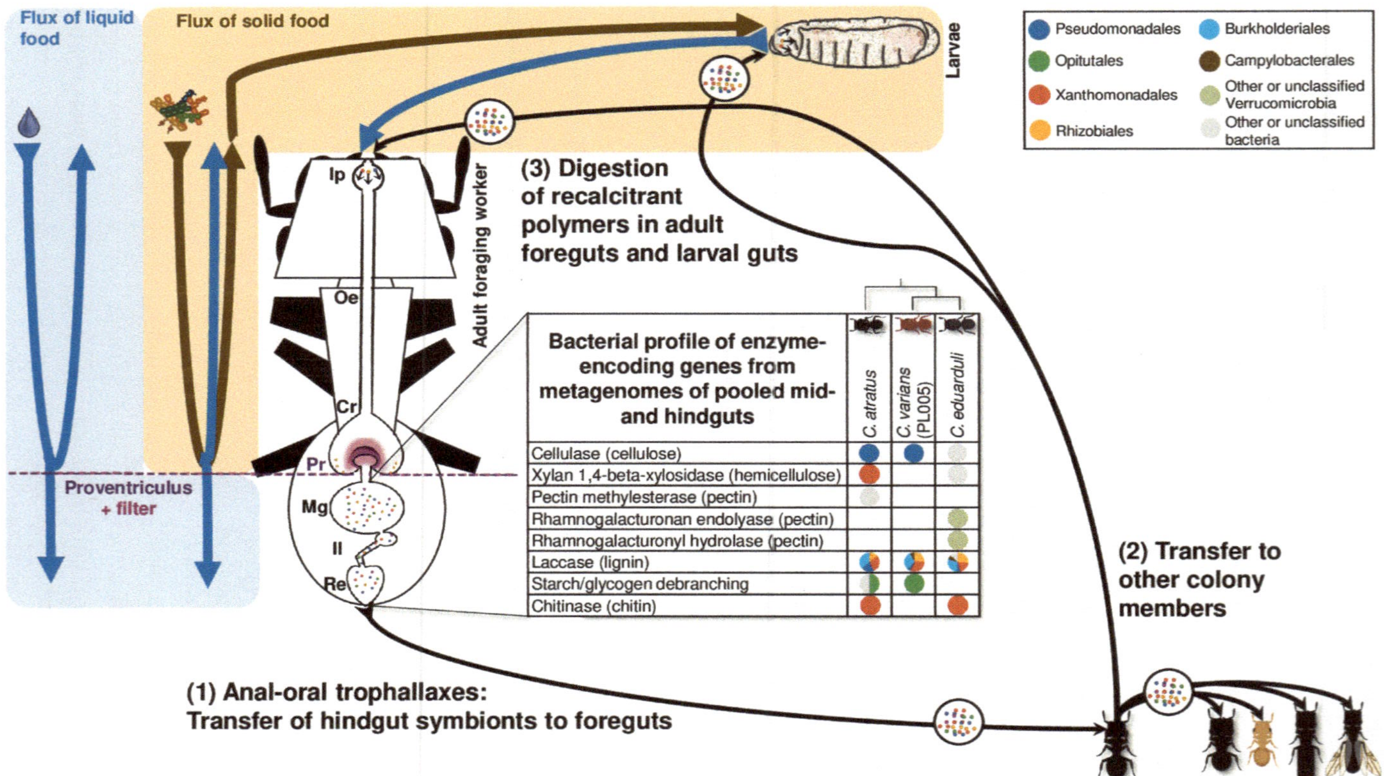

Fig. 5 See legend on opposite page.

digestive caste of ant colonies (Cassill et al., 2005; Sorensen et al., 1983), the guts of these early life stages could be predominant sites of recalcitrant polymer degradation (Lanan et al., 2016), potentially facilitated by larval gut symbionts, which are thought to include at least a subset of the adult worker's microbiome (Russell et al., 2009). (ii) As shown by Rodrigues (2016), adult *Cephalotes* ants may transfer hindgut bacterial symbionts to the infrabuccal pocket of other workers in the colony. Thus, the *Cephalotes* infrabuccal pocket may be a site of recalcitrant polymer degradation where hindgut bacterial symbionts secrete digestive enzymes (Rodrigues, 2016).

4.4 Summary

Fungus-growing ants have been studied over the past century for their association with a fungal digestive symbiont, which are now known to themselves harbour a suite of digestive bacteria. In leaf-cutter ants, recent studies suggest that gut bacteria could be involved in the processing of chitin derivatives and citrate, which could supplement the energy budget of the ant host. In *Cephalotes* ants, the dense and diverse bacterial gut community could play a crucial role in the degradation of food sources from their plant-surface

Fig. 5 Hypotheses on digestive processes and fluxes in *Cephalotes* ant colonies. Blue arrows represent liquid food, brown arrows represent solid food and black arrows represent transfer of adult gut symbionts. Adult *Cephalotes* workers forage for liquid (exudate) or solid (plant-surface foraging) food, but the proventriculus (in purple) prevents solid food (and recalcitrant polymers) from reaching the midgut. Liquid food (and simple carbohydrates) can pass through the proventriculus and be digested in the midgut. The proventriculus filter (in pink) prevents microorganisms from accessing the midgut and disturbing the established midgut and hindgut bacterial communities. The midgut and hindgut bacterial symbionts colonize these gut chambers when the newly eclosed workers feed on the rectal fluids of mature ants (see Lanan et al., 2016). The taxonomic profile of *Cephalotes* sp. midgut and hindgut bacterial genes encoding enzymes to degrade diverse plant and fungal/arthropod polymers is shown in the middle (metagenomes from Hu et al., 2018). Altogether, the *Cephalotes* gut symbionts encode a variety of enzymes to catabolize recalcitrant polymers. Rodrigues (2016) showed that anal-oral trophallaxes also occur among mature *Cephalotes* workers. (1) Anal-oral trophallaxes could thus be a way to transfer hindgut symbionts to the foregut of other workers which would propagate the bacterial symbionts via serial oral-oral trophallaxes and (2) reach non-worker ants, including larvae. (3) Once located in the foregut (and/or the larval gut), the gut bacterial symbionts would enter in contact with recalcitrant polymers where they could synthesize digestive enzymes, participating in their catabolism. Gut sections are as follow: Ip, infrabuccal pocket; Oe, oesophagus; Cr, crop; Pr, proventriculus; Mg, midgut; Il', ileum; Re, rectum.

foraging habits and in the processing of exudate nutrients. Genomic approaches provide promising starting points to generate hypotheses on the role of ant gut microbes in carbohydrate utilization. Nevertheless, ultimate questions on digestive symbioses in ants remain unanswered. Are bacterial products of carbohydrate digestion directly available for ant host assimilation and how do hosts benefit from this? Can ants digest the bacteria themselves, and how much does this contribute to the carbon needs of the insect? Tools such as metabolomics and labelled isotope tracing can help to answer these questions and improve our understanding of ant-microbe digestive symbioses.

5. Conclusion

The strong microbe-host fidelity, enabled by social interactions among insects described in this chapter, has aligned host and symbiont interests, allowing high degrees of specialization and, in some cases, coevolution. Gut microbes, and external fungal symbionts, have shaped the enzymatic repertoire of their associated social insects, supporting existing endogenous functions and adding new ones. This suite of enzymes is capable of digesting carbon-rich molecules found in their diets that include wood, pollen and fungi. Through social transmission of their microbiomes, both simple and recalcitrant carbon sources can be utilized by the colony, sustaining the energy needs of the system. The diversity and ecological dominance of social termites, bees, and ants speak to their evolutionary success. And in this success, microbially conferred abilities to feed on carbon sources not available to other animals—and to, hence, sustain both foragers and non-foragers—have played a significant role.

References

Alaux, C., Dantec, C., Parrinello, H., Le Conte, Y., 2011. Nutrigenomics in honey bees: digital gene expression analysis of pollen's nutritive effects on healthy and varroa-parasitized bees. BMC Genomics 12, 496.

Ali, H.R.K., Hemeda, N.F., Abdelaliem, Y.F., 2019. Symbiotic cellulolytic bacteria from the gut of the subterranean termite *Psammotermes hypostoma* Desneux and their role in cellulose digestion. AMB Express 9, 111.

Anderson, K.E., Russell, J.A., Moreau, C.S., Kautz, S., Sullam, K.E., Hu, Y., Basinger, U., Mott, B.M., Buck, N., Wheeler, D.E., 2012. Highly similar microbial communities are shared among related and trophically similar ant species. Mol. Ecol. 21, 2282–2296.

Anderson, K.E., Carroll, M.J., Sheehan, T., Mott, B.M., Maes, P., Corby-Harris, V., 2014. Hive-stored pollen of honey bees: many lines of evidence are consistent with pollen preservation, not nutrient conversion. Mol. Ecol. 23, 5904–5917.

Andrews, J.H., Harris, R.F., 2000. The ecology and biogeography of microorganisms on plant surfaces. Annu. Rev. Phytopathol. 38, 145–180.
Andrews, J.H., Kenerley, C.M., Nordheim, E.V., 1980. Positional variation in phylloplane microbial populations within an apple tree canopy. Microb. Ecol. 6, 71–84.
Aylward, F.O., Burnum, K.E., Scott, J.J., Suen, G., Tringe, S.G., Adams, S.M., Barry, K.W., Nicora, C.D., Piehowski, P.D., Purvine, S.O., Starrett, G.J., Goodwin, L.A., Smith, R.D., Lipton, M.S., Currie, C.R., 2012. Metagenomic and metaproteomic insights into bacterial communities in leaf-cutter ant fungus gardens. ISME J. 6, 1688–1701.
Aylward, F.O., Khadempour, L., Tremmel, D.M., McDonald, B.R., Nicora, C.D., Wu, S., Moore, R.J., Orton, D.J., Monroe, M.E., Piehowski, P.D., 2015. Enrichment and broad representation of plant biomass-degrading enzymes in the specialized hyphal swellings of *Leucoagaricus gongylophorus*, the fungal symbiont of leaf-cutter ants. PLoS One 10, e0134752.
Babendreier, D., Kalberer, N.M., Romeis, J., Fluri, P., Mulligan, E., Bigler, F., 2005. Influence of Bt-transgenic pollen, Bt-toxin and protease inhibitor (SBTI) ingestion on development of the hypopharyngeal glands in honeybees. Apidologie 36, 585–594.
Babendreier, D., Joller, D., Romeis, J., Bigler, F., Widmer, F., 2007. Bacterial community structures in honeybee intestines and their response to two insecticidal proteins. FEMS Microbiol. Ecol. 59, 600–610.
Baker, H.G., Baker, I., Hodges, S.A., 1998. Sugar composition of nectars and fruits consumed by birds and bats in the tropics and subtropics. Biotropica 30, 559–586.
Ballor, N., Paulson, I., Leadbetter, J., 2012. Genomic analysis reveals multiple [FeFe] hydrogenases and hydrogen sensors encoded by Treponemes from the H_2-rich termite gut. Microb. Ecol. 63, 282–294.
Bárány, I., Fadón, B., Risueño, M.C., Testillano, P.S., 2010. Cell wall components and pectin esterification levels as markers of proliferation and differentiation events during pollen development and pollen embryogenesis in *Capsicum annuum* L. J. Exp. Bot. 61, 1159–1175.
Barker, R.J., 1977. Some carbohydrates found in pollen and pollen substitutes are toxic to honey bees. J. Nutr. 107, 1859–1862.
Barker, R.J., Lehner, Y., 1974. Acceptance and sustenance value of naturally occurring sugars fed to newly emerged adult workers of honey bees (*Apis mellifera* L.). J. Exp. Zool. 187, 277–285.
Baroni Urbani, C., De Andrade, M., 1997. Pollen eating, storing, and spitting by ants. Naturwissenschaften 84, 256–258.
Bass, M., Cherrett, J.M., 1995. Fungal hyphae as a source of nutrients for the leaf-cutting ant *Atta sexdens*. Physiol. Entomol. 20, 1–6.
Bauer, S., Tholen, A., Overmann, J., Brune, A., 2000. Characterization of abundance and diversity of lactic acid bacteria in the hindgut of wood-and soil-feeding termites by molecular and culture-dependent techniques. Arch. Microbiol. 173, 126–137.
Belchior, C., Sendoya, S.F., Del-Claro, K., 2016. Temporal variation in the abundance and richness of foliage-dwelling ants mediated by extrafloral nectar. PLoS One 11, e0158283.
Belt, T., 1874. The Naturalist in Nicaragua. Murray, London.
Bignell, D.E., 1984. Direct potentiometric determination of redox potentials of the gut contents on the termites *Zootermopsis nevadensis* and *Cubitermes severus* and in three other arthropods. J. Insect Physiol. 30, 169–174.
Blatrix, R., Djiéto-Lordon, C., Mondolot, L., La Fisca, P., Voglmayr, H., McKey, D., 2012. Plant-ants use symbiotic fungi as a food source: new insight into the nutritional ecology of ant–plant interactions. Proc. R. Soc. B Biol. Sci. 279, 3940–3947.
Blochmann, F., 1882. Über das Vorkommen bakterienähnlicher Gebilde in den Geweben und Eiern verschiedener Insekten. Zbl. Bakt. 11, 234–240.

Blüthgen, N., Fiedler, K., 2004. Competition for composition: lessons from nectar-feeding ant communities. Ecology 85, 1479–1485.
Blüthgen, N., Gebauer, G., Fiedler, K., 2003. Disentangling a rainforest food web using stable isotopes: dietary diversity in a species-rich ant community. Oecologia 137, 426–435.
Bonnin, E., Garnier, C., Ralet, M.-C., 2014. Pectin-modifying enzymes and pectin-derived materials: applications and impacts. Appl. Microbiol. Biotechnol. 98, 519–532.
Botta, L., Brunori, F., Tulimieri, A., Piccinino, D., Meschini, R., Saladino, R., 2017. Laccase-mediated enhancement of the antioxidant activity of propolis and poplar bud exudates. ACS Omega 2, 2515–2523.
Bowman, S.M., Free, S.J., 2006. The structure and synthesis of the fungal cell wall. Bioessays 28, 799–808.
Boyd, N.D., Martin, M.M., 1975. Faecal proteinases of the fungus-growing ant, *Atta texana*: properties, significance and possible origin. Insect Biochem. 5, 619–635.
Breznak, J.A., Brune, A., 1994. Role of microorganisms in the digestion of lignocellulose by termites. Annu. Rev. Entomol. 39, 453–487.
Breznak, J.A., Prankratz, S., 1977. In situ morphology of the gut microbiota of wood-eating termites [*Reticulitermes flavipes* (Kollar) and *Coptotermes formosanus* (Shiraki)]. Appl. Environ. Microbiol. 33, 406–426.
Breznak, J.A., Switzer, J.M., 1986. Acetate synthesis from H_2 plus CO_2 by termite gut microbes. Appl. Environ. Microbiol. 52, 623–630.
Brightwell, R.J., Silverman, J., 2009. Effects of honeydew-producing hemipteran denial on local argentine ant distribution and boric acid bait performance. J. Econ. Entomol. 102, 1170–1174.
Brown, B.P., Wernegreen, J.J., 2019. Genomic erosion and extensive horizontal gene transfer in gut-associated Acetobacteraceae. BMC Genomics 20, 472.
Brune, A., 1998. Termite guts: the world's smallest bioreactors. Trends Biotechnol. 16, 16–21.
Brune, A., 2004. Symbiotic associations between termites and prokaryotes. In: Dworkin, M. (Ed.), The Prokaryotes: An Evolving Electronic Resource for the Microbiological Community. Springer-Verlag, NY.
Brune, A., 2014. Symbiotic digestion of lignocellulose in termite guts. Nat. Rev. Microbiol. 12, 168–180.
Brune, A., Deitrich, C., 2015. The gut microbiota of termites: digesting the diversity in the light of ecology and evolution. Annu. Rev. Microbiol. 69, 145–166.
Brune, A., Emerson, D., Breznak, J., 1995a. The termite gut microflora as an oxygen sink: microelectrode determination of oxygen and pH gradients in guts of lower and higher termites. Appl. Environ. Microbiol. 61, 2681–2687.
Brune, A., Miambi, E., Breznak, J., 1995b. Roles of oxygen and the intestinal microflora in the metabolism of lignin-derived phenylpropanoids and other monoaromatic compounds by termites. Appl. Environ. Microbiol. 61, 2688–2695.
Buckley, R., 1987. Interactions involving plants, homoptera, and ants. Annu. Rev. Ecol. Syst. 18, 111–135.
Bugg, T.D.H., Ahmad, M., Hardiman, E.M., Singh, R., 2011. The emerging role for bacteria in lignin degradation and bio-product formation. Curr. Opin. Biotechnol. 22, 394–400.
Bution, M.L., Caetano, F.H., 2008. Ileum of the *Cephalotes* ants: a specialized structure to harbor symbionts microorganisms. Micron 39, 897–909.
Byk, J., Del-Claro, K., 2010. Nectar- and pollen-gathering *Cephalotes* ants provide no protection against herbivory: a new manipulative experiment to test ant protective capabilities. Acta Ethol. 13, 33–38.
Byk, J., Del-Claro, K., 2011. Ant–plant interaction in the neotropical savanna: direct beneficial effects of extrafloral nectar on ant colony fitness. Popul. Ecol. 53, 327–332.

Caetano, F., da Cruz-Landim, C., 1985. Presence of microorganisms in the alimentary canal of ants of the tribe Cephalotini (Myrmicinae): location and relationship with intestinal structures. Naturalia 10, 37–47.
Carbajo, J.M., Junca, H., Terrón, M.C., González, T., Yagüe, S., Zapico, E., González, A.E., 2002. Tannic acid induces transcription of laccase gene cglcc1 in the white-rot fungus *Coriolopsis gallica*. Can. J. Microbiol. 48, 1041–1047.
Cassill, D.L., Butler, J., Vinson, S.B., Wheeler, D.E., 2005. Cooperation during prey digestion between workers and larvae in the ant, *Pheidole spadonia*. Insect. Soc. 52, 339–343.
Cembrowski, A.R., Reurink, G., Arcila Hernandez, L.M., Sanders, J.G., Youngerman, E., Frederickson, M.E., 2015. Sporadic pollen consumption among tropical ants. Insect. Soc. 62, 379–382.
Chen, I.-M.A., Chu, K., Palaniappan, K., Pillay, M., Ratner, A., Huang, J., Huntemann, M., Varghese, N., White, J.R., Seshadri, R., 2019. IMG/M v. 5.0: an integrated data management and comparative analysis system for microbial genomes and microbiomes. Nucleic Acids Res. 47, D666–D677.
Cleveland, L.R., 1923. Symbiosis between termites and their intestinal protozoa. Proc. Natl. Acad. Sci. 9, 424–428.
Cleveland, L.R., 1925a. Toxicity of oxygen for protozoa in vivo and in vitro: animals defaunated without injury. Biol. Bull. 48, 455–468.
Cleveland, L.R., 1925b. The physiological and symbiotic relationships between the intestinal protozoa of termites and their host with special reference to *Reticulitermes flavipes* Kollar. Biol. Bull. Mar. Biol. Woods Hole 46, 289–293.
Cleveland, L.R., 1926. Symbiosis among animals with special reference to termites and their intestinal flagellates. Q. Rev. Biol. 1, 51–64.
Cole, B.J., 1980. Repertoire convergence in two Mangrove ants, *Zacryptocerus varians* and *Camponotus* (Colobopsis) sp. Insect. Soc. 27, 265–275.
Cole, B.J., 1983. Assembly of mangrove ant communities: colonization abilities. J. Anim. Ecol. 52, 349–355.
Coley, P.D., Kursar, T.A., Machado, J.-L., 1993. Colonization of tropical rain forest leaves by epiphylls: effects of site and host plant leaf lifetime. Ecology 74, 619–623.
Cook, S.C., Davidson, D.W., 2006. Nutritional and functional biology of exudate-feeding ants. Entomol. Exp. Appl. 118, 1–10.
Costa, R.A.C., da Cruz-Landim, C., 2005. Hydrolases in the hypopharyngeal glands of workers of *Scaptotrigona postica* and *Apis mellifera* (Hymenoptera, Apinae). Genet. Mol. Res. 4, 616–623.
Cox-Foster, D.L., Conlan, S., Holmes, E.C., Palacios, G., Evans, J.D., Moran, N.A., Quan, P.-L., Briese, T., Hornig, M., Geiser, D.M., Martinson, V., van Engelsdorp, D., Kalkstein, A.L., Drysdale, A., Hui, J., Zhai, J., Cui, L., Hutchison, S.K., Simons, J.F., Egholm, M., Pettis, J.S., Lipkin, W.I., 2007. A metagenomic survey of microbes in honey bee colony collapse disorder. Science 318, 283–287.
Crailsheim, K., 1988. Intestinal transport of sugars in the honeybee (*Apis mellifera* L.). J. Insect Physiol. 34, 839–845.
Crotti, E., Rizzi, A., Chouaia, B., Ricci, I., Favia, G., Alma, A., Sacchi, L., Bourtzis, K., Mandrioli, M., Cherif, A., Bandi, C., Daffonchio, D., 2010. Acetic acid bacteria, newly emerging symbionts of insects. Appl. Environ. Microbiol. 76, 6963.
Czechowski, W., Markó, B., Erős, K., Csata, E., 2011. Pollenivory in ants (Hymenoptera: Formicidae) seems to be much more common than it was thought. Ann. Zool. 61, 519–525.
Dáttilo, W., Marquitti, F.M.D., Guimarães Jr., P.R., Izzo, T.J., 2014. The structure of ant–plant ecological networks: is abundance enough? Ecology 95, 475–485.
Davidson, D.W., 1997. The role of resource imbalances in the evolutionary ecology of tropical arboreal ants. Biol. J. Linn. Soc. 61, 153–181.

Davidson, D.W., 1998. Resource discovery versus resource domination in ants: a functional mechanism for breaking the trade-off. Ecol. Entomol. 23, 484–490.
Davidson, D.W., Cook, S.C., Snelling, R.R., Chua, T.H., 2003. Explaining the abundance of ants in lowland tropical rainforest canopies. Science 300, 969–972.
Davidson, D.W., Cook, S.C., Snelling, R.R., 2004. Liquid-feeding performances of ants (Formicidae): ecological and evolutionary implications. Oecologia 139, 255–266.
De Fine Licht, H.H., Boomsma, J.J., 2014. Variable interaction specificity and symbiont performance in Panamanian *Trachymyrmex* and *Sericomyrmex* fungus-growing ants. BMC Evol. Biol. 14, 244. https://doi.org/10.1186/s12862-014-0244-6.
De Fine Licht, H.H., Schiøtt, M., Mueller, U.G., Boomsma, J.J., 2010. Evolutionary transitions in enzyme activity of ant fungus gardens. Evolution 64, 2055–2069.
De Fine Licht, H.H., Schiøtt, M., Rogowska-Wrzesinska, A., Nygaard, S., Roepstorff, P., Boomsma, J.J., 2013. Laccase detoxification mediates the nutritional alliance between leaf-cutting ants and fungus-garden symbionts. Proc. Natl. Acad. Sci. 110, 583–587.
den Besten, G., van Eunen, K., Groen, A.K., Venema, K., Reijngoud, D.-J., Bakker, B.M., 2013. The role of short-chain fatty acids in the interplay between diet, gut microbiota, and host energy metabolism. J. Lipid Res. 54, 2325–2340.
Dillon, R.J., Dillon, V.M., 2004. The gut bacteria of insects: nonpathogenic interactions. Annu. Rev. Entomol. 49, 71–92.
Do, T.H., Nguyen, T.T., Nguyen, T.N., Le, Q.G., Nguyen, C., Kimura, K., Truong, N.H., 2014. Mining biomass-degrading genes through Illumina-based de novo sequencing and metagenomic analysis of free-living bacteria in the gut of the lower termite *Coptotermes gestroi* harvested in Vietnam. J. Biosci. Bioeng. 118, 665–671.
Donohoe, D.R., Garge, N., Zhang, X., Sun, W., O'Connell, T.M., Bunger, M.K., Bultman, S.J., 2011. The microbiome and butyrate regulate energy metabolism and autophagy in the mammalian colon. Cell Metab. 13, 517–526.
Douglas, A.E., 2013. Microbial brokers of insect-plant interactions revisited. J. Chem. Ecol. 39, 952–961.
Dussutour, A., Simpson, S., 2008. Carbohydrate regulation in relation to colony growth in ants. J. Exp. Biol. 211, 2224–2232.
Dussutour, A., Simpson, S.J., 2009. Communal nutrition in ants. Curr. Biol. 19, 740–744.
Ebert, A., Brune, A., 1997. Hydrogen concentration profiles at the oxic-anoxic interface: a microsensor study of the hindgut of the wood-feeding lower termite *Reticulitermes flavipes* (Kollar). Appl. Environ. Microbiol. 63, 4039–4046.
Eilmus, S., Heil, M., 2009. Bacterial associates of arboreal ants and their putative functions in an obligate ant-plant mutualism. Appl. Environ. Microbiol. 75, 4324.
Eisner, T., 1957. A comparative morphological study of the proventriculus of ants (Hymenoptera: Formicidae). Bull. Mus. Comp. Zool. Harv. Coll. 116, 437–490.
Ellegaard, K.M., Engel, P., 2019. Genomic diversity landscape of the honey bee gut microbiota. Nat. Commun. 10, 1–13.
Engel, P., Martinson, V.G., Moran, N.A., 2012. Functional diversity within the simple gut microbiota of the honey bee. Proc. Natl. Acad. Sci. 109, 11002–11007.
Engel, P., Kwong, W.K., Moran, N.A., 2013. *Frischella perrara* gen. nov., sp. nov., a gammaproteobacterium isolated from the gut of the honeybee, *Apis mellifera*. Int. J. Syst. Evol. Microbiol. 63, 3646–3651.
Eutick, M.L., O'Brien, R.W., Slaytor, M., 1978. Bacteria from the guts of Australian termites. Appl. Environ. Microbiol. 35, 823–828.
Evison, S.E.F., Ratnieks, F.L.W., 2007. New role for majors in Atta leafcutter ants. Ecol. Entomol. 32, 451–454.
Fagundes, R., Dáttilo, W., Ribeiro, S.P., Rico-Gray, V., Del-Claro, K., 2016. Food source availability and interspecific dominance as structural mechanisms of ant-plant-hemipteran multitrophic networks. Arthropod Plant Interact. 10, 207–220.

Febvay, G., Decharme, M., Kermarrec, A., 1984. Digestion of chitin by the labial glands of *Acromyrmex octospinosus* Reich (Hymenoptera: Formicidae). Can. J. Zool. 62, 229–234.
Feng, T., Liu, H., Xu, Q., Sun, J., Shi, H., 2015. Identification and characterization of two endogenous β-glucosidases from the termite *Coptotermes formosanus*. Appl. Biochem. Biotechnol. 176, 2039–2052.
Fiedler, K., 2007. Stable N-isotope signatures of central European ants—assessing positions in a trophic gradient. Insect. Soc. 54, 393–402.
Fischer, K.M., Volkl, W., Hoffmann, H.K., 2005. Honeydew production and honeydew sugar composition of polyphagous black bean aphid, *Aphis fabae* (Hemiptera: Aphididae) on various host plants and implications for ant-attendance. Eur. J. Entomol. 102, 155–160.
Floren, A., Biun, A., Linsenmair, E.K., 2002. Arboreal ants as key predators in tropical lowland rainforest trees. Oecologia 131, 137–144.
Fujita, A., Hojo, M., Aoyagi, T., Hayashi, Y., Arakawa, G., Tokuda, G., Watanabe, H., 2010. Details of the digestive system in the midgut of *Coptotermes formosanus* Shiraki. J. Wood Sci. 56, 222–226.
Gardner, P.T., White, T.A.C., McPhail, D.B., Duthie, G.G., 2000. The relative contributions of vitamin C, carotenoids and phenolics to the antioxidant potential of fruit juices. Food Chem. 68, 471–474.
Gilliam, M., 1979. Microbiology of pollen and bee bread: the genus *Bacillus*. Apidologie 10, 269–274.
Gilliam, M., 1997. Identification and roles of non-pathogenic microflora associated with honey bees. FEMS Microbiol. Lett. 155, 1–10.
Gilliam, M., Lorenz, B.J., Richardson, G.V., 1988. Digestive enzymes and micro-organisms in honey bees, *Apis mellifera*: influence of streptomycin, age, season and pollen. Microbios 223, 95–114.
González-Peña, M.d.C., Anderson, A.J., Smith, D.M., Moreira, G.S., 2002. Effect of dietary cellulose on digestion in the prawn *Macrobrachium rosenbergii*. Aquaculture 211, 291–303.
Gordon, D.M., 2012. The dynamics of foraging trails in the tropical arboreal ant *Cephalotes goniodontus*. PLoS One 7, e50472.
Gow, N.A.R., Latge, J.-P., Munro, C.A., 2017. The fungal cell wall: structure, biosynthesis, and function. Microbiol. Spectr. 5, 1–25.
Graber, J.R., Breznak, J.A., 2004. Physiology and nutrition of *Treponema primitia*, an H_2/CO_2-acetogenic spirochete from termite hindguts. Appl. Environ. Microbiol. 70, 1307–1314.
Gutierrez, J., 1956. The metabolism of cellulose-digesting, symbiotic flagellates of the genus *Trichonympha* from the termite *Zootermopsis*. J. Protozool. 3, 39–42.
Heil, M., Baumann, B., Krüger, R., Linsenmair, K.E., 2004. Main nutrient compounds in food bodies of Mexican acacia ant-plants. Chem 14, 45–52.
Hogan, M.E., Slaytor, M., Obrien, R.W., 1985. Transport of volatile fatty acids across the hindgut of the cockroach *Panesthia cribrata* Saussure and the termite, *Mastotermes darwiniensis* Froggatt. J. Insect Physiol. 31, 587–591.
Hölldobler, B., Wilson, E.O., 1990. The Ants. Harvard University Press.
Hu, Y., Lukasik, P., Moreau, C.S., Russell, J.A., 2014. Correlates of gut community composition across an ant species (*Cephalotes varians*) elucidate causes and consequences of symbiotic variability. Mol. Ecol. 23, 1284–1300.
Hu, Y., Sanders, J.G., Łukasik, P., D'Amelio, C.L., Millar, J.S., Vann, D.R., Lan, Y., Newton, J.A., Schotanus, M., Kronauer, D.J., 2018. Herbivorous turtle ants obtain essential nutrients from a conserved nitrogen-recycling gut microbiome. Nat. Commun. 9, 964.
Hungate, R.E., 1938. Studies on the nutrition of *Zootermopsis*. II. The relative importance of the termite and the protozoa in wood digestion. Ecology 19, 1–25.

Hungate, R.E., 1939. Experiments on the nutrition of *Zootermopsis*. III. The anaerobic carbohydrate dissimilation by the intestinal protozoa. Ecology 20, 230–245.

Hungate, R.E., 1943. Quantitative analysis on the cellulose fermentation by termite protozoa. Ann. Entomol. Soc. Am. 36, 730–739.

Hyodo, F., Azuma, J., Abe, T., 2005. Estimation of effect of passage through the gut of a lower termite, *Coptotermes formosanus* Shiraki, on lignin by solid-state CP/MAS ^{13}C NMR. Holzforschung 53, 244–246.

Inoue, T., Murashima, K., Azuma, J.-I., Sugimoto, A., Slaytor, M., 1997. Cellulose and xylan utilisation in the lower termite *Reticulitermes speratus*. J. Insect Physiol. 43, 235–242.

James, E.R., Okamoto, N., Burki, F., Scheffrahn, R.H., Keeling, P.J., 2013. *Cthulhu macrofasciculumque* n. g., n. sp. and *Cthylla microfasciculumque* n. g., n. sp., a newly identified lineage of parabasalian termite symbionts. PLoS One 8, e58509.

Janusz, G., Pawlik, A., Sulej, J., Świderska-Burek, U., Jarosz-Wilkołazka, A., Paszczyński, A., 2017. Lignin degradation: microorganisms, enzymes involved, genomes analysis and evolution. FEMS Microbiol. Rev. 41, 941–962.

Jeyaprakash, A., Hoy, M.A., Allsopp, M.H., 2003. Bacterial diversity in worker adults of *Apis mellifera* capensis and *Apis mellifera* scutellata (Insecta: Hymenoptera) assessed using 16S rRNA sequences. J. Invertebr. Pathol. 84, 96–103.

Juturu, V., Wu, J.C., 2012. Microbial xylanases: engineering, production and industrial applications. Biotechnol. Adv. 30, 1219–1227.

Katznelson, H., 1950. *Bacillus pulvifaciens* (n. sp.), an organism associated with powdery scale of honey-bee larvae. J. Bacteriol. 59, 153–155.

Kaya, M., Halıcı, M.G., Duman, F., Erdoğan, S., Baran, T., 2015. Characterisation of α-chitin extracted from a lichenised fungus species *Xanthoria parietina*. Nat. Prod. Res. 29, 1280–1284.

Ke, J., Sun, J.-Z., Nguyen, H.D., Singh, D., Lee, K.C., Beyenal, H., Chen, S.-L., 2010. In-situ oxygen profiling and lignin modification in guts of wood-feeding termites. Insect Sci. 17, 277–290.

Ke, J., Laskar, D.D., Singh, D., Chen, S., 2011a. In situ lignocellulosic unlocking mechanism for carbohydrate hydrolysis in termites: crucial lignin modification. Biotechnol. Biofuels 4, 17.

Ke, J., Singh, D., Yang, X., Chen, S., 2011b. Thermal characterization of softwood lignin modification by termite *Coptotermes formosanus* (Shiraki). Biomass Bioenergy 35, 3617–3626.

Kešnerová, L., Moritz, R., Engel, P., 2016. *Bartonella apis* sp. nov., a honey bee gut symbiont of the class Alphaproteobacteria. Int. J. Syst. Evol. Microbiol. 66, 414–421.

Kešnerová, L., Mars, R.A., Ellegaard, K.M., Troilo, M., Sauer, U., Engel, P., 2017. Disentangling metabolic functions of bacteria in the honey bee gut. PLoS Biol. 15, e2003467.

Kitade, O., Hayashi, Y., Takatsuto, K., Matsumoto, T., 2012. Variation and diversity of symbiotic protist composition in the damp wood termite *Hodotermopsis sjoestedti*. Jpn. J. Protozool. 45, 29–36.

Koch, H., Schmid-Hempel, P., 2011. Socially transmitted gut microbiota protect bumble bees against an intestinal parasite. Proc. Natl. Acad. Sci. 108, 19288–19292.

König, H., Li, L., Fröhlich, J., 2013. The cellulolytic system of the termite gut. Appl. Microbiol. Biotechnol. 97, 7943–7962.

Kuhnigk, T., Borst, E.-M., Ritter, P., Kampfer, A., Graf, H., Hertel, H., Konig, H., 1994. Degradation of lignin monomers by the hindgut flora of xylophagous termites. Syst. Appl. Microbiol. 17, 76–85.

Kuhnigk, T., Borst, E.M., Breunig, A., König, H., Collins, M.D., Hutson, R.A., Kämpfer, P., 1995. *Bacillus oleronius* sp.nov., a member of the hindgut flora of the termite *Reticulitermes santonensis* (Feytaud). Can. J. Microbiol. 41, 699–706.

Kwong, W.K., Moran, N.A., 2013. Cultivation and characterization of the gut symbionts of honey bees and bumble bees: description of *Snodgrassella alvi* gen. nov., sp. nov., a member of the family Neisseriaceae of the Betaproteobacteria, and *Gilliamella apicola* gen. nov., sp. nov., a member of Orbaceae fam. nov., Orbales ord. nov., a sister taxon to the order "Enterobacteriales" of the Gammaproteobacteria. Int. J. Syst. Evol. Microbiol. 63, 2008–2018.

Kwong, W.K., Moran, N.A., 2016a. Gut microbial communities of social bees. Nat. Rev. Microbiol. 14, 374.

Kwong, W.K., Moran, N.A., 2016b. *Apibacter adventoris* gen. nov., sp. nov., a member of the phylum Bacteroidetes isolated from honey bees. Int. J. Syst. Evol. Microbiol. 66, 1323–1329.

Kwong, W.K., Engel, P., Koch, H., Moran, N.A., 2014. Genomics and host specialization of honey bee and bumble bee gut symbionts. Proc. Natl. Acad. Sci. U. S. A. 111, 11509–11514.

Kwong, W.K., Medina, L.A., Koch, H., Sing, K.W., Soh, E.J.Y., Ascher, J.S., Jaffe, R., Moran, N.A., 2017. Dynamic microbiome evolution in social bees. Sci. Adv. 3, e1600513.

Lach, L., 2005. Interference and exploitation competition of three nectar-thieving invasive ant species. Insect. Soc. 52, 257–262.

Lach, L., Hobbs, R.J., Majer, J.D., 2009. Herbivory-induced extrafloral nectar increases native and invasive ant worker survival. Popul. Ecol. 51, 237–243.

Lanan, M., 2015. Spatiotemporal resource distribution and foraging strategies of ants (Hymenoptera: Formicidae). Myrmecol. News 20, 53–70.

Lanan, M.C., Rodrigues, P.A., Agellon, A., Jansma, P., Wheeler, D.E., 2016. A bacterial filter protects and structures the gut microbiome of an insect. ISME J. 10, 1866–1876.

Lapolla, J.S., 2009. Taxonomic revision of the Southeast Asian ant genus *Euprenolepis*. Zootaxa 2046, 1–25.

Larson, H.K., Goffredi, S.K., Parra, E.L., Vargas, O., Pinto-Tomas, A.A., McGlynn, T.P., 2014. Distribution and dietary regulation of an associated facultative Rhizobiales-related bacterium in the omnivorous giant tropical ant, *Paraponera clavata*. Naturwissenschaften 101, 397–406.

Leadbetter, J., Breznak, J., 1996. Physiological ecology of *Methanobrevibacter cuticularis* sp. nov. and *Methanobrevibacter curvatus* sp. nov., isolated from the hindgut of the termite *Reticulitermes flavipes*. Appl. Environ. Microbiol. 62, 3620–3631.

Leadbetter, J.R., Schmidt, T.M., Graber, J.R., Breznak, J.A., 1999. Acetogenesis from H_2 plus CO_2 by spirochetes from termite guts. Science 283, 686–689.

Leidy, J., 1881. The parasites of the termites. J. Acad. Nat. Sci. Phila. 8, 425–447.

Lin, J.Y., Hobson, W.J., Wertz, J.T., 2016a. *Ventosimonas gracilis* gen. nov., sp. nov., a member of the Gammaproteobacteria isolated from *Cephalotes varians* ant guts representing a new family, Ventosimonadaceae fam. nov., within the order 'Pseudomonadales'. Int. J. Syst. Evol. Microbiol. 66, 2869–2875.

Lin, J.Y., Russell, J.A., Sanders, J.G., Wertz, J.T., 2016b. *Cephaloticoccus* gen. nov., a new genus of 'Verrucomicrobia' containing two novel species isolated from *Cephalotes* ant guts. Int. J. Syst. Evol. Microbiol. 66, 3034–3040.

Lin, J.Y., Hobson, W.J., Wertz, J.T., 2017. *Saccharedens versatilis* gen. nov., sp. nov., a sugar-degrading member of the Burkholderiales isolated from *Cephalotes rohweri* ant guts. Int. J. Syst. Evol. Microbiol. 67, 447–453.

Littledyke, M., Cherrett, J.M., 1976. Direct ingestion of plant sap from cut leaves by the leaf-cutting ants *Atta cephalotes* (L.) and *Acromyrmex octospinosus* (reich) (Formicidae, Attini). Bull. Entomol. Res. 66, 205–217.

Lortzing, T., Calf, O.W., Böhlke, M., Schwachtje, J., Kopka, J., Geuß, D., Kosanke, S., van Dam, N.M., Steppuhn, A., 2016. Extrafloral nectar secretion from wounds of *Solanum dulcamara*. Nat. Plants 2, 16056.

Martins, C., Moreau, C.S., 2020. Influence of host phylogeny, geographical location and seed harvesting diet on the bacterial community of globally distributed *Pheidole* ants. PeerJ 8, e8492.

Martinson, V.G., Danforth, B.N., Minckley, R.L., Rueppell, O., Tingek, S., Moran, N.A., 2011. A simple and distinctive microbiota associated with honey bees and bumble bees. Mol. Ecol. 20, 619–628.

Mauldin, J.K., Smythe, R.V., Baxter, C.C., 1972. Cellulose catabolism and lipid synthesis by the subterranean termite, *Coptotermes formosanus*. Insect Biochem. 2, 209–217.

McEwen, S.E., Slaytor, M., O'Brien, R.W., 1980. Cellobiase activity in three species of Australian termites. Insect Biochem. 10, 563–567.

Mikaelyan, A., Thompson, C., Meuser, K., Zheng, H., Rani, P., Plarre, R., Brune, A., 2017. High-resolution phylogenetic analysis of Endomicrobia reveals multiple acquisitions of endosymbiotic lineages by termite gut flagellates. Environ. Microbiol. Rep. 9, 477–483.

Mohr, K.I., Tebbe, C.C., 2006. Diversity and phylotype consistency of bacteria in the guts of three bee species (Apoidea) at an oilseed rape field. Environ. Microbiol. 8, 258–272.

Moran, N.A., Hansen, A.K., Powell, J.E., Sabree, Z.L., 2012. Distinctive gut microbiota of honey bees assessed using deep sampling from individual worker bees. PLoS One 7, e36393.

Moreira-Soto, R.D., Sanchez, E., Currie, C.R., Pinto-Tomás, A.A., 2017. Ultrastructural and microbial analyses of cellulose degradation in leaf-cutter ant colonies. Microbiology 163, 1578–1589.

Mullin, C.A., Frazier, M., Frazier, J.L., Ashcraft, S., Simonds, R., Vanengelsdorp, D., Pettis, J.S., 2010. High levels of miticides and agrochemicals in North American apiaries: implications for honey bee health. PLoS One 5, e9754.

Nakashima, K., Watanabe, H., Saitoh, H., Tokuda, G., Azuma, J.-I., 2002. Dual cellulose-digesting system of the wood-feeding termite, *Coptotermes formosanus* Shiraki. Insect Biochem. Mol. Biol. 32, 777–784.

Nalepa, C.A., 2015. Origin of termite eusociality: trophallaxis integrates the social, nutritional, and microbial environments. Ecol. Entomol. 40, 323–335.

Nelsen, M.P., Ree, R.H., Moreau, C.S., 2018. Ant–plant interactions evolved through increasing interdependence. Proc. Natl. Acad. Sci. 115, 12253–12258.

Ni, J., Tokuda, G., 2013. Lignocellulose-degrading enzymes from termites and their symbiotic microbiota. Biotechnol. Adv. 31, 838–850.

Noda, S., Mantini, C., Meloni, D., Inoue, J.-I., Kitade, O., 2012. Molecular phylogeny and evolution of parabasia with improved taxon sampling and new protein markers of actin and elongation factor 1-a. PLoS One 7, e29938.

Noda, S., Shimizu, D., Yuki, M., Kitade, O., Ohkuma, M., 2018. Host-symbiont cospeciation of termite-gut cellulolytic protists of the genera *Teranympha* and *Eucomonympha* and their *Treponema* endosymbionts. Microbes Environ. 33, 26–33.

Noirot, C., Noirot-Thimothee, C., 1969. The digestive system. In: Krishna, K., Weesner, F.M. (Eds.), Biology of Termites. Academic Press, New York, NY, pp. 49–88.

O'Brien, G.W., Veivers, P.C., McEwen, S.E., Slaytor, M., O'Brien, R.W., 1979. The origin and distribution of cellulase in the termites, *Nasutitermes exitiosus* and *Coptotermes lacteus*. Insect Biochem. 9, 619–625.

Odelson, D.A., Breznak, J.A., 1983. Volatile fatty acid production by the hindgut microbiota of xylophagous termites. Appl. Environ. Microbiol. 45, 1602–1613.

Odelson, D.A., Breznak, J.A., 1985. Nutrition and growth characteristics of *Trichomitopsis termopsidis*, a cellulolytic protozoan from termites. Appl. Environ. Microbiol. 49, 614–621.

Ohashi, K., Sawata, M., Takeuchi, H., Natori, S., Kubo, T., 1996. Molecular cloning of cDNA and analysis of expression of the gene for alpha-glucosidase from the hypopharyngeal gland of the honeybee *Apis mellifera* L. Biochem. Biophys. Res. Commun. 221, 380–385.

Ohkuma, M., Brune, A., 2011. Diversity, structure, and evolution of the termite gut microbial community. In: Bignell, D.E., Roisin, Y., Lo, N. (Eds.), Biology of Termites: A Modern Synthesis. Springer, Dordrecht, Netherlands, pp. 413–438.

Olofsson, T.C., Vasquez, A., 2008. Detection and identification of a novel lactic acid bacterial flora within the honey stomach of the honeybee *Apis mellifera*. Curr. Microbiol. 57, 356–363.

Onchuru, T.O., Javier Martinez, A., Ingham, C.S., Kaltenpoth, M., 2018. Transmission of mutualistic bacteria in social and gregarious insects. Curr. Opin. Insect Sci. 28, 50–58.

Ottonetti, L., Tucci, L., Chelazzi, G., Santini, G., 2008. Stable isotopes analysis to assess the trophic role of ants in a Mediterranean agroecosystem. Agric. For. Entomol. 10, 29–36.

Paoli, P.P., Donley, D., Stabler, D., Saseendranath, A., Nicolson, S.W., Simpson, S.J., Wright, G.A., 2014. Nutritional balance of essential amino acids and carbohydrates of the adult worker honeybee depends on age. Amino Acids 46, 1449–1458.

Passos, L., Oliveira, P.S., 2004. Interaction between ants and fruits of *Guapira opposita* (Nyctaginaceae) in a Brazilian sandy plain rainforest: ant effects on seeds and seedlings. Oecologia 139, 376–382.

Pasti, M.B., Belli, M.L., 1985. Cellulolytic activity of *Actinomycetes* isolated from termites (Termitidae) gut. FEMS Microbiol. Lett. 26, 107–112.

Paul, J., Saxena, S., Varma, A., 1993. Ultrastructural studies of the termite (*Odontotermes obesus*) gut microflora and its cellulolytic properties. World J. Microbiol. Biotechnol. 9, 108–112.

Peterson, B.F., Scharf, M.E., 2018. Metatranscriptomic techniques for identifying cellulases in termites and their symbionts. Methods Mol. Biol. Clifton NJ 1796, 85–101.

Peterson, B.F., Stewart, H.L., Scharf, M.E., 2015. Quantification of symbiotic contributions to lower termite lignocellulose digestion using antimicrobial treatments. Insect Biochem. Mol. Biol. 59, 80–88.

Podterob, A.P., 2008. Chemical composition of lichens and their medical applications. Pharm. Chem. J. 42, 582–588.

Poissonnier, L.-A., Simpson, S.J., Dussutour, A., Buhl, J., 2020. Regulation of macronutrient intake in termites: a dietary self-selection experiment. J. Insect Physiol. 120, 103983.

Pollegioni, L., Tonin, F., Rosini, E., 2015. Lignin-degrading enzymes. FEBS J. 282, 1190–1213.

Powell, S., 2008. Ecological specialization and the evolution of a specialized caste in *Cephalotes* ants. Funct. Ecol. 22, 902–911.

Powell, S., Franks, N.R., 2006. Ecology and the evolution of worker morphological diversity: a comparative analysis with *Eciton* army ants. Funct. Ecol. 20, 1105–1114.

Pringle, E.G., 2019. Convergence, constraint and the potential for mutualism between ants and gut microbes. Mol. Ecol. 28, 699–702.

Quinlan, R.J., Cherrett, J.M., 1979. The role of fungus in the diet of the leaf-cutting ant *Atta cephalotes* (L.). Ecol. Entomol. 4, 151–160.

Radek, R., Strassert, J.F.H., Krüger, J., Meuser, K., Scheffrahn, R.H., Brune, A., 2014. Phylogeny and ultrastructure of *Oxymonas jouteli*, a Rostellum-free species, and *Opisthomitus longiflagellatus* sp. nov., Oxymonadid flagellates from the gut of *Neotermes jouteli*. Protist 165, 384–399.

Ricigliano, V.A., Fitz, W., Copeland, D.C., Mott, B.M., Maes, P., Floyd, A.S., Dockstader, A., Anderson, K.E., 2017. The impact of pollen consumption on honey bee (Apis mellifera) digestive physiology and carbohydrate metabolism. Arch. Insect Biochem. Physiol. 96, e21406. https://doi.org/10.1002/arch.21406.

Roche, R.K., Wheeler, D.E., 1997. Morphological specializations of the digestive tract of *Zacryptocerus rohweri* (Hymenoptera: Formicidae). J. Morphol. 234, 253–262.

Rodrigues, P.A.D.P., 2016. Bacterial symbionts at the colony and individual levels: integration through behavior and morphology an a social insect. PhD. The University of Arizona, Ann Arbor.

Rodríguez-Zaragoza, S., 1994. Ecology of free-living amoebae. Crit. Rev. Microbiol. 20, 225–241.
Romero Rodriguez, M.A., Vazquez Oderiz, M.L., Lopez Hernandez, J., Lozano, J.S., 1992. Determination of vitamin C and organic acids in various fruits by HPLC. J. Chromatogr. Sci. 30, 433–437.
Roulston, T.H., Cane, J.H., 2000. Pollen nutritional content and digestibility for animals. Plant Syst. Evol. 222, 187–209.
Rubin, B.E., Kautz, S., Wray, B.D., Moreau, C.S., 2019. Dietary specialization in mutualistic acacia-ants affects relative abundance but not identity of host-associated bacteria. Mol. Ecol. 28, 900–916.
Russell, J.A., Moreau, C.S., Goldman-Huertas, B., Fujiwara, M., Lohman, D.J., Pierce, N.E., 2009. Bacterial gut symbionts are tightly linked with the evolution of herbivory in ants. Proc. Natl. Acad. Sci. 106, 21236–21241.
Sabree, Z.L., Hansen, A.K., Moran, N.A., 2012. Independent studies using deep sequencing resolve the same set of core bacterial species dominating gut communities of honey bees. PLoS One 7, e41250.
Saha, B.C., 2003. Hemicellulose bioconversion. J. Ind. Microbiol. Biotechnol. 30, 279–291.
Salem, H., Florez, L., Gerardo, N., Kaltenpoth, M., 2015. An out-of-body experience: the extracellular dimension for the transmission of mutualistic bacteria in insects. Proc. Biol. Sci. 282, 20142957.
Sanders, J.G., Powell, S., Kronauer, D.J.C., Vasconcelos, H.L., Frederickson, M.E., Pierce, N.E., 2014. Stability and phylogenetic correlation in gut microbiota: lessons from ants and apes. Mol. Ecol. 23, 1268–1283.
Sapountzis, P., Zhukova, M., Hansen, L.H., Sørensen, S.J., Schiøtt, M., Boomsma, J.J., 2015. Acromyrmex leaf-cutting ants have simple gut microbiota with nitrogen-fixing potential. Appl. Environ. Microbiol. 81, 5527.
Sapountzis, P., Zhukova, M., Shik, J.Z., Schiott, M., Boomsma, J.J., 2018. Reconstructing the functions of endosymbiotic Mollicutes in fungus-growing ants. eLife 7, e39209.
Sapountzis, P., Nash, D.R., Schiøtt, M., Boomsma, J.J., 2019. The evolution of abdominal microbiomes in fungus-growing ants. Mol. Ecol. 28, 879–899.
Saran, R.K., Rust, M.K., 2005. Feeding, uptake, and utilization of carbohydrates by western subterranean termite (Isoptera: Rhinotermitidae). J. Econ. Entomol. 98, 1284–1293.
Sato, T., Kuwahara, H., Fujita, K., Noda, S., Kihara, K., Yamada, A., Ohkuma, M., Hongoh, Y., 2014. Intranuclear verrucomicrobial symbionts and evidence of lateral gene transfer to the host protist in the termite gut. ISME J. 8, 1008–1019.
Schäfer, A., Konrad, R., Kuhnigk, T., Kämpfer, P., Hertel, H., König, H., 1996. Hemicellulose-degrading bacteria and yeasts from the termite gut. J. Appl. Bacteriol. 80, 471–478.
Scheller, H.V., Ulvskov, P., 2010. Hemicelluloses. Annu. Rev. Plant Biol. 61, 263–289.
Schiott, M., Rogowska-Wrzesinska, A., Roepstorff, P., Boomsma, J.J., 2010. Leaf-cutting ant fungi produce cell wall degrading pectinase complexes reminiscent of phytopathogenic fungi. BMC Biol. 8, 156.
Schmitt-Wagner, D., Friedrich, M.W., Wagner, B., Brune, A., 2003a. Axial dynamics, stability, and interspecies similarity of bacterial community structure in the highly compartmentalized gut of soil-feeding termites (*Cubitermes* spp.). Appl. Environ. Microbiol. 69, 6018–6024.
Schmitt-Wagner, D., Friedrich, M.W., Wagner, B., Brune, A., 2003b. Phylogenetic diversity, abundance, and axial distribution of bacteria in the intestinal tract of two soil-feeding termites (*Cubitermes* spp.). Appl. Environ. Microbiol. 69, 6007–6017.
Schultz, T.R., Brady, S.G., 2008. Major evolutionary transitions in ant agriculture. Proc. Natl. Acad. Sci. 105, 5435.

Schultz, J.E., Breznak, J.A., 1978. Heterotrophic bacteria present in hindguts of wood-eating termites [*Reticulitermes flavipes* (Kollar)]. Appl. Environ. Microbiol. 35, 930–936.

Schultz, J.E., Breznak, J.A., 1979. Cross-feeding of lactate between *Streptococcus lactis* and *Bacteroides* sp. isolated from termite hindguts. Appl. Environ. Microbiol. 37, 1206–1210.

Shi, J., Cui, M., Yang, L., Kim, Y.J., Zhang, D., 2015a. Genetic and biochemical mechanisms of pollen wall development. Trends Plant Sci. 20, 741–753.

Shi, Y., Huang, Z., Han, S., Fan, S., Yang, H., 2015b. Phylogenetic diversity of Archaea in the intestinal tract of termites from different lineages. J. Basic Microbiol. 55, 1021–1028.

Shik, J.Z., Rytter, W., Arnan, X., Michelsen, A., 2018. Disentangling nutritional pathways linking leafcutter ants and their co-evolved fungal symbionts using stable isotopes. Ecology 99, 1999–2009.

Silva, A., Bacci Jr., M., Gomes de Siqueira, C., Correa Bueno, O., Pagnocca, F.C., Aparecida Hebling, M.J., 2003. Survival of *Atta sexdens* workers on different food sources. J. Insect Physiol. 49, 307–313.

Sinotte, V.M., Renelies-Hamilton, J., Taylor, B.A., Ellegaard, K.M., Sapountzis, P., Vasseur-Cognet, M., Poulsen, M., 2020. Synergies between division of labor and gut microbiomes of social insects. Front. Ecol. Evol. 7, 503. https://doi.org/10.3389/fevo.2019.00503.

Slaytor, M., 2000. Energy metabolism in the termite and its gut microbiota. In: Abe, T., Bignell, M., Higashi, E. (Eds.), Termite Evolution Sociality Symbiosis and Ecology. Kluwer Academic Publishers, Norwell, MA, pp. 307–332.

Söllinger, A., Tveit, A.T., Poulsen, M., Noel, S.J., Bengtsson, M., Bernhardt, J., Frydendahl Hellwing, A.L., Lund, P., Riedel, K., Schleper, C., Højberg, O., Urich, T., 2018. Holistic assessment of rumen microbiome dynamics through quantitative metatranscriptomics reveals multifunctional redundancy during key steps of anaerobic feed degradation. mSystems 3, e00038-18.

Sols, A., Cadenas, E., Alvarado, F., 1960. Enzymatic basis of mannose toxicity in honey bees. Science 131, 297–298.

Sorensen, A.A., Kamas, R.S., Vinson, S.B., 1983. The influence of oral secretions from larvae on levels of proteinases in colony members of *Solenopsis invicta* Buren (Hymenoptera: Formicidae). J. Insect Physiol. 29, 163–168.

Speranza, A., Calzoni, G.L., Pacini, E., 1997. Occurrence of mono- or disaccharides and polysaccharide reserves in mature pollen grains. Sex. Plant Reprod. 10, 110–115.

Su, L., Yang, L., Huang, S., Su, X., Li, Y., Wang, F., Wang, E., Kang, N., Xu, J., Song, A., 2016. Comparative gut microbiomes of four species representing the higher and the lower termites. J. Insect Sci. 16, 97. https://doi.org/10.1093/jisesa/iew081.

Tamschick, S., Radek, R., 2013. Colonization of termite hindgut walls by oxymonad flagellates and prokaryotes in Incisitermes tabogae, *I. marginipennis* and *Reticulitermes flavipes*. Eur. J. Protistol. 49, 1–14.

Tarmadi, D., Tobimatsu, Y., Yamamura, M., Miyamoto, T., Miyagawa, Y., Umezawa, T., Yoshimura, T., 2018. NMR studies on lignocellulose deconstructions in the digestive system of the lower termite *Coptotermes formosanus* Shiraki. Sci. Rep. 8, 1290.

Thayer, D.W., 1976. Facultative wood-digesting bacteria from the hindgut of the termite *Reticulitermes hesperus*. J. Gen. Microbiol. 96, 287–296.

Tholen, A., Brune, A., 2000. Impact of oxygen on metabolic fluxes and in situ rates of reductive acetogenesis in the hindgut of the wood-feeding termite *Reticulitermes flavipes*. Environ. Microbiol. 2, 436–449.

Tholen, A., Schink, B., Brune, A., 1997. The gut microflora of *Reticulitermes flavipes*, its relation to oxygen, and evidence for oxygen-dependent acetogenesis by the most abundant *Enterococcus* sp. FEMS Microbiol. Ecol. 24, 137–149.

Tokuda, G., Lo, N., Watanabe, H., Slaytor, M., Matsumoto, T., Noda, H., 1999. Metazoan cellulase genes from termites: intron/exon structures and sites of expression. Biochim. Biophys. Acta 1447, 146–159.

Tokuda, G., Tsuboi, Y., Kihara, K., Saitou, S., Moriya, S., Lo, N., Kikuchi, J., 2014. Metabolomic profiling of ^{13}C-labelled cellulose digestion in a lower termite: insights into gut symbiont function. Proc. R. Soc. B Biol. Sci. 281, 20140990.

Tokuda, G., Mikaelyan, A., Fukui, C., Matsuura, Y., Watanabe, H., Fujishima, M., Brune, A., 2018. Fiber-associated spirochetes are major agents of hemicellulose degradation in the hindgut of wood-feeding higher termites. Proc. Natl. Acad. Sci. U. S. A. 115, e11996–e12004.

Trager, W., 1934. The cultivation of a cellulose-digesting flagellate, *Trichomonas termopsidis* and of certain other termite protozoa. Biol. Bull. Mar. Biol. Woods Hole 66, 182–190.

Utami, Y.D., Kuwahara, H., Igai, K., Murakami, T., Sugaya, K., Morikawa, T., Nagura, Y., Yuki, M., Deevong, P., Inoue, T., Kihara, K., Lo, N., Yamada, A., Ohkuma, M., Hongoh, Y., 2019. Genome analyses of uncultured TG2/ZB3 bacteria in 'Margulisbacteria' specifically attached to ectosymbiotic spirochetes of protists in the termite gut. ISME J. 13, 455–467.

van Velzen, E., Etienne, R.S., 2013. The evolution and coexistence of generalist and specialist herbivores under between-plant competition. Theor. Ecol. 6, 87–98.

Veivers, P.C., Obrien, R.W., Slaytor, M., 1982. Role of bacteria in maintaining the redox potential in the hindgut of termites and preventing entry of foreign bacteria. J. Insect Physiol. 28, 947–951.

Viljakainen, S., Visti, A., Laakso, S., 2002. Concentrations of organic acids and soluble sugars in juices from nordic berries. Acta Agric. Scand. Sect. B Soil Plant Sci. 52, 101–109.

Watanabe, H., Noda, H., Tokuda, G., Lo, N., 1998. A cellulase gene of termite origin. Nature 394, 330–331.

Weber, N.A., 1966. Fungus-growing ants. Science 153, 587–604.

Weber, N.A., 1972. The fungus-culturing behavior of ants. Am. Zool. 12, 577–587.

Wenzel, M., Schonig, I., Berchtold, M., Kampfer, P., Konig, H., 2002. Aerobic and facultatively anaerobic cellulolytic bacteria from the gut of the termite *Zootermopsis angusticollis*. J. Appl. Microbiol. 92, 32–40.

Wertz, J.T., Breznak, J.A., 2007a. *Stenoxybacter acetivorans* gen. nov., sp. nov., an acetate-oxidizing obligate microaerophile among diverse O_2-consuming bacteria from termite guts. Appl. Environ. Microbiol. 73, 6819–6828.

Wertz, J.T., Breznak, J.A., 2007b. Physiological ecology of *Stenoxybacter acetivorans*, an obligate microaerophile in termite guts. Appl. Environ. Microbiol. 73, 6829–6841.

Wertz, J.T., Kim, E., Breznak, J.A., Schmidt, T.M., Rodrigues, J.L.M., 2011. Genomic and physiological characterization of the Verrucomicrobia isolate *Diplosphaera colotermitum* gen. nov., sp. nov. reveals microaerophily and nitrogen fixation genes. Appl. Environ. Microbiol. 78, 1544–1555. https://doi.org/10.1128/AEM.06466-11.

Wertz, J.T., Kim, E., Breznak, J.A., Schmidt, T.M., Rodrigues, J.L.M., 2018. Genomic and physiological characterization of the verrucomicrobia isolate *Geminisphaera colitermitum* gen. nov., sp. nov., reveals microaerophily and nitrogen fixation genes. Appl. Environ. Microbiol. 84, 13. Second correction for Wertz et al.

Wheeler, W.M., 1925. A new guest-ant and other new Formicidae from Barro Colorado Island, Panama. Biol. Bull. 49, 150–181.

Wills, B.D., Chong, C.D., Wilder, S.M., Eubanks, M.D., Holway, D.A., Suarez, A.V., 2015. Effect of carbohydrate supplementation on investment into offspring number, size, and condition in a social insect. PLoS One 10, e0132440.

Wilson, E.O., 1976. A social ethogram of the neotropical arboreal ant *Zacryptocerus varians* (Fr. Smith). Anim. Behav. 24, 354–363.

Witte, V., Maschwitz, U., 2008. Mushroom harvesting ants in the tropical rain forest. Naturwissenschaften 95, 1049–1054.
Wright, G.A., Nicolson, S.W., Shafir, S., 2018. Nutritional physiology and ecology of honey bees. Annu. Rev. Entomol. 63, 327–344.
Wu, W., Gu, D., Yan, S., Li, Z., 2019. RNA interference of endoglucanases in the formosan subterranean termite *Coptotermes formosanus* shiraki (Blattodea: Rhinotermitidae) by dsRNA injection or ingestion. J. Insect Physiol. 112, 15–22.
Xie, L., Zhang, L., Zhong, Y., Liu, N., Long, Y., Wang, S., Zhou, X., Zhou, Z., Huang, Y., Wang, Q., 2012. Profiling the metatranscriptome of the protistan community in *Coptotermes formosanus* with emphasis on the lignocellulolytic system. Genomics 99, 246–255.
Yamin, M.A., 1978. Axenic cultivation of the cellulolytic flagellate *Trichomitopsis termopsidis* (Cleveland) from the termite *Zootermopsis*. J. Protozool. 25, 535–538.
Yamin, M.A., 1980. Cellulose metabolism by the termite flagellate *Trichomitopsis termopsidis*. Appl. Environ. Microbiol. 39, 859–863.
Yamin, M.A., 1981. Cellulose metabolism by the flagellate *Trichonympha* from a termite is independent of endosymbiotic bacteria. Science 211, 58–59.
Yuki, M., Kuwahara, H., Shintani, M., Izawa, K., Sato, T., Starns, D., Hongoh, Y., Ohkuma, M., 2015. Dominant ectosymbiotic bacteria of cellulolytic protists in the termite gut also have the potential to digest lignocellulose. Environ. Microbiol. 17, 4942–4953.
Zavarzina, A.G., Romankevich, E.A., Peresypkin, V.I., Ulyantzev, A.S., Belyaev, N.A., Zavarzin, A.A., 2015. Lignin phenols derivatives in lichens. Dokl. Biochem. Biophys. 465, 394–397.
Zheng, H., Nishida, A., Kwong, W.K., Koch, H., Engel, P., Steele, M.I., Moran, N.A., 2016. Metabolism of toxic sugars by strains of the bee gut symbiont *Gilliamella apicola*. mBio 7, e01326-16.
Zheng, H., Powell, J.E., Steele, M.I., Dietrich, C., Moran, N.A., 2017. Honeybee gut microbiota promotes host weight gain via bacterial metabolism and hormonal signaling. Proc. Natl. Acad. Sci. U. S. A. 114, 4775–4780.
Zheng, H., Perreau, J., Powell, J.E., Han, B., Zhang, Z., Kwong, W.K., Tringe, S.G., Moran, N.A., 2019. Division of labor in honey bee gut microbiota for plant polysaccharide digestion. Proc. Natl. Acad. Sci. 116, 25909–25916.
Zhou, X., Wheeler, M.M., Oi, F.M., Scharf, M.E., 2008. RNA interference in the termite *Reticulitermes flavipes* through ingestion of double-stranded RNA. Insect Biochem. Mol. Biol. 38, 805–815.

Further reading

Alippi, A.M., López, A.C., Aguilar, O.M., 2002. Differentiation of *Paenibacillus* larvae subsp. larvae, the cause of American foulbrood of honeybees, by using PCR and restriction fragment analysis of genes encoding 16S rRNA. Appl. Environ. Microbiol. 68, 3655–3660.
Bourguignon, T., Lo, N., Dietrich, C., Šobotník, J., Sidek, S., Roisin, Y., Brune, A., Evans, T.A., 2018. Rampant host switching shaped the termite gut microbiome. Curr. Biol. 28. 649–654.e2.
Ravenscraft, A., Berry, M., Hammer, T., Peay, K., Boggs, C., 2019. Structure and function of the bacterial and fungal gut microbiota of neotropical butterflies. Ecol. Monogr. 89, e01346. https://doi.org/10.1002/ecm.1346.

CHAPTER FOUR

Microbial symbionts of herbivorous species across the insect tree

Enric Frago[a,b], Sharon E. Zytynska[c], Nina E. Fatouros[d]
[a]CIRAD, CBGP, Montpellier, France
[b]CBGP, CIRAD, INRA, IRD, Montpellier SupAgro, University Montpellier, Montpellier, France
[c]Institute of Integrative Biology, Department of Evolution, Ecology and Behaviour, University of Liverpool, Liverpool, United Kingdom
[d]Biosystematics Group, Wageningen University and Research, Wageningen, The Netherlands

Contents

Abstract

Microbes play crucial roles in the biology of herbivorous insects, and the last decade has provided exciting new evidence for a prominent role of microbial symbiosis in detoxification of plant toxins, manipulation of plant defences and defence against natural enemies. We provide an order by order update of symbioses across herbivorous insects, particularly focusing on recent published evidence, and on how symbionts interact

Advances in Insect Physiology, Volume 58
ISSN 0065-2806
https://doi.org/10.1016/bs.aiip.2020.04.002

with the defensive system of the plant. While the hemimetabolous Hemiptera order largely relies on obligatory microbial symbioses, we did not find such a close relationship between symbionts and hosts in the other three orders Orthoptera, Phasmatodea and Thysanoptera. These three orders mostly harbour transient gut symbionts and/or rely on laterally transferred genes from microbes. Despite the radical changes and harsh conditions during metamorphosis, numerous holometabolous species transmit symbionts vertically and show close associations with both intra- and extracellular symbionts. The last section of this book chapter discusses the role that symbionts will play in future scenarios of global warming, but also their implications for the transmission of plant viruses and modern agriculture.

1. Introduction

The natural world is full of examples of intricate symbiotic interactions between species like the coevolution of orchid bees and orchids, or ants and acacia trees. Many interactions are not seen by the naked eye but can have dramatic impacts for the individuals involved. Insects have evolved obligate (required) and facultative (helpful, but not required) symbiotic associations with bacteria, fungi and protozoans. Symbiotic interactions in insect herbivores have provided key knowledge into the ecology and evolution of animal symbioses (Douglas, 2011; McCutcheon et al., 2019), exemplifying how such interactions can influence animal diet and defence against natural enemies. Such unseen interactions are the focus of this chapter.

Technological advances, and molecular biology in particular, have made the study of insect-microbe interactions more open and accessible across the world. This has resulted in a quick increase in the number of research papers describing different systems, yet there is still a bias towards model systems that have been studied for longer. Unsurprisingly, this has also led to a rise in the number of reviews and forward-looking perspectives on this topic that consider broader groups (Douglas, 2015; Flórez et al., 2015; Hammer and Bowers, 2015; Moran et al., 2019), yet many are taxon-focused (Biedermann and Vega, 2020; Dicke et al., 2020; Duplouy and Hornett, 2018; Kaltenpoth and Engl, 2014; Mason et al., 2019a; McLean et al., 2016; Paniagua-Voirol et al., 2018; Sudakaran et al., 2017). We present a systematic overview of symbionts within the different orders of herbivorous insects (those that feed on living tissue of vascular plants). In each section we present an overview of symbiosis in the group and highlight relevant recent papers together with any previous reviews that go into more detailed specifics for that group. Our aim for this chapter is to present a broad overview of insect-symbiont-plant interactions, and an appreciation of the potential impacts these can have in the future.

Insects have evolved herbivory several times, and particularly within eight orders of winged insects classified together within the subclass Pterygota (McKenna et al., 2019; Wiens et al., 2015). Herbivorous species-rich clades, i.e. Hemiptera (stink, true and shield bugs, leaf and tree hoppers, psyllids, whiteflies, aphids, and cicadas), Lepidoptera (moths and butterflies) and Coleoptera (beetles) make up the vast majority of the approximately 450,000 described herbivore species (Stork, 2018). Feeding on plants is challenging, often because of high levels of toxins or low levels of nutrients, and due to this many insect herbivores have evolved symbiotic relationships with microbes that have enabled them to feed on this low-quality food source (Hammer and Bowers, 2015). In some cases, herbivorous insects have taken up beneficial genes from the microbes (horizontal gene transfer) leaving the microbe itself obsolete to the insect (McKenna et al., 2019; Shelomi et al., 2016; Wybouw et al., 2016). About half of all described insect species are herbivores. So the question arises whether this great success of insect herbivory is largely due to vertically-transmitted beneficial symbionts and/or ancient lateral gene transfers from bacteria, viruses and/or fungi. Alternatively, opportunistic microbes, co-opted from the plants' phyllosphere or the soil, could further benefit these herbivores. For example, there are Lepidopteran species whose larvae have been shown to largely consist of transient gut bacterial communities rather than resident symbionts (Hammer et al., 2019b; Mason et al., 2019a; Moran et al., 2019; Paniagua-Voirol et al., 2018).

Plants have a complex defensive system. These defences are considered constitutive when they are always present, or induced when they are triggered upon insect attack. Constitutive defences often come from the secondary metabolism of the plant and include a whole suite of compounds that exert toxic effects on insects (Schoonhoven et al., 2005). These defences, however, can be costly and most plants have also evolved a defensive layer that is only induced once the plant perceives the attack by the insect, and which is similar to the immune system of animals. Induced defences can be either direct or indirect. Direct ones are aimed at killing, poisoning or repelling herbivores, whereas indirect ones enhance the effectiveness of the attacker's natural enemies (Dicke and Baldwin, 2010; Hilker and Fatouros, 2015; Pieterse et al., 2013). Plant volatiles are important in mediating induced indirect defences as they are capable of making often inconspicuous herbivores more easily located by predators and parasitic wasps and flies (Turlings and Erb, 2018). One way that plants fine-tune their antiherbivore defences is via phytohormones, which are induced differently depending, among other factors, on the specific stressor that attacks them (Stam et al., 2013). While the jasmonic acid (JA) pathway is mainly induced through attack by leaf-chewing insects and some phloem feeders, the salicylic acid (SA) pathway mainly acts upon induction by phloem feeders and plant pathogens. Both pathways 'crosstalk' so that upregulation of one pathway can downregulate the other, especially under multiple attack, i.e. JA-SA crosstalk (Erb et al., 2012; Lazebnik et al., 2014). Some insects that trigger induced plant defences upon feeding have evolved counter defensive strategies that down-regulate these responses. These strategies can be achieved using the intrinsic metabolic repertoire of the insect (Erb and Reymond, 2019), or by co-opting this function from a microbial symbiont. During the last few years, numerous studies have indicated symbiont roles in modulating plant hormones and antiherbivore defences, and, in particular, altering crosstalk between JA and SA pathways. These discoveries motivated several review papers, some of them very recent (Casteel and Hansen, 2014; Engel and Moran, 2013; Frago et al., 2012; Giron et al., 2013, 2017; Hammer and Bowers, 2015; Mason et al., 2019a; Oliver and Martinez, 2014; Shikano et al., 2017; Sugio et al., 2015).

1.1 Scope of this review

This review will focus on microbial symbionts of herbivorous species, and particularly their effects on host plant use. There have been some recent reviews on this topic, and this is why we decided to perform an order by order revision, focusing on recent publications: we hereby focus on four hemimetabolous and four holometabolous insect orders altogether containing most herbivorous insect species. The last decade has seen an important amount of research on defensive insect symbioses. This topic has also been extensively reviewed in recent years (Flórez et al., 2015; Kaltenpoth and Engl, 2014; Monticelli et al., 2019; Oliver et al., 2010; Zytynska and Meyer, 2019), and will not be the main focus of this book chapter. However, the most well studied examples will be discussed, particularly when defensive symbioses may help understanding host plant use by herbivores. Manipulation of plant defences might imply symbiont colonization of the plant and replication in it, and this is commonly the case of plant pathogens that are vectored by insects. Although in these systems the insect has often been considered a passive vector, on many occasions the pathogen becomes a mutualist because the insect obtains a benefit from feeding on a diseased plant (Frago et al., 2012). This type of symbiosis has been largely documented in phloem feeding Hemiptera, particularly between aphids and thrips that vector viruses, but also between psyllids and bacteria. Since these types of interactions have also been recently reviewed by Eigenbrode et al. (2018), we will discuss here cases in which symbionts help their hosts manipulate plant physiology without systemically colonizing the plant. A last section is also included where the implications of symbionts in insect herbivores in a changing world are discussed.

2. Hemimetabola

The development of hemimetabolous insects is characterized by a series of moults, leading to an adult stage with wings. From the 18 hemi metabolous insect orders, four of them largely contain herbivorous species, namely Orthoptera (grasshoppers and crickets), Phasmatodea (walking sticks), Thysanoptera (thrips) and the largest one, Hemiptera (including, among others, aphids, shield bugs, tree and plant hoppers and whiteflies) (Fig. 1). Hemimetabolous insects are more likely to harbour stable symbiotic associations than holometabolous ones because their body plan is more stable over development (Hammer and Moran, 2019). Many species are associated with obligate symbionts that allow them to exploit otherwise unusable feeding niches, as seen with hemipterans in the suborder Sternorrhyncha that are specialized on plant phloem sap (see below) (Sudakaran et al., 2017; Wernegreen, 2004). Yet, other hemimetabolous lineages such as the Phasmatodea seem to have evolved a lifestyle largely independent of microbial symbionts (Shelomi et al., 2013, 2016).

2.1 Orthoptera

The few studies dissecting the microbial community of grasshoppers and crickets mainly focus on edible species sold for human consumption (Stoops et al., 2016) or pest species (Idowu et al., 2009). Orthoptera are chewing insects that macerate foliar tissue using strong mandibles. Evidence shows resident gut microbial communities in Orthoptera are rare, and that the microbiome is mostly acquired from the environment (Mason et al., 2019a). For example, guts of first instar variegated grasshoppers, *Zonocerus variegatus*, hatching from eggs were found to be sterile (Ademolu and Idowu, 2011). Further, Dillon & Charnley (2002) found that the gut microbiota of desert locusts, *Schistocerca gregaria* (Fig. 2A) was dominated by Enterobacteriaceae found in

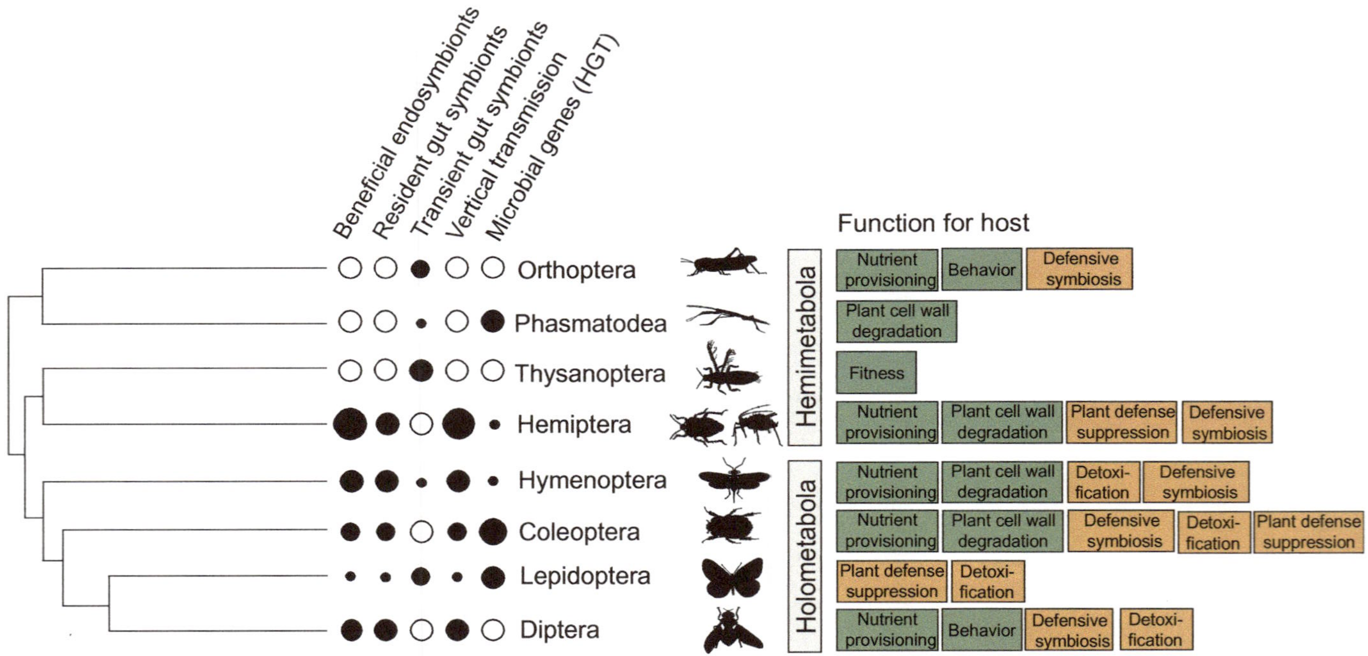

Fig. 1 Different types of symbionts, transmission routes and functions for hosts across herbivorous insect orders using the phylogenetic tree from Misof et al. (2014). Herbivorous insects are distributed across four hemimetabolous and four holometabolous orders in total consisting of about 450,000 species (Wiens et al., 2015). Traits can be very common (large black circle), less common (small black circle), or absent or not known/studied (white circle). Green boxes represent traits that are essential for the host, while orange ones represent benefits against natural enemies or plants defences.

Fig. 2 Examples of herbivorous insects and interactions with microbial symbionts: (A) through gut microbial metabolism, the desert locust *Schistocerca gregaria* produces phenolics used as protection against pathogens and/or components of the locusts' aggregation pheromone, (B) phasmids like *Clonopsis gallica* acquire plant cell wall degrading enzymes from bacteria by horizontal gene transfer, (C) western flower thrips *Frankliniella occidentalis* are associated with γ-proteobacteria like *Erwinia* and *Pantoea*, (D) aphids host up to nine common endosymbionts, and the symbiont *Regiella insecticola* protects the green peach aphid *Myzus persicae* from parasitoids, (E) whiteflies like *Bemisia tabaci* are reported to host seven facultative symbionts namely *Rickettsia*, *Wolbachia*, *Hamiltonella*, *Arsenophonus*, *Cardinium*, *Fritschea*, and *Hemipteriphilus*, (F) psyllids like *Psylla pyri* are associated with *Carsonella ruddii* symbionts, (G) the bean bug *Riptortus pedestris* (Vengolis, https://commons.wikimedia.org/wiki) harbours *Burkholderia* gut symbionts that confers them with insecticide resistance, (H) termites host microbiomes that are predominantly composed of resident, beneficial microbes, (I) fungus growing leaf-cutter ants keep vertically transmitted *Streptomyces* bacteria on their cuticle that produce antibiotics to suppress fungus garden-parasites, (J) solitary bees show a great diversity

(Continued)

the environment, suggesting they are acquired after hatching. Despite the transient nature of these associations, the relationship between some of the acquired bacteria and locusts seems to be mutualistic. Secondary plant chemicals like phenolics, are degraded by microbial metabolism in the locusts' gut and the products used as protection against pathogens and/or components of the locusts' aggregation pheromone (Dillon and Charnley, 2002; Dillon et al., 2000). Previous digestion of plant material in the locusts' gut, however, appeared to be a prerequisite for the production of phenolics by the gut bacteria. Microbial transformation of plant secondary metabolites may be widespread in this order, as other Orthoptera species were also shown to have antimicrobial phenolics in their gut fluid (Dillon and Charnley, 2002).

2.2 Phasmatodea

To break down ingested foliar tissue or to facilitate plant penetration, herbivorous insects depend on plant cell wall degrading enzymes (PCWDEs) like cellulases and pectinases, that are often produced by symbiotic bacteria. However, recent studies suggest that phasmids (stick insects) do not depend on gut symbionts for plant digestion, but potentially on transient bacteria

Fig. 2—Cont'd and variability in their gut microbiome depending on the environment and the way microbes are transmitted, (K) tenthredinid sawfly larvae show a low apparent diversity of gut bacteria mainly attributed to the host plant, (L) the tortoise beetle *Cassida rubiginosa* acquired an extracellular bacterium, *Stammera,* that took over pectinase activities to digest plant cell wall components, (M) some bark and ambrosia beetles like the European Shot-hole Borer *Anisandrus dispar* are fungal farmers and transport their symbiotic fungi in mycangia, (N) oral secretions of Colorado potato beetle *Leptinotarsa decemlineata,* contain gut bacteria deposited into wounds that suppress plant defences, (O) extracellular *Burkholderia gladioli* bacteria were shown to protect eggs of Lagriinae beetles (Tenebrionidae) from pathogenic microbes, (P) the microbiome of caterpillars of the cabbage moth *Mamestra brassicae* is modulated by bacteria and fungi present in the soil and host plant, (Q) eggs and hatching caterpillars of the cabbage white butterfly *Pieris brassicae* contain low amounts of bacteria which do not affect egg-mediated priming of plant defences, (R) the microbiome of different wild mushroom-feeding *Drosophila* species are very similar, probably reflecting similarities in the ecological niche exploited, (S) tephritids like the Cherry fruit fly *Rhagoletis cerasi* carry in their guts yeasts and bacteria that are transferred to the fruit by females during oviposition. *Part A: credits: Antoine Foucart; Part B: credits: Nina E. Fatouros; Part C: credits: Tibor Bukovinszky; Part D: credits: Nina E. Fatouros; Part E: credits: Tibor Bukovinszky; Part F: credits: Tibor Bukovinszky; Part H: credits: Nina E. Fatouros; Part I: credits: Nina E. Fatouros; Part J: credits: Jitte Groothuis; Part K: credits: Nina E. Fatouros; Part L: credits: Nina E. Fatouros; Part M: credits: Peter H.W. Biedermann; Part N: credits: Nina E. Fatouros; Part O: credits: Martin Kaltenpoth; Part P: credits: Tibor Bukovinszky; Part Q: credits: Tibor Bukovinszky; Part R: credits: Jitte Groothuis; Part S: credits: Tibor Bukovinszky.*

obtained from the diet or environment (Shelomi et al., 2015). Similar to caterpillars and Orthoptera, the gut of phasmids is a straight and narrow tube lacking chambers that could harbour obligate symbionts. In addition, egg-smearing, coprophagy or other known modes of transmission of resident symbionts that are found in other groups like cockroaches (Blattodea) and termites (Isoptera) are absent. This evidence suggests that vertical transmission of stable symbiont communities is unlikely in this group (Salem et al., 2015). Next generation sequencing and microscopy of two phasmid species revealed that the only heritable symbionts were *Spiroplasma*, which can manipulate the reproduction of parthenogenetic species (Shelomi et al., 2013).

The lack of PCWDE-producing heritable symbionts is likely due to ancient horizontal transfer of microbial genes. A major clade of phasmids (the Euphasmatodea that includes all lineages but Timematodea) produces pectinases endogenously without the need for a microbe (Fig. 2B). This evolutionary novelty was likely acquired from bacteria by horizontal gene transfer because the most similar homologues of these genes were found in γ-proteobacteria (Shelomi et al., 2016). Horizontal gene transfer of pectinase genes has also been shown for beetles, butterflies and moths (see below), but the phasmatodean genes are not related to the latter two (Shelomi et al., 2016). Whether the pectinase genes identified in Euphasmatodea were acquired from stable symbiotic associations or from transient microbes occurring in ancestral stick insect species is a question that still needs to be clarified.

2.3 Thysanoptera

The role of microbial symbionts has been investigated only in a few species of thrips, mainly pest species. Some herbivorous thrips seem to harbour gut bacteria, consisting largely of Enterobacteriaceae. In a recent review, Schausberger (2018) presents an overview of the endosymbiotic and gut/saliva bacteria from the thrips family Thripidae, which contains most pest species. Of particular interest are two of the most widespread pest species, the western flower thrips, *Frankliniella occidentalis* (Fig. 2C) and the onion thrips, *Thrips tabaci*, both associated with γ-proteobacteria, including *Erwinia* and *Pantoea* or *Pantoea*-like bacteria. Empirical studies have shown *Erwinia* to positively affect thrips fitness, yet the transmission of these gut bacteria is likely not vertical and they are probably acquired from the plant surface (De Vries et al., 2004). *Pantoea* bacteria are also likely to be transmitted horizontally among thrips via the plant. This bacterium causes plant diseases like centre rot in onion, and is vectored by the onion-infesting thrips *F. fusca*. The bacteria are transmitted to the plant and other thrips through faeces and persist in thrips through several life stages (Dutta et al., 2016). It is not known, however, if this bacterium confers any fitness advantage to its host, so its condition of insect mutualist remains to be demonstrated. A more recent study on *T. tabaci* identified a gut bacterial community mainly consisting of Proteobacteria (Gawande et al., 2019). In this study, the diversity and structure of the bacterial community depended on the environment and habitat where thrips lived. Based on this evidence, thrips seem to primarily acquire microbes from their environment (Dickey et al., 2014), and evidence for beneficial endosymbionts and vertically transmitted resident gut bacteria is thus scarce in this group. This contrasts with the closely-related order Hemiptera, probably reflecting the difference in feeding mode (i.e. cell-content feeding vs phloem feeding).

2.4 Hemiptera

Phloem-feeding Hemiptera, and aphids in particular, have long been used as model systems to study nutritional symbioses. Plant phloem is of poor nutritional quality and feeding exclusively on this substrate is challenging because many essential amino-acids that are usually needed for animal development are lacking. To complement this poor diet, many phloem-feeding Hemiptera have engaged in obligatory associations with nutritional bacteria (Douglas, 1998). Acquisition of these symbionts was at the core of the diversification of many groups, including *Buchnera aphidicola* in aphids, *Portiera aleyrodidarum* in whiteflies and *Carsonella ruddii* in psyllids (Dolling, 1991; Moran et al., 2008) (Fig. 2D–F). There are many reviews on the evolution and mechanistic functioning of obligatory symbioses in hemipterans (Douglas, 1998, 2015; Moran et al., 2005; Sudakaran et al., 2017) so we will cover this topic here in less detail, and focus on the facultative associations that are not required for survival. The role of hemipteran facultative endosymbionts in host plant use, adaptation to abiotic conditions and interactions with natural enemies has been studied for more than a decade, particularly in well-established model species like aphids and whiteflies (Brownlie and Johnson, 2009; Frago et al., 2012; Guo et al., 2017; Kanakala and Ghanim, 2019; Moran et al., 2008; Oliver et al., 2010, 2014; Vorburger, 2018; Zytynska and Meyer, 2019; Zytynska and Weisser, 2016). Studies on these types of symbioses are mostly limited to bacteria, with some fungi examples like that of the brown planthopper, *Nilaparvata lugens*

(Chen et al., 1981). Little effort has been made to identify the importance of other microbial groups, including protozoa, which may be reasonably suspected to play important roles as well.

2.4.1 Aphids

Aphid endosymbionts are predominantly vertically transmitted from mother to offspring during clonal reproduction, with some reported cases of transmission failure in the field depending on the symbiont species, other hosted symbionts, and host genotype (Rock et al., 2018). Horizontal transfer of symbionts among aphids can also occur during sexual reproduction, via the host plant, by parasitic wasps when ovipositing eggs into the aphids, or even through infected honeydew (reviewed in Chrostek et al., 2017 and Zytynska and Weisser, 2016). Thus, on average, only 40–60% of aphids in a population are infected by even the most prevalent facultative symbiont.

Aphid facultative symbionts have been particularly well studied in the pea aphid, *Acyrthosiphon pisum*, and have been found to provide a variety of services to their hosts including protection against parasitic wasps, predators, fungal pathogens and heat shock (reviewed by Guo et al., 2017; Oliver et al., 2010, 2014). Whether facultative symbionts play a role in host-plant specialization in this species is a question that has long been debated. This taxon is considered as a complex of species because it has a continuum of populations that are specialized on various plant species in the Fabaceae family, and strong associations between particular species of symbionts and host plants are observed (Oliver et al., 2010; Peccoud et al., 2009; Peccoud and Simon, 2010; Tsuchida et al., 2002; Via et al., 2000). Tsuchida et al. (2004) provided the first experimental evidence of a nutritional role of facultative symbionts in *A. pisum*. This study showed that the facultative γ-proteobacterium, called pea aphid U-type symbiont (PAUS; now known as *Regiella insecticola*) had a positive effect on host fitness when feeding on white clover, *Trifolium repens*, and this effect could be transferred to another aphid *Megoura crassicauda* by artificially transferring the symbiont. This result was not recovered, however, in a later study by Leonardo (2004) revealing that the effect of the symbiont depended on the specific combination of aphid and symbiont genotypes. A more recent study that tracked the evolutionary history of the pea aphid taxon using phylogenetic techniques revealed that acquisition of particular symbiont species coincided with the colonization of novel host plants (Henry et al., 2013). Facultative symbionts thus played a role in the colonization of new host plants, but whether this was due to nutritional provisioning is a question that cannot be answered through phylogenetic inference. The main reason is that, in *A. pisum* some facultative symbionts are known to increase their hosts' resistance against natural enemies and environmental stressors (Flórez et al., 2015; Oliver et al., 2010, 2014), which are factors that may modulate colonization of new host plants. As suggested by McLean et al. (2016), for example, *A. pisum* aphids collected on clover often carry the symbiont *Regiella insecticola*, which is also known to confer resistance against the specialist aphid fungal pathogen *Pandora*. Is it possible that acquisition of this symbiont allowed aphids to colonize clover, but that this was selected because pressure by the natural enemy was stronger on this plant? This idea has been explored in a few studies, often with mixed results (Hrček et al., 2016; Sochard et al., 2019) as not all symbiont genotypes protect likewise, and there are strong interactions between the protection conferred by the symbiont and the intrinsic resistance that symbiont-free aphids may also possess.

Two closely related symbiont species *Hamiltonella defensa* and *Regiella insecticola* stand out as defensive symbionts that protect their hosts against parasitic wasps and fungal pathogens, respectively (Flórez et al., 2015; Oliver et al., 2010, 2014). Facultative symbionts are not usually considered as nutritional, but in some species they have been found to replace, or complement the nutritional services of the primary symbiont *Buchnera*, thus becoming obligatory (Sudakaran et al., 2017). In *Cinara cedri*, *Serratia symbiotica* was the first example of a facultative symbiont taking over the synthesis of essential nutrients (tryptophan and riboflavin) from *Buchnera* (Lamelas et al., 2011). A more recent study also found that many *Cinara* aphids also carry *Erwinia* symbionts, which live within their own bacteriocytes near to those of *Buchnera* (Manzano-Marin et al., 2020). These authors show that this symbiont has vitamin-synthesizing genes that compensate for deficiencies in the *Cinara-Buchnera* interaction, and even provide a new function potentially synthesizing biotin and thiamine for the aphid. Moreover, in one aphid lineage a subset of genes was found to have been transferred to a different symbiont (*H. defensa*) suggesting that lateral gene transfers could drive the establishment and dynamics of multi-symbiont communities.

2.4.2 Whiteflies

Whiteflies and psyllids are sister lineages with a common ancestor that established a long-term association with a bacterium, which diversified and provided each group their obligate symbiont (Santos-Garcia et al., 2018). Whiteflies are reported to host seven facultative symbionts around the world namely *Rickettsia*, *Wolbachia*, *Hamiltonella*, *Arsenophonus*, *Cardinium*, *Fritschea*, and *Hemipteriphilus* (Kanakala and Ghanim, 2019). The whitefly *Bemisia tabaci* forms a cryptic species complex, with sibling species now considered distinct at 4% genetic divergence. However, across these species genetically-similar facultative symbionts (i.e. same genetic cluster) were identified—with symbionts from the same genetic cluster hosted by highly-divergent host species (Kanakala and Ghanim, 2019). Relative to aphids, whitefly facultative symbionts seem thus to be more readily transferred from one individual to another by horizontal transmission (Ahmed et al., 2013; Qi et al., 2019). The transfer by parasitoid wasps has been shown to occur rather frequently, but only for specific symbiont species. For example, a study in China identified five symbiont species in the whitefly (*Hamiltonella*, *Arsenophonus*, *Cardinium*, *Rickettsia* and *Wolbachia*) and only the latter two were found in a common parasitic wasp of the whitefly (Qi et al., 2019). In this study symbiont horizontal transmission was confirmed because of shared symbiont haplotypes. Similar to *Cinara* aphids, a co-obligate relationship between whitefly primary and various facultative symbionts has been shown to provide nutritional benefits to the host. For example, the whitefly facultative symbiont *Hamiltonella*, which is often found sharing bacteriocytes with the obligate *Portiera*, is thought to produce several B vitamins (Rao et al., 2015). While research on symbiosis in whiteflies and other non-aphid hemipterans is a little behind that on aphids, we find that there are some similarities (e.g. types of symbiont, or action of effect) but also many differences (e.g. horizontal transmission rates) between the systems.

2.4.3 Stinkbugs, true bugs and shield bugs

The hemipteran suborder Heteroptera includes herbivorous families like Pentatomidae, Acanthosomatidae, Alydidae and Scutelleridae (i.e. stinkbugs, true bugs and shield bugs). In these groups, nutritional bacteria are not endosymbionts that live within host tissues, but microbes hosted in specialized gut structures. Removal of these bacteria using antibiotics often leads to dramatic increases in insect mortality, which suggests that in most species these microbes are obligatory (Zytynska et al., 2019). In some species these symbionts have also been found to confer other important services. For example, the bean bug (*Riptortus pedestris*) (Fig. 2G) harbours *Burkholderia* gut symbionts that the nymphs acquire every generation from the soil. The bacteria were shown to confer insecticide resistance to the pest insects, quickly established within a single generation and potentially transferred horizontally to other pests (Kikuchi et al., 2012).

2.4.4 Symbionts modulate induced plant defences in Hemiptera

As far as we are aware, endosymbiont manipulation of plant defences was first reported in the whitefly *B. tabaci*. When infected with the symbiont *Hamiltonella*, the insect downregulated tomato defences for its own benefit (Su et al., 2015). The authors of this report also showed that the effect was lost in plant mutants that were deficient in salicylic acid dependent responses, which revealed that this strategy is based on the manipulation of the balance between opposing phytohormonal routes (i.e. the aforementioned crosstalk). This result evidenced a mechanism of symbiont manipulation of plant defences that has been found in other insect groups (see below), and which consists in cheating the plant by inducing salicylic acid defences that are usually triggered against pathogens, leading to a down-regulation of insect-specific defences that depend on jasmonic acid. In the grain aphid *Sitobion miscanthi*, the symbiont *Hamiltonella* has also been found to down-regulate wheat defences, although not by altering the balance between the two phytohormones mentioned (Li et al., 2019). In this study, the symbiont down-regulated all defences that the authors measured, including jasmonic and salicylic routes, but also polyphenol oxidase (PPO) and peroxidase (POD). Although plant defences in grasses are less well understood than in tomato, this study presents an alternative strategy in which the symbiont does not target a specific phytohormonal route, but probably a more global defensive mechanism found downstream of the phytohormonal routes mentioned.

Plant defences do not only act against herbivores by directly affecting their fitness, but also indirectly by attracting herbivores' natural enemies, for example via the emission of attractive plant volatiles. As mentioned before, in the pea aphid *A. pisum*, the symbiont *H. defensa* is known to protect aphids against parasitic wasps and predators (Flórez et al., 2015) by preventing the development of the wasp's larva or by reducing the fitness of predators. In a recent study, this defensive effect has been found to extend to parasitic wasp recruitment (Frago et al., 2017). *Hamiltonella defensa* reduced plant volatile emissions, rendering plants less attractive to the wasp *Aphidius ervi*, and ultimately reducing attacks on young aphids. Based on this and previous studies, *H. defensa* thus stands out as a symbiont capable of manipulating plant defences upon aphid and whitefly feeding. Surprisingly, however, the study by Frago et al. (2017) revealed that the symbionts *Regiella*, *Spiroplasma*, *Serratia* and *Rickettsiella* also manipulated the emission of plant volatiles and plant attraction to the wasp. This result suggests that many symbionts, and not only *H. defensa*, may play a role in manipulation of plant physiology in aphids.

The mechanism by which hemipteran endosymbionts challenge induced plant defences is so far little understood. Three potential (and non-exclusive) explanations are possible: (1) injection of salivary effectors of bacterial origin into the plant, (2) effectors of bacterial origin delivered to the plant via honeydew, or (3) behavioural changes in the insect. The first mechanism is the most likely occurring in the whitefly example discussed above (Su et al., 2015) because the authors succeeded in emulating manipulation of plant defences by artificially injecting the saliva of the insect into the plant. Although in this example the symbiont was the facultative *H. defensa*, this mechanism can potentially involve obligatory symbionts, as *Buchnera* proteins have been found in the saliva of the aphids *A. pisum* and *Megoura viciae* (Vandermoten et al., 2014). Plant sap is rich in sugars but poor in proteins, and for this reason phloem feeders excrete the excess sugar and produce honeydew. Working with the pea aphid *A. pisum*, Schwartzberg and Tumlinson (2014) demonstrated for the first time that experimental honeydew deposition suppressed jasmonic acid in plants while increasing levels of salicylic acid. Although the honeydew also contained the latter phytohormone, the levels in plants were larger, so that the plant played a role in such an increase. This study did not look for hidden microbial players in mediating plant responses, even if many proteins of symbiont origin are found in the honeydew of this same aphid species (Sabri et al., 2013). Although it is possible that symbionts (or proteins of symbiont origin) in the honeydew manipulate plant responses for the benefit of the insect hosts, most research pinpoints honeydew as the Achilles heel in these associations. For instance, in the aphid *Macrosiphum euphorbiae*, GroEL proteins that originate from the obligatory symbiont *Buchnera* are responsible for triggering plant defences (Chaudhary et al., 2014). In the brown planthopper, *Nilaparvata lugens*, which is one of the most important pests of rice, honeydew-associated microbes elicited plant defences and even the release of plant volatiles (Wari et al., 2019). More precisely, the authors 'cured' insects from bacterial associates using antibiotics and demonstrated that bacteria amplified antiherbivore defences, specifically phytoalexins. By culturing microbes obtained from honeydew, the authors also identified the species *Acinetobacter soli* and *Serratia marcescens* as potential suspects that the plant uses as clues of insect attack. Symbionts in aphid honeydew can also betray their hosts by increasing attraction of their natural enemies as found in *A. pisum* aphids whose gut resident *Staphylococcus sciuri* directly produces volatiles that attract aphid natural enemies (Leroy et al., 2011).

The last potential mechanism whereby symbionts may allow insects to manipulate plant defences is by altering their host's behaviour. Aphids need to penetrate plant tissues with their mouthparts, and before starting to suck plant sap they often perform several punctures known as probing behaviours. While doing so, the plant suffers from mechanical damage and the insect releases salivary effectors into plant tissues, which are two important elements for the plant to recognize the attacker, but also for the insect to avoid detection. Symbionts are increasingly recognized as having an influence on animal behaviour (Rohrscheib and Brownlie, 2013; Su et al., 2013b), and a recent study demonstrated that the facultative symbiont *H. defensa* alters probing behaviour in the aphid *Rhopalosipum padi* (Leybourne et al., 2020). The change in feeding behaviour due to symbiont carrying led to a less efficient nutritional intake and reduced insect fitness. It is possible, however, that in other hemipteran species, or host plants, these behavioural alterations weaken plant induced defences.

3. Holometabola

Holometabola is the most species-rich animal lineage comprising about 850,000 described species (i.e. 50% of all animal species) (Beutel and Pohl, 2006). Development through a complete metamorphosis is seen as its key success (Beutel et al., 2011). Yet, the radical change in form and function from larval to adult life stage is a constraint for the acquisition of vertically transmitted symbionts (Hammer and Moran, 2019). A complete metamorphosis implies that larval organs break down and may not be reformed during the adult stage: symbionts in the gut or bacteriomes need to reallocate and recolonize newly-formed structures. Moreover, symbionts need to withstand severe conditions during metamorphosis in the pupal phase. This has led to the evolution of alternative transmission routes that bypass metamorphosis. These alternatives mostly include externally transferred symbionts via inoculation of the substrate (in *Drosophila*, for instance), or social transmission (known from ants and bees, and in hemimetabolous insects such as cockroaches and termites) (Salem et al., 2015) (Fig. 2H). Unlike in hemimetabolous insects, microbiota of Holometabola can change dramatically between larvae and adults as they often exploit different feeding niches. This diversion in feeding niches is especially apparent in Lepidoptera and herbivorous Hymenoptera (sawflies and gall wasps) because larvae are folivorous or xylophagous while adults feed on nectar or pollen (Hammer and Moran, 2019). Despite these difficulties, many holometabolous insects possess vertically transmitted symbionts that have found ways to counter the constraints imposed by metamorphosis (Fig. 1).

3.1 Hymenoptera

About 15% of the ca. 150,000 hymenopteran species are phytophagous. This estimate includes bees that feed on nectar and pollen, ants that feed on plant exudates, fungus-growing ants (including leaf-cutter ants) (Fig. 2I), plant-galling wasps, and the paraphyletic group of basal hymenopteran lineages (previously named Symphyta) that includes sawflies, horntails, and wood wasps. Many hymenopterans have established intimate nutritional symbioses with fungi. As reviewed by Biedermann and Vega (2020), herbivorous hymenopteran groups having symbiotic associations with fungi include fungus-growing ants that harvest plants upon which fungi are cultivated and then fed upon, wood wasps that inoculate fungi into the xylem to feed their solitary larvae, and stingless bees that feed their larvae with a culture of fungi growing in nectar. These fungi mostly facilitate digestion of low-quality diets like wood or leaves, but they also degrade and detoxify plant defensive compounds. Regarding bacterial mutualists, the honey bee gut microbiome has become a model of host-microbe interactions and has been intensely discussed in previous reviews (Douglas, 2019; Engel et al., 2016; Zheng et al., 2018). Distinctive gut bacterial lineages are vertically transmitted through social contacts likely facilitating host-symbiont coevolution (Engel and Moran, 2013; Moran et al., 2019). Yet, this social-bee model is not representative for the vast majority of bees that are solitary (>90% of the >17,500 species) (Fig. 2J). Relative to social bees, solitary ones show a greater diversity and variability in their gut microbiome, which strongly depends on the environment and the way microbes are transmitted among individual insects (Voulgari-Kokota et al., 2019). A recent study has revealed that the microbiome of megachilid bees, for instance, depends on the flowers they visit (McFrederick et al., 2017), and Kim et al. (2019) found that pollinating Hymenoptera can even act as agents of horizontal transmission of plant symbionts. In this latter study, a *Streptomyces* strain moves from the rhizosphere into strawberry roots up to their flowers and is then transferred by pollinating honey bees to other flowers. Like in fungus-growing ants (see below), the *Streptomyces* strain studied protected the bees but also the plants from pathogens. Symbionts in the remaining herbivorous hymenopteran groups, mainly sawflies, ants and gall wasps have been largely studied too, and in our review, we detail results from recent studies.

Many ant species have acquired vertically-transmitted bacterial symbionts along their evolutionary history (Moreau, 2020). Most herbivorous ants are considered as canopy foragers that feed on plant exudates, insect honeydew, pollen and vertebrate waste, and they obtain little nitrogen even when they complement their diet via predation. In *Cephalotes, Dolichoderus* and *Camponotus* ants, genomic evidence suggests that bacteria located in the gut may perform useful nitrogen-metabolic services for their hosts (Bisch et al., 2018; Gil et al., 2003; Hu et al., 2018). Due to symbiont enrichment in herbivorous ants, and relatedness of some gut bacteria to nitrogen-fixing rhizobia that are mutualistic with leguminous plants, some authors have speculated that endosymbionts have facilitated the origins and maintenance of the 'herbivorous' lifestyle across this insect family (Kaltenpoth and Flórez, 2020; Russell et al., 2009; Stoll et al., 2007). Nitrogen fixation was originally posited as a mutualistic service, but in vivo demonstration of this activity by internally harboured ant symbionts has proven elusive, and shotgun (meta) genomic sequencing has similarly failed to identify nitrogen fixation genes in abundant endosymbionts. Such efforts have, however, implicated symbionts in mutualistic nitrogen recycling, with studies in both the *Cephalotes* and *Camponotus* systems combining in vivo experiments with genomics to support such roles (Feldhaar et al., 2007; Gil et al., 2003; Hu et al., 2018).

Beyond these taxa, fungus-growing leaf-cutting ants (subtribe Attina, genus *Atta* and *Acromyrmex*) have established intimate associations with the fungus they cultivate on harvested plant material (reviewed in Moreau, 2020). In addition to these associations, however, leaf-cutting ants have established symbiosis with bacteria. Although the gut microbiome is quite simple in attine ants, these species keep vertically-transmitted Actinobacteria (e.g. *Streptomyces*) on their cuticle that produce antibiotics to suppress fungus garden-parasites, an important service to ensure that only the right fungi grows on harvested plant material (Currie et al., 1999). These cuticular actinobacterial biofilms are vertically transmitted by many fungus-growing ant genera, which makes these species exceptional in keeping distinct microbiomes externally and internally. Notably, the composition of the gut microbiomes appears to be affected by the presence/absence of these other symbionts on the cuticle: gut microbiomes of ant species that carry cuticular Actinobacteria tend to be more similar than those without, especially in the ant lineages arising later in the history of this taxon (Sapountzis et al., 2019).

Besides interactions with cuticular symbionts, gut microbial symbionts can affect the social dynamics between leaf-cutter ants by changing the cuticular chemicals that ants use as recognition cues. In the leaf-cutting ant *Acromyrmex echinatior* antibiotic treatment led to a more aggressive behaviour against nestmates, which in turn correlated with a decrease in the abundance of two antifungal compounds, that are produced against fungus garden-parasites (Teseo et al., 2019). These compounds are produced by exocrine metapleural glands that are unique to ants and which also produce secretions with antibiotic properties that modulate Actinobacteria communities present on the ant cuticle (Poulsen et al., 2003). After antibiotic treatment, when ants were fed with faecal droplets the gut bacterial community was partly restored together with normal behaviour (Teseo et al., 2019).

Sawflies are a group of herbivorous Hymenoptera, comprising some species that are serious pests of wheat. A first comprehensive microbiota screening of six sawfly species representing four different Symphyta families (Agridae, Diprionidae, Pamphiliidae and Tenthredinidae) (Fig. 2K) revealed a low apparent diversity of gut bacteria. The bacteria found were mainly α- or γ-proteobacteria that were mainly attributed to the host plant (Graham et al., 2008). For example, *Rhanella* sp. found in half of the screened species may have been acquired from the host plant as they were isolated from foliage in other studies (e.g. Hashidoko et al., 2002). A more recent study revealed that these sawflies are colonized by a novel *Spiroplasma* species, detected in both adults and larvae that is likely to be either transmitted vertically or horizontally by larval feeding on the inner wheat stem (Yeoman et al., 2019). This symbiont carries several genes encoding for carbohydrate-metabolism as well as biosynthetic pathways of essential B-vitamins. Furthermore, *Spiroplasma* genes encode for cardiolipin synthase and chitinase, both potentially involved in insect defence, which would add sawflies as another insect group using *Spiroplasma* as a defensive symbiont (Ballinger and Perlman, 2019).

The hymenopteran lineage Cynipoidea comprises both plant and insect parasites. Within this lineage, the Cynipidae (about 1400 spp.) family is entirely specialized in forming galls mainly on oak trees and rose bushes (Ronquist et al., 2015). Gall wasps induce plants to modify their host physiology and develop complex gall structures that often resemble novel plant organs. Wasp larvae feed within the protective gall. A transcriptional analysis of ovaries and venom glands of two gall wasp species revealed that unlike many hymenopteran parasitoid species, gall wasps seem to be deprived of viral genes or virus particles in the venom glands, which could aid at reprograming plant cells for gall development (Cambier et al., 2019). However, similar to insect herbivores in other orders, it is likely that cellulase genes expressed in the venom glands and/or ovaries are of bacterial origin. The acquisition of such horizontally transferred genes might have been an important adaptation in the evolution of Cynipidae to become plant parasites.

Yet, a phylogenomic approach on the whole lineage is needed to further elucidate whether cellulase genes originate from horizontal gene transfer events as was shown for the Phasmatodea discussed above (Shelomi et al., 2016).

Defensive symbioses have been identified in herbivorous Hymenoptera, particularly against pathogens that threaten fungal gardens in leaf-cutter ants (as mentioned above) or bee larvae, but also against eukaryotic parasites (reviewed in Flórez et al., 2015 and Kaltenpoth and Engl, 2014). In the bumblebee *Bombus terrestis*, for instance, gut symbionts reduced infection rates by the trypanosomatid parasite *Crithidia bombi* (Koch and Schmid-Hempel, 2011).

3.2 Coleoptera

The enormous diversity of beetles is especially attributed to the adaptive radiation of specialized herbivorous beetles feeding on angiosperms. A recent phylogenetic study revealed that specialization mainly in leaf and seed mining, and stem and wood boring in the highly diverse Phytophaga lineage (longhorn beetles, leaf beetles, and weevils; >125,000 described species) was enabled by plant cell wall-degrading enzymes (PCWDEs) obtained from bacteria and fungi via horizontal gene transfer events (Kirsch et al., 2014; McKenna et al., 2019). The origin of genes associated with PCWDEs were found to be phylogenetically and temporally linked to significant increases in diversification rates some 200 million years ago. These enzymes obtained via horizontal gene transfer are therefore proposed as a key innovation that facilitated the evolution of specialized plant-feeding niches in herbivorous beetles, which account for almost half of all described beetle species (Coleoptera >400,000 species). Despite the importance of horizontal gene transfer for the success of this group, stable associations with microbes are also common in phytophagous beetles. This suggests an ongoing process of loss and replacement of genes related to cell-wall degradation, but also of microbes capable of performing this function (Kirsch et al., 2014; Salem et al., 2017).

Co-speciation between bacterial symbionts and leaf beetles (Chrysomelidae) has been shown for several species belonging to different subfamilies (Fukumori et al., 2017; Kölsch and Synefiaridou, 2012; Salem et al., 2017). In *Sitophilus* grain weevils, for example, intracellular symbionts provide essential amino acids (Vigneron et al., 2014). A more recent example has revealed that the tortoise leaf beetle *Cassida rubiginosa* (Fig. 2L) acquired an obligate symbiont that took over pectinase activities to digest plant cell wall components (Salem et al., 2017). This extracellular bacterium, *Stammera*, has a reduced genome, a common characteristic of endosymbionts in stable symbiotic associations (Moran et al., 2008). In this example, the bacteria are localized in sac-like structures connected to the foregut as well as in the females' reproductive tract, both evidence of stable associations. They are vertically transferred through the egg, which is topped with a cap-like structure enclosing spheres densely populated with *Stammera*. Such complex structures guarantee that the symbiont is encapsulated throughout the hosts' development and are more typical for partnerships where symbionts are transferred intracellularly during oogenesis (Salem et al., 2017). Even if horizontally acquired PCWDEs are widespread in beetles (McKenna et al., 2019), co-opting symbionts that produce these same molecules may be more common than currently thought particularly when means of vertical transmission and morphological adaptations to accommodate the microbes in the intestine have evolved (Mason et al., 2019a). However, how often a similar type of symbiosis such as with *Stammera* exists in other beetle lineages remains to be investigated.

Gut microbes have also been found to provide their hosts with essential amino acids, and this seems particularly common in species that feed on nutrient-limited food sources like wood (Biedermann and Vega, 2020). The Asian longhorned beetle, *Anoplophora glabripennis*, for instance, uses gut bacteria and fungi to obtain these organic compounds (Ayayee et al., 2016). Microbial co-option may not only be common to perform universal enzymatic reactions like pectinase activities, but also to digest specific plant secondary metabolites. For example, one of the most devastating pests of coffee, the coffee berry borer, *Hypothenemus hampei*, carries a *Pseudomonas* bacterium in the gut that degrades caffeine and allows insect survival on coffee fruits (Ceja-Navarro et al., 2015). Plant secondary metabolites can even lead to the convergence of gut microbial communities, particularly in well-defended host plants. This has been demonstrated in Cycads, which are a group of plants that have evolved tight interactions with the insects that pollinate them, as they interact through larval feeding on the plant as well as through pollination services. Cycads contain carcinogenic and neurotoxic compounds in their tissues and it is thought that insect development on them is facilitated by bacterial symbionts. Salzman et al. (2018) compared the microbiomes of three beetles and two butterflies that thrive on Cycads, and found that they share at least two bacteria that are absent in non-cycad feeding close relatives.

One of the coleopteran groups where symbiosis has been more extensively studied includes bark and ambrosia beetles (reviewed in Biedermann and Vega, 2020 and Hulcr and Stelinski, 2017) (Fig. 2M), probably because some species are devastating forest pests. Bark and ambrosia beetles have evolved tight associations with fungi. In this group, fungi are usually vertically transmitted and carried between plants in specialized structures in the beetles' elytra, and the fungi have even lost their capacity to reproduce sexually. Both adults and larvae of the beetle feed on fungal gardens, and the fungus can also physically block resin ducts, which can act as a plant defence against herbivores. The use of symbiotic fungi has also evolved in other beetle families including lizard beetles, ship-timber beetles and leaf-rolling weevils (reviewed in Biedermann and Vega, 2020). Some bark beetles like pine engravers and *Dendroctonus* spp. (both comprising severe pests of coniferous trees) also contain bacteria that decrease defensive monoterpenes produced by their host plants (Howe et al., 2018). As found in the cycad example mentioned above (Salzman et al., 2018), beetles seem to have convergent microbiomes strongly influenced by the host plants' defensive chemistry (Howe et al., 2018; Mason et al., 2019a).

Many beetles use microbes to defend themselves, particularly against pathogens as found in the bark beetle *Dendroctonus frontalis* and in leaf-rolling weevils in the genus *Euops* (reviewed in Biedermann and Vega, 2020 and Flórez et al., 2015). In Lagriinae beetles (Fig. 2O), extracellular bacteria were shown to protect eggs from pathogenic microbes (Kaltenpoth and Flórez, 2020). Symbiotic *Burkholderia gladioli* are present in the females' accessory reproductive glands and are smeared onto the eggs during oviposition. The bacteria then enter the egg and colonize the embryo. The symbionts produce an antifungal polyketide that protects the eggs from fungi present in the soil environment. Conceivably these symbionts were originally acquired from plants as they can cause disease in the beetles' host, and they can also be transmitted horizontally among beetles via the plant (Flórez et al., 2017). Besides lagriine beetles, multiple other insect taxa rely on the symbiosis with *Burkholderia* bacteria, which have been found to provide defensive and nutritional services, and also detoxification of insecticides (reviewed in Kaltenpoth and Engl, 2014). As suggested by these authors, it is likely that insects often acquire symbionts from the environment because these bacteria are widespread in the soil and on plants.

3.2.1 Symbionts modulate induced plant defences in Coleoptera

Bark and ambrosia beetles farm mutualistic fungi to feed both larvae and adults. These species attack conifers, which prevent insect colonization using resins enriched with toxic compounds like terpenes. At least in some conifer species, resin production is an induced defence as its production is increased after wounding (e.g. Lombardero et al., 2000). Beetle fungal symbionts can impair the efficiency of these defences by actively degrading the toxic compounds (e.g. Boone et al., 2013; Hammerbacher et al., 2013). In the pine weevil, *Hylobius abietis*, although diterpene acids had no toxic effect against the beetle, gut bacteria were found to be able to degrade these compounds and to provide the resulting, catabolized molecules as a nutritional food source for the insect (Berasategui et al., 2017). Beetle symbionts have also been found to manipulate plant defences downstream of the production of toxic compounds via altering defence-related genes and phytohormones. The Western corn rootworm, *Diabrotica virgifera*, is one of the most important pests of maize. A transcriptomic study found that beetles carrying the endosymbiont *Wolbachia* suppressed the induction of defence-related genes in the plant (Barr et al., 2010). This finding, however, was later contradicted by Robert et al. (2013) who showed that the symbiont did not alter the induction of specific markers of defence genes, or the emission of defensive plant volatiles. Bacterial symbionts not only colonize guts but find their way to salivary glands, enabling their occurrence in oral secretions. As also found in whiteflies and lepidopterans, oral secretions of two chrysomelids, the Colorado potato beetle (*Leptinotarsa decemlineata*) (Fig. 2N) and false potato beetle (*L. juncta*) are deposited into wounds during feeding to downregulate plant defences with a concomitant increase in insect performance (Chung et al., 2013, 2017; Wang et al., 2016). Based on a sequencing analysis of beetle regurgitates, similar roles have been suggested in the soybean pest *Epilachna varivestis* (Gedling et al., 2018). A detailed revision of these types of strategies can be found in Mason et al. (2019a). Symbionts in oral secretions benefit their hosts because they manipulate crosstalk between phytohormonal routes. While they still trigger the expression of salicylic acid-dependent genes (usually ineffective against insect attackers), they downregulate jasmonic acid-dependent responses that are specific against insect attackers. The first evidence of this strategy in beetles identified the symbionts belonging to the genera *Stenotrophomonas*, *Pseudomonas* and *Enterobacter* as responsible for this effect (Chung et al., 2013). In species using this strategy, the bacterial community in oral secretions may influence plant specialization because the microbiome of oral secretions depends on the host plant, and also because manipulation of plant defences depends on both the host plant and the insect species (Chung et al., 2017; Wang et al., 2016). Whether suppression of plant defences by symbiotic bacteria in oral secretions is widespread in chewing insects, and whether this mechanism helps them to conquer new host plants is a question that needs to be further investigated.

3.3 Lepidoptera

Although Lepidoptera is one of the most species-rich orders with almost exclusively folivorous feeding caterpillars, the role of microbial symbionts as nutritional partners remains ambiguous (Duplouy and Hornett, 2018; Paniagua-Voirol et al., 2018). Many recent studies provide evidence that caterpillars collected from the wild lack a resident gut microbiome, so that their guts are mostly colonized by transient and often diet-associated species (Hammer et al., 2017; Jones et al., 2019; Phalnikar et al., 2019; Staudacher et al., 2016; Szenteczki et al., 2019; Whitaker et al., 2016). As found in Coleoptera, evidence accumulates that ancient horizontal gene transfer events from microbes may have played a major role in the evolution of herbivory in Lepidoptera (Sun et al., 2013; Wybouw et al., 2014, 2016). The caterpillars' gut commonly harbours very low numbers of microbes, which is probably due to its harsh conditions of fast food passage and high pH levels, especially in the midgut (Engel and Moran, 2013; Paniagua-Voirol et al., 2018). Moreover, specialized tissues or structures that contain microbes, as are commonly found in other groups like Coleoptera or Hemiptera, have so far not been found in Lepidoptera. Gut microbial symbiosis is therefore unlikely to contribute to the diversification and evolution of specialization on Angiosperm families in this taxon, even if gut microbes may provide important nutritional services. For example, Ravenscraft et al. (2019) sampled adults of almost 300 tropical butterfly species, and found that communities of gut bacteria and fungi were a subset of those found in the food. This suggests that the adult gut acquires microbes from the plant but selects those that may aid in digestion. This speculation was indirectly supported by means of culture-based assays that revealed that the microbiomes of frugivores and nectarivores differed in their catabolic abilities. Microbial symbionts may also detoxify plant secondary metabolites in larvae as found, for example, for the degradation of phenolic glycosides in the gypsy moth *Lymantria dispar* (Mason et al., 2014), proteinase inhibitors in the velvet bean caterpillar *Anticarsia gemmatalis* (Pilon et al., 2013; Visôtto et al., 2009), and potentially in cycad feeding specialists as discussed in the Coleoptera section (Salzman et al., 2018).

Relative to caterpillars, adult butterflies were also shown to have richer bacterial microbiomes (Hammer et al., 2019a; Ravenscraft et al., 2019). A large screening of microbiome from wild-caught butterfly species across the Heliconiini revealed that closely related butterfly lineages harboured similar microbiomes (Hammer et al., 2019a). A correlative analysis between microbial community similarities and host phylogenetic distances showed moderate congruent topologies of the butterfly phylogeny with the microbiome dendrogram. For instance, pollen-feeding *Heliconius* had distinct, albeit low diversity bacterial communities relative to non-feeding butterflies. Although host-microbe co-diversification could be a plausible explanation for these observations, Hammer et al. (2019a) proposed that these differences were due to insect guts filtering out bacterial lineages depending on the diet of the host. Similar results were shown by Ravenscraft et al. (2019) in a study comprising over 50 different neotropical butterfly species. The role of microbes in adult butterflies thus remains little understood, but it can range from preventing pathogen colonization to direct antiparasite interactions. These studies highlight that in Lepidoptera, microbiomes shift in their composition and diversity across the different life stages and emphasizes the need to separately look at the community composition in relation to diet and development.

If Lepidopteran microbiomes are assembled from microbes in the local environment, from which microhabitats do these microbes arise? Soil microbes are among the most abundant and diverse group of organisms on Earth (Delgado-Baquerizo et al., 2018), and Hannula et al. (2019) showed that the entire microbiome of caterpillars of the cabbage moth, *Mamestra brassicae* (Fig. 2P), was affected by bacteria and fungi largely present in the soil and not in the host plant they were feeding on. These findings support prior discoveries that soil legacy effects have impacts on plant growth and plant-insect interactions (Pineda et al., 2017). Yet, Hannula et al. (2019) conducted their experiments under artificial conditions in a greenhouse with a moth strain reared for several years under laboratory conditions. Further studies need to address whether soil microbes affect aboveground insect microbiomes under more natural conditions, and how widespread such soil legacy effects are. This may be potentially more common in polyphagous caterpillars like *M. brassicae* that are often in direct contact with the soil because they often move up and down the plant to switch host plants until they pupate. On the contrary, specialist butterfly species like *Pieris brassicae* (Fig. 2Q) usually stay on the same plant throughout most of larval development eventually moving up to feed upon flowers (Lucas-Barbosa et al., 2013). It would be interesting to understand whether such butterfly species acquire microbes from the soil as well but maybe indirectly via the plant as found in the honeybee example discussed above (Kim et al., 2019).

3.3.1 Symbionts modulate induced plant defences in Lepidoptera

Probably one of the first examples of plant physiology manipulation by insect symbionts was found in the Gracillariidae leaf miner moth *Phyllonorycter blancardella*. This moth carries the bacterium *Wolbachia*, which aids its host by manipulating plant hormones called Cytokinins (Giron et al., 2013). In autumn, such phytohormone manipulation keeps the leaf surrounding the insect photosynthetically active and green, in otherwise decaying leaves (i.e. the 'green island effect'). A more recent study revealed that acquisition of *Wolbachia* in two lepidopteran lineages (including 60 species from the Gracillariidae family) was associated with the green island phenotype (Gutzwiller et al., 2015). This study is one of the few exploring the role of symbionts in manipulating plant physiology through use of a phylogenetic framework.

Similar to the whitefly *B. tabaci* and some beetle species, oral secretions of the fall armyworm, *Spodoptera frugiperla*, and corn earworm, *Helicoverpa zea* (both Noctuidae), have also been found to modulate plant defences. While for the Colorado potato beetle the effect found was clearly beneficial, for these moths the benefit is less clear. Acevedo et al. (2017) identified a whole suite of bacteria from oral secretions of *S. frugiperla*, and using culture-dependent methods tested their effect on plant defences. Results did not point to a clear benefit for the insect because different bacterial species had varying effects on different markers of defence-related genes. A similar result was found in a later study exploring the presence of phytohormones in the saliva of this same moth (Acevedo et al., 2019). Studies with *H. zea*, however, clearly showed that gut-associated bacteria often betray their host as they triggered insect-specific plant defences (i.e. enzyme polyphenol oxidase and genes regulated by jasmonic acid) (Wang et al., 2017). Bacteria that reside in the gut can also become the target of plant defences. When feeding on highly resistant maize plants with increased concentrations of chitinases and large trichomes (a type of physical plant defence), *S. frugiperda* usually shows reduced fitness. A study by Mason et al. (2019b) revealed that these effects may be partially explained by plant defences weakening protective layers of the insect gut thus allowing resident bacteria to colonize the hemolymph and become pathogenic.

Defences of plants can be primed by a preceding environmental stimulus, such as herbivore egg deposition, reliably indicating upcoming herbivorous feeding damage (Hilker et al., 2015). Paniagua-Voirol et al. (2020) investigated whether bacteria present on eggs of cabbage white butterflies, *Pieris brassicae*, (Fig. 2Q) are involved in eliciting an egg-mediated enhancement of plant defences against caterpillars. When treating butterflies with antibiotics, the egg-mediated plant response diminished: both caterpillar performances and transcriptional responses of the plant were similar to control plants that did not receive egg deposition. However, culture-independent quantification of bacteria revealed very low and inconsistent bacterial abundances on the eggs. Despite these results, bacteria were four times more abundant in female butterflies than in eggs, thus confirming that adult butterflies have very diverse microbiomes (Hammer et al., 2019a; Ravenscraft et al., 2019). The present study was done with butterflies cultured for numerous generations under laboratory and therefore more sterile environmental conditions. Previous studies on different moth species comparing field and laboratory populations have shown that laboratory reared insects are depauperate and largely dominated by a few bacterial strains (Belda et al., 2011; Staudacher et al., 2016; Xiang et al., 2006). More studies with Lepidoptera directly obtained from the field are thus likely to provide novel and exciting discoveries in this group.

3.4 Diptera

One of the best examples of co-speciation between dipterans and symbionts is that of ambrosia gall midges (Cecidomyiidae) and their fungal associate *Botryosphaeria dothidea*. In this group, fungal spores are deposited into plant tissues along with insect eggs, and hyphae line the inner walls of the gall providing food for the larvae and protection against natural enemies (e.g. Rohfritsch, 2008). Joy (2013) performed a large phylogenetic analysis of midge-plant associations comprising a total of 351 genera of gall-inducing midges and found that an association with the fungus accelerated diversification in this group and broadened the number of host plants exploited. In particular, symbiotic lineages were found to be more than 17 times more diverse than non-symbiotic ones, and had undergone a sevenfold increase in the number of host plant species used.

The number of species of herbivorous dipterans that make galls, however, is a minority, as most species feed on decaying plant material. Symbiosis with bacteria and yeasts in this guild has been studied in pest species in the families Drosophilidae and Tephritidae (Fig. 2R and S), and particularly in the model species *Drosophila melanogaster*. In laboratory-reared *D. melanogaster*, even if some microbes are resident (Ma and Leulier, 2018), the composition of microbial communities (both bacteria and yeasts) is usually similar between the food and the insect (reviewed in Broderick and Lemaitre, 2012). Bacterial symbionts are likely to be acquired by the larvae from the environment, and the microbes introduced to larval food sources by adult flies where they metabolize or detoxify products that are then fed upon by developing larvae. In wild *Drosophila*, however, gut microbes are usually different from that of the food source (e.g. Wong et al., 2015), which questions whether some resident microbes exist in this group. *Drosophila* is a diverse genus of flies, and whether symbionts have played a role in their diversification is a question that has been recently debated. Using six species obtained from laboratory cultures Brooks et al. (2016) pinpointed microbes as the driving force of *Drosophila* evolution because the authors found a strong congruence between the fly phylogeny and the dissimilarity matrices of bacterial communities. In a more recent study with four mushroom-feeding species of wild *Drosophila*, Martinson et al. (2017b) challenged this view. In this later report, the authors found that the microbiota of the different species were very similar, probably reflecting an effect of the exploited ecological niche and not the phylogeny of the host. This same result was found in a group of cactus-feeding *Drosophila* species that are endemic in the Sonoran desert (Martinson et al., 2017a; Starmer and Fogleman, 1986).

Tephritids are a group of dipterans comprising important pests like the oriental fruit fly (*Bactrocera dorsalis*) and the Mediterranean fruit fly (*Ceratitis capitata*), the latter species being the model species for the study of symbiosis in the Tephritidae family. As with *Drosophila*, tephritids carry yeasts and bacteria in their guts that are transferred to the fruit by females during oviposition where they proliferate eventually causing decay (reviewed in Lauzon, 2003). The bacteria found in *C. capitata* mostly comprise *Enterobacter*, *Klebsiella* and *Pectobacterium*, some of them with the ability to fix atmospheric nitrogen and to degrade pectin, thereby playing an important role in fly nutrition. The tight interaction between the fly and some of these bacteria is revealed by the presence of an oesophageal bulb in the fly capable of hosting the microbes (Sacchetti et al., 2008). Some Tephritidae pests are currently controlled using the sterile insect technique and mass releases of parasitoids. These techniques require rearing large numbers of flies and research on fly symbionts aims at improving this method (Ras et al., 2017).

Many fly species are pests that cause sour rot in fruits. An important question in these species is whether microbes that are carried by adults and deposited in fruits during oviposition are the main players inducing fruit decay. In an experiment inoculating grape fruits with different microbial symbionts of *Drosophila* sp. it was found that sour rot requires both the insect and its associated consortium of microbes, particularly yeasts and acetic acid bacteria (Barata et al., 2012). Another important question in this guild of flies is how rotting microbes penetrate the fruit. Some species can only colonize and spread rotting in already damaged fruits, while others are capable of attacking healthy fruits. This is the case of the spotted wing drosophila, *D. suzukii*, an emerging global pest. Although this fly can also cause sour rot in fruits (Ioriatti et al., 2018), whether vertically-transmitted symbionts play a role in allowing it to attack healthy fruits is not yet known. A recent study, however, demonstrated that even if *D. suzukii* is not capable of producing extensive fruit rot in vineyards, the wounds it produces to healthy fruits opens a window of susceptibility to more damaging drosophilids like *D. melanogaster* (Rombaut et al., 2017). This study highlights the importance of the community context to understand the role of fly symbionts in the biology of key pests.

4. The impact of insect symbionts in a changing world: Outlook and perspective

4.1 Insect symbionts in a warmer world

There is now unequivocal evidence that the climate is changing. One of the clearest trends is that mean global temperatures increased on average 0.6 °C during the last century, and particularly during the second half (IPCC, 2019). Recent reviews suggest that symbionts will critically modulate insect responses to global warming by either exacerbating extinctions or by buffering them (Corbin et al., 2017; Renoz et al., 2019; Thierry et al., 2019). Global warming may disrupt the symbiosis between insect hosts and their symbionts (symbiosis meltdown) or symbionts may buffer any negative effects of warming on the host (adaptation).

The phenomenon of symbiosis meltdown has been well described in some taxa like corals (Toby Kiers et al., 2010). In insect herbivores, there are few good examples in which increased temperatures lead to the death of the insect via negative effects on obligatory bacterial symbionts. These examples include pea aphids, the stink bug *Nezara*, the whitefly *Bemisia tabaci*, mealybugs and weevils (recently reviewed in Corbin et al., 2017; Renoz et al., 2019; Thierry et al., 2019; Wernegreen, 2012). In all these examples, experiments have shown that experimentally increasing temperatures led to symbiont mortality and concomitantly to the death of the insect host (i.e. the symbiont is the thermal 'weak link', Corbin et al., 2017).

A symbiosis meltdown does not necessarily result in extinction of both species. Renoz et al. (2019) provide several examples in which insects have lost a weak symbiotic partner by incorporating the microbial function into the host's genetic repertoire via horizontal gene transfer. Moreover, if the symbiont is facultative, the symbiont partner could be temporarily lost and subsequently re-acquired from the environment. This may be possible in groups like stink bugs that commonly acquire mutualistic bacterial symbionts from the soil each generation (e.g. Kikuchi et al., 2012). The plant host can also influence these interactions, for example by protecting insect symbionts from heat stress on their surfaces or even inside their tissues. The colonization of plant tissues by insect symbionts has been recently demonstrated in phloem-feeding insects as a means for horizontal transmission (reviewed in Chrostek et al., 2017) and can have additional benefits with increasing temperatures. A recent study showed that an important part of the microbiota of the moth *M. brassicae* is acquired from the soil (Hannula et al., 2019). Since soil may buffer temperature variations, plant roots could be thus an important reservoir of insect symbionts during heat waves. A few facultative endosymbionts have been found to protect their hosts against heat shocks (e.g. above 35 °C), and it is likely that associations with these bacteria will be favoured on a warmer planet. The aphid symbiont *Serratia symbiotica* was the first found to confer heat resistance, but similar roles have been reported for *H. defensa* in pea aphids and the whitefly *B. tabaci*, and also for *Rickettsia* and *Fukatsuia* in pea aphids (Oliver et al., 2010; Thierry et al., 2019). As far as we are aware, gut symbionts with the ability to protect insects against heat shock have not been identified.

Global warming may also make insects less protected by symbionts because at increased temperatures defensive symbionts may be lost, or become more costly. For example, the symbiont *H. defensa* protects its aphid hosts from parasitoids under normal temperatures, but fails at doing so when the host is exposed to heat shock (Bensadia et al., 2006; Doremus et al., 2018). The host plant can again alter these interactions. A recent study by Sochard et al. (2019) showed that the defensive symbiont *H. defensa* defended aphids against a parasitic wasp when these aphids were originally collected from *Medicago sativa* plants, but not from *Ononis spinosa* host plants. It is thus feasible that global warming alters not only aphid persistence on particular host plants, via effects on the symbionts, but also the associated community of natural enemies. Some natural enemies like parasitic wasps also carry symbionts that are needed for successful parasitism (Dicke et al., 2020). Future studies are therefore needed to explore this question, as symbiosis meltdown in natural enemies may release herbivore populations from top-down control and trigger insect outbreaks with cascading effects on plant communities.

4.2 Insect symbionts mediate the transmission of plant viruses

Many insects are important vectors of plant viruses and the first study to associate insect symbionts with virus transmission was in the early 1990s by Van den Heuvel et al. (1994). They were studying the transmission of potato leafroll virus by the aphid *Myzus persicae*, and found the virus had a high affinity for the protein symbionin (now named GroEL) that is produced by the aphid primary symbiont (*Buchnera aphidicola*). Subsequent research has questioned its role for circulating viruses, as *Buchnera*-GroEL was localized in the bacteriocytes rather than the haemolymph, gut or fat body, which would be needed for circulative virus transmission (reviewed by Pinheiro et al., 2015). Yet, the observation of a similar effect from a GroEL protein of the secondary symbiont *H. defensa* in whiteflies, and compelling experimental data that *H. defensa* enhances viral transmission (Bello et al., 2019; Guo et al., 2014; Hong et al., 2017; Su et al., 2013a), still shows we do not yet have a full understanding of this.

Direct interactions between insect symbionts and viruses are, in no doubt, important for circulative and persistent viruses that stay with the insect its whole life. However, all viruses (non-persistent, semi-persistent, or persistent) can also manipulate the insect to increase the chance of transmission. Such indirect interactions can occur through influencing host-plant selection (Ingwell et al., 2012), the production of winged individuals for dispersal (Hodge and Powell, 2010), and increased feeding on virus-infected plants (Angelella et al., 2018). All of these viral effects can also be influenced by insect symbionts (Frago et al., 2012). For example, aphids probe plants as a 'taste test' when finding new suitable host plants, and this probing activity can be increased both by *H. defensa* symbiont infection and by the *watermelon mosaic virus* infecting pumpkin plants and cowpea aphids (*Aphis craccivora*) feeding on it (Angelella et al., 2018). For the symbiont effect, an increased amount of probing when infected by *H. defensa* could act to further reduce plant defences which can also benefit successful virus transmission. A plant's induced defence response to virus infection can act to reduce the fitness of both virus and insect, but it is in the virus' best interests to maximize the fitness of the insect host. In this way persistent viruses and insect symbionts could be both considered as mutualists to the insect because they fight the plants immune system, minimize pathogenic effects in the vector and ultimately increase insect fitness (Pinheiro et al., 2015). Another challenge where both virus and bacterial symbionts need to benefit the host is in response to abiotic stress. Hong et al. (2017) showed that increasing ozone concentrations further enhanced *tomato yellow leaf curl virus* transmission among plants by whiteflies (beyond any symbiont effect). They concluded that the effect occurred through ozone-stress acting to decrease plant resistance and immune responses to the virus, and hypothesized that environmental stress, mediated through changes in the insect protective microbiome, resulted in indirect effects on insect 'endurance'.

Transmission of viruses among insect hosts does not just occur as a vector moves from plant to plant (horizontal transmission), but also from parent to offspring through vertical transmission (Jia et al., 2018). This is also the main way in which bacterial symbionts are transmitted across insect hosts, and therefore this route is important when considering the interactions between viruses and insect symbionts. Green leafhoppers (*Nephotettix cincticeps*) host two obligate symbionts (*Sulcia* and *Nasuia*) and are common pests of rice fields, transmitting various plant viruses. Two recent studies have investigated the role of these symbionts on the vertical transmission of *rice dwarf virus* (Jia et al., 2017; Wu et al., 2019). Both studies show that the virus uses the symbionts as a vehicle to transfer to the next generation, by binding to the outer envelope of *Sulcia* and moving through the envelope to enter the periplasmic space of *Nasuia*. For both, as the symbiont is transmitted to the offspring, so is the virus. It is unknown if similar mechanisms may occur in other commonly studied symbiont systems like whiteflies and aphids.

4.3 Insect symbionts in agriculture, beyond effects on natural enemies

Future outlooks for managed systems under climate change predict increased impacts of pests and disease (IPCC, 2019). Many insect symbionts act to benefit the survival (nutrition, resistance to biotic and abiotic stress) and dispersal (host-plant use or production of dispersal morphs) of their insect hosts. The impact of insect-symbiont interactions on pest biocontrol via natural enemies has been recently explored in the realm of food web interactions (McLean et al., 2016), the spread of symbiont-mediated natural enemy resistance in closed greenhouse systems (Vorburger, 2018) and how plant biodiversity can mitigate symbiont-mediated resistance in open crop fields (Zytynska and Meyer, 2019).

Beyond nutritional and protective services, insect symbionts in agriculture can also benefit their host by detoxifying and degrading chemical pesticides. In natural systems, a host's bacterial community can help them feed on highly-defended plant hosts by transforming secondary plant metabolites (Dillon and Charnley, 2002; Howe et al., 2018). Adaptation of this to combat the effects of sprayed chemical pesticides leads to symbiont-mediated pesticide resistance (van den Bosch and Welte, 2017) and has been demonstrated in a number of pest species. For example, a *Burkholderia* symbiont of the bean bug *Riptortus pedestris* (Fig. 2G), a major pest of soybean in Japan, can degrade the organophosphate fenitrothion (Kikuchi et al., 2012), *Bacillus cereus* in the guts of diamondback moths, *Plutella xylostella*, can breakdown indoxacarb and acephate (Ramya et al., 2016; van den Bosch and Welte, 2017), and the gut symbiont *Citrobacter* of tephritid fruit fly pests (Fig. 2S) can degrade trichlorphon (Cheng et al., 2017). The increasing focus on biopesticides should therefore take into consideration the adaptive potential not only of the host insect being targeted but also of the symbiotic bacteria that are hosted by the insect. Moreover, horizontal gene transfer can also transfer resistance traits from bacteria to the host (Husnik and McCutcheon, 2018), which means that even if the symbiotic relationship breaks down the insect will still maintain the resistance trait.

Targeting insect symbionts for pest biocontrol has strong potential and would be particularly valuable against symbionts synthesizing essential nutrients for their host, e.g. obligate symbionts of insects feeding on vertebrate blood, plant sap, and deadwood (Arora and Douglas, 2017; Popa et al., 2012). If an obligate symbiont can be eliminated from the insect host, the insect will suffer from reduced nutrition leading to reduced reproduction, population growth and survival. Research to identify appropriate genes that can be targeted for deletion by RNA interference (RNAi) to reduce pest populations is ongoing (Niu et al., 2018). For example, the deletion of *amiD* and *ldcA1* genes caused the symbionts to be attacked as another pathogen, hindering symbiont acceptance by the insect host with knock-on effects for host fitness (Chung et al., 2018). However, there will be a selective pressure for insect host to evolve to regain this symbiosis or to evolve a different strategy to compensate for this loss (e.g. with other bacterial symbionts) and over time we might expect such targets to reduce in effectiveness. Despite these issues, in the short-term and with sufficient management control for the use of such methods this is a promising area for pest control. In the United-States, four transgenic insecticidal proteins are registered for corn rootworm management and the latest involves use of

RNAi to silence important genes influencing cornworm development (reviewed by Fishilevich et al., 2016). The use of RNAi allows a large number of different genes to be targeted, and as we learn more about the molecular basis of ecological interactions more targets will be uncovered. Targeting a larger number of genes with RNAi lowers the probabilities that insects adapt to and overcome such transcriptional inhibitions. In mosquitoes, the deletion of an outer-membrane gene (ompA) of a dominant gut bacterial symbiont using CRISPR/Cas9 impaired biofilm formation and reduced gut colonization (Hegde et al., 2019). While this mutation did not influence host fitness, it is thought to have implications for vector competency where this method of gene editing can be used for site-specific integration of genes into the symbiont that reduce the spread of viral diseases. Targeting insect symbionts that are also involved in virus transmission could benefit plant production through simultaneous reduction of pests, vectors and disease (Chuche et al., 2017). All these methods would need to be implemented as part of an integrated pest management plan that involves a multifaceted approach to pest control using a combination of pest abundance surveys, increased recruitment of natural enemies, alongside use of symbiont- or insect-targeted biopesticides.

5. Conclusion

Our systematic review of microbial symbiosis in herbivorous insects shows that not all insect orders rely equally on microbial symbionts to thrive on plants, and that there is much variation with respect to symbiont localization in hosts (e.g. guts versus hemocoel). Symbiont-dependent insect species are particularly present in Hemiptera, which evolved to feed on low-nutrient plant sap with the help of obligate endosymbionts that have been vertically transmitted for millions of years. Other orders seem to rely on genes acquired from microbes via ancient lateral gene transfers, and have developed distinct feeding niches without depending on costly symbioses. When present in these lineages, gut microbes were often shown to be transient and taken up from the environment, such as the host plant or the soil. In accordance with Hammer et al. (2019b) who states that animals vary greatly in their degree of reliance on microbial symbiosis, we see this diversity of interactions, particularly in some diverse orders like Lepidoptera or Coleoptera. In these orders, some species use the highly diverse microbiomes of their food plants to filter them into their guts and to harbour them for a certain period of time rather than let them become resident.

Further large-scale screenings of herbivorous species and their food plants should shed light on how common this rather loose type of association between microbes and their insect hosts is. Although the hemimetabolous Hemiptera order is largely relying on obligatory microbial symbiosis, we did not find such a close relationship between symbionts and hosts in the other three orders Orthoptera, Phasmatodea and Thysanoptera. Instead, these harbour transient gut symbionts and/or rely on laterally transferred genes from microbes. Despite the radical changes and harsh conditions during metamorphosis numerous holometabolous species vertically transmit their symbionts and show close associations with both intra- and extracellular symbionts. Some species maintain the same symbiont in larvae and adults but harbour them in different tissues, while in other species, communities differ between larvae and adults, largely because of variations in habitat use or diet.

The role that symbionts can play in future scenarios of global warming has a strong potential to change insect population dynamics and ultimately the persistence of insect populations, the structure of their communities, and the services they provide. Global warming and habitat loss are triggering massive species extinctions, and these extinctions maybe either exacerbated or buffered by symbionts as they can be the weaker link in insects, or protect against environmental stress. Symbiont effects can change the pest status of some insects, triggering outbreaks in some species, but also reducing the densities of others. However, this also gives rise to the potential to better control these pest populations by manipulating the symbiont relationship with their host. The last few years have revealed an unprecedented decline in insect abundance in many natural ecosystems (Hallmann et al., 2017). It is likely that a subset of this loss has been mediated by impacts on insect symbionts. Populations of rare or endangered species, for example, can be reduced if these species carry symbionts that are susceptible to increased warming. Extinctions can also be triggered if communities become dominated by few species that carry symbionts that allow them to better sustain increased temperatures. Habitat loss is an important element challenging insect populations. When the amount of habitat is reduced, species persistence is jeopardized because resources are of lower quality and critical resources can be missing (Samways et al., 2010). These critical resources can be symbionts, an effect that can be particularly dramatic for species that rely upon environmental symbionts, acquired anew at every generation. In the last 20 years, the steadily growing literature investigating the ecology and evolutionary potential of insect symbionts has enabled the development of strategies to transfer basic science to commercialisation. We are now at an age where we can bring together ecology, evolution and biotechnology to benefit the management of cropping systems, and to better understand the role of microbes in natural insect populations to protect endangered or beneficial species.

Acknowledgements

We are grateful to Emmanuelle Jousselin, Piotr Łukasik, Simon Fellows and Peter Biedermann for insightful discussions, and to Jacob Russell, Kerry M. Oliver and two anonymous reviewers for their comments on an earlier version of this manuscript. We are also grateful to Antoine Foucart, Tibor Bukovinszky, Jitte Groothuis, Peter Biedermann and Martin Kaltenpoth for sharing with us insect pictures. Sharon Zytynska is funded by a BBSRC (UKRI) David Phillips Fellowship BB/S010556/1, Nina Fatouros is funded by the Dutch Research Council (NWO/TTW Vidi grant no. 14854), and Enric Frago is funded by CIRAD and by the Agence Nationale de la Recherche (ANR) via the PRIMA S2 2018 project INTOMED.

References

Acevedo, F.E., Peiffer, M., Tan, C.-W., Stanley, B.A., Stanley, A., Wang, J., Jones, A.G., Hoover, K., Rosa, C., Luthe, D., Felton, G., 2017. Fall armyworm-associated gut bacteria modulate plant defense responses. Mol. Plant Microbe Interact. 30, 127–137. https://doi.org/10.1094/mpmi-11-16-0240-r.

Acevedo, F.E., Smith, P., Peiffer, M., Helms, A., Tooker, J., Felton, G.W., 2019. Phytohormones in fall armyworm saliva modulate defense responses in plants. J. Chem. Ecol. 45 (7), 598–609. https://doi.org/10.1007/s10886-019-01079-z.

Ademolu, K.O., Idowu, A.B., 2011. Occurrence and distribution of microflora in the gut regions of the variegated grasshopper *Zonocerus variegatus* (Orthoptera: Pyrgomorphidae) during development. Zool. Stud. 50, 409–415.

Ahmed, M.Z., de Barro, P.J., Ren, S.X., Greeff, J.M., Qiu, B.L., 2013. Evidence for horizontal transmission of secondary endosymbionts in the *Bemisia tabaci* cryptic species complex. PLoS One 8, e53084. https://doi.org/10.1371/journal.pone.0053084.

Angelella, G., Nalam, V., Nachappa, P., White, J., Kaplan, I., 2018. Endosymbionts differentially alter exploratory probing behavior of a nonpersistent plant virus vector. Microb. Ecol. 76, 453–458. https://doi.org/10.1007/s00248-017-1133-5.

Arora, A.K., Douglas, A.E., 2017. Hype or opportunity? Using microbial symbionts in novel strategies for insect pest control. J. Insect Physiol. 103, 10–17. https://doi.org/10.1016/j.jinsphys.2017.09.011.

Ayayee, P.A., Larsen, T., Rosa, C., Felton, G.W., Ferry, J.G., Hoover, K., 2016. Essential amino acid supplementation by gut microbes of a wood-feeding cerambycid. Environ. Entomol. 45, 66–73. https://doi.org/10.1093/ee/nvv153.

Ballinger, M.J., Perlman, S.J., 2019. The defensive spiroplasma. Curr. Opin. Insect Sci. 32, 36–41. https://doi.org/10.1016/j.cois.2018.10.004.

Barata, A., Santos, S.C., Malfeito-Ferreira, M., Loureiro, V., 2012. New insights into the ecological interaction between grape berry microorganisms and *Drosophila* flies during the development of sour rot. Microb. Ecol. 64, 416–430. https://doi.org/10.1007/s00248-012-0041-y.

Barr, K.L., Hearne, L.B., Briesacher, S., Clark, T.L., Davis, G.E., 2010. Microbial symbionts in insects influence down-regulation of defense genes in maize. PLoS One 5, e11339. https://doi.org/10.1371/journal.pone.0011339.

Belda, E., Pedrola, L., Peretó, J., Martínez-Blanch, J.F., Montagud, A., Navarro, E., Urchueguía, J., Ramón, D., Moya, A., Porcar, M., 2011. Microbial diversity in the midguts of field and lab-reared populations of the European corn borer *Ostrinia nubilalis*. PLoS One 6, e21751. https://doi.org/10.1371/journal.pone.0021751.

Bello, V.H., Watanabe, L.F.M., Santos, B.R., Marubayashi, J.M., Yuki, V.A., De Marchi, B.R., Pavan, M.A., Krause-Sakate, R., 2019. Evidence for increased efficiency of virus transmission by populations of Mediterranean species of *Bemisia tabaci* with high *Hamiltonella* prevalence. Phytoparasitica 47, 293–300.

Bensadia, F., Boudreault, S., Guay, J.F., Michaud, D., Cloutier, C., 2006. Aphid clonal resistance to a parasitoid fails under heat stress. J. Insect Physiol. 52, 146–157. https://doi.org/10.1016/j.jinsphys.2005.09.011.

Berasategui, A., Salem, H., Paetz, C., Santoro, M., Gershenzon, J., Kaltenpoth, M., Schmidt, A., 2017. Gut microbiota of the pine weevil degrades conifer diterpenes and increases insect fitness. Mol. Ecol. 26, 4099–4110. https://doi.org/10.1111/mec.14186.

Beutel, R.G., Pohl, H., 2006. Endopterygote systematics—where do we stand and what is the goal (Hexapoda, Arthropoda)? Syst. Entomol. 31, 202–219. https://doi.org/10.1111/j.1365-3113.2006.00341.x.

Beutel, R.G., Friedrich, F., Hörnschemeyer, T., Pohl, H., Hünefeld, F., Beckmann, F., Meier, R., Misof, B., Whiting, M.F., Vilhelmsen, L., 2011. Morphological and molecular evidence converge upon a robust phylogeny of the megadiverse Holometabola. Cladistics 27 (4), 341–355. https://doi.org/10.1111/j.1096-0031.2010.00338.x.

Biedermann, P.H.W., Vega, F.E., 2020. Ecology and evolution of insect-fungus mutualisms. Annu. Rev. Entomol. 65, 431–455.

Bisch, G., Neuvonen, M.M., Pierce, N.E., Russell, J.A., Koga, R., Sanders, J.G., Łukasik, P., Andersson, S.G.E., 2018. Genome evolution of Bartonellaceae symbionts of ants at the opposite ends of the trophic scale. Genome Biol. Evol. 10, 1687–1704. https://doi.org/10.1093/gbe/evy126.

Boone, C.K., Keefover-Ring, K., Mapes, A.C., Adams, A.S., Bohlmann, J., Raffa, K.F., 2013. Bacteria associated with a tree-killing insect reduce concentrations of plant defense compounds. J. Chem. Ecol. 39, 1003–1006. https://doi.org/10.1007/s10886-013-0313-0.

Broderick, N.A., Lemaitre, B., 2012. Gut-associated microbes of *Drosophila melanogaster*. Gut Microbes 3, 307–321.

Brooks, A.W., Kohl, K.D., Brucker, R.M., van Opstal, E.J., Bordenstein, S.R., 2016. Phylosymbiosis: relationships and functional effects of microbial communities across host evolutionary history. PLoS Biol. 14, e2000225. https://doi.org/10.1371/journal.pbio.2000225.

Brownlie, J.C., Johnson, K.N., 2009. Symbiont-mediated protection in insect hosts. Trends Microbiol. 17, 348–354.

Cambier, S., Ginis, O., Moreau, S.J.M., Gayral, P., Hearn, J., Stone, G.N., Giron, D., Huguet, E., Drezen, J.M., 2019. Gall wasp transcriptomes unravel potential effectors involved in molecular dialogues with oak and rose. Front. Physiol. 10, 926. https://doi.org/10.3389/fphys.2019.00926.

Casteel, C.L., Hansen, A.K., 2014. Evaluating insect-microbiomes at the plant-insect interface. J. Chem. Ecol. 40, 836–847. https://doi.org/10.1007/s10886-014-0475-4.

Ceja-Navarro, J.a., Vega, F.E., Karaoz, U., Hao, Z., Jenkins, S., Lim, H.C., Kosina, P., Infante, F., Northen, T.R., Brodie, E.L., 2015. Gut microbiota mediate caffeine detoxification in the primary insect pest of coffee. Nat. Commun. 6, 7618. https://doi.org/10.1038/ncomms8618.

Chaudhary, R., Atamian, H.S., Shen, Z., Briggs, S.P., Kaloshian, I., 2014. GroEL from the endosymbiont *Buchnera aphidicola* betrays the aphid by triggering plant defense. Proc. Natl. Acad. Sci. U. S. A. 111, 8919–8924. https://doi.org/10.1073/pnas.1407687111.

Chen, C.-C., Cheng, L.-L., Hou, R.F., 1981. Studies on the intracellular yeast-like symbiote in the Brown Planthopper, *Nilaparvata lugens* StÅl: II. Effects of antibiotics and elevated temperature on the symbiotes and their host. Z. Angew. Entomol. 92, 440–449. https://doi.org/10.1111/j.1439-0418.1981.tb01694.x.

Cheng, D.F., Guo, Z.J., Riegler, M., Xi, Z.Y., Liang, G.W., Xu, Y.J., 2017. Gut symbiont enhances insecticide resistance in a significant pest, the oriental fruit fly *Bactrocera dorsalis* (Hendel). Microbiome 5, 13. https://doi.org/10.1186/s40168-017-0236-z.

Chrostek, E., Pelz-Stelinski, K., Hurst, G.D.D., Hughes, G.L., 2017. Horizontal transmission of intracellular insect symbionts via plants. Front. Microbiol. 8, 2237. https://doi.org/10.3389/fmicb.2017.02237.

Chuche, J., Auricau-Bouvery, N., Danet, J.L., Thiéry, D., 2017. Use the insiders: could insect facultative symbionts control vector-borne plant diseases? J. Pest Sci. 90, 51–68. m. https://doi.org/10.1007/s10340-016-0782-3.

Chung, S.H., Rosa, C., Scully, E.D., Peiffer, M., Tooker, J.F., Hoover, K., Luthe, D.S., Felton, G.W., 2013. Herbivore exploits orally secreted bacteria to suppress plant defenses. Proc. Natl. Acad. Sci. U. S. A. 110, 15728–15733. https://doi.org/10.1073/pnas.1308867110.

Chung, S.H., Scully, E.D., Peiffer, M., Geib, S.M., Rosa, C., Hoover, K., Felton, G.W., 2017. Host plant species determines symbiotic bacterial community mediating suppression of plant defenses. Sci. Rep. 7, 39690. https://doi.org/10.1038/srep39690.

Chung, S.H., Jing, X., Luo, Y., Douglas, A.E., 2018. Targeting symbiosis-related insect genes by RNAi in the pea aphid-*Buchnera* symbiosis. Insect Biochem. Mol. Biol. 95, 55–63.

Corbin, C., Heyworth, E.R., Ferrari, J., Hurst, G.D.D., 2017. Heritable symbionts in a world of varying temperature. Heredity (Edinb). 118, 10–20. https://doi.org/10.1038/hdy.2016.71.

Currie, C.R., Scott, J.A., Summerbell, R.C., Malloch, D., 1999. Fungus-growing ants use antibiotic-producing bacteria to control garden parasites. Nature 398, 701–704. https://doi.org/10.1038/19519.

De Vries, E.J., Jacobs, G., Sabelis, M.W., Menken, S.B.J., Breeuwer, J.A.J., 2004. Diet-dependent effects of gut bacteria on their insect host: the symbiosis of *Erwinia* sp. and western flower thrips. Proc. R. Soc. B Biol. Sci. 271, 2171–2178. https://doi.org/10.1098/rspb.2004.2817.

Delgado-Baquerizo, M., Oliverio, A.M., Brewer, T.E., Benavent-González, A., Eldridge, D.J., Bardgett, R.D., Maestre, F.T., Singh, B.K., Fierer, N., 2018. A global atlas of the dominant bacteria found in soil. Science 359, 320–325. https://doi.org/10.1126/science.aap9516.

Dicke, M., Baldwin, I.T., 2010. The evolutionary context for herbivore-induced plant volatiles: beyond the "cry for help" Trends Plant Sci. 15, 167–175. https://doi.org/10.1016/j.tplants.2009.12.002.

Dicke, M., Cusumano, A., Poelman, E.H., 2020. Microbial symbionts of parasitoids. Annu. Rev. Entomol. 65, 1–20. https://doi.org/10.1146/annurev-ento-011019-024939.

Dickey, A.M., Trease, A.J., Jara-Cavieres, A., Kumar, V., Christenson, M.K., Potluri, L.-P., Morgan, J.K., Shatters, R.G., Mckenzie, C.L., Davis, P.H., Osborne, L.S., 2014. Estimating bacterial diversity in *Scirtothrips dorsalis* (Thysanoptera: Thripidae) via next generation sequencing. Fla. Entomol. 97, 362–366. https://doi.org/10.1653/024.097.0204.

Dillon, R., Charnley, K., 2002. Mutualism between the desert locust *Schistocerca gregaria* and its gut microbiota. Res. Microbiol. 153, 503–509.

Dillon, R.J., Vennard, C.T., Charnley, A.K., 2000. Pheromones: exploitation of gut bacteria in the locust. Nature 403, 851. https://doi.org/10.1038/35002669.

Dolling, W.R., 1991. Hemiptera. Oxford University Press.

Doremus, M.R., Smith, A.H., Kim, K.L., Holder, A.J., Russell, J.A., Oliver, K.M., 2018. Breakdown of a defensive symbiosis, but not endogenous defences, at elevated temperatures. Mol. Ecol. 27, 2138–2151. https://doi.org/10.1111/mec.14399.

Douglas, A.E., 1998. Nutritional interactions in insect-microbial symbioses: aphids and their symbiotic bacteria Buchnera. Annu. Rev. Entomol. 43, 17–37.

Douglas, A.E., 2011. Lessons from studying insect symbioses. Cell Host Microbe 10, 359–367. https://doi.org/10.1016/j.chom.2011.09.001.

Douglas, A.E., 2015. Multiorganismal insects: diversity and function of resident microorganisms. Annu. Rev. Entomol. 60, 17–34. https://doi.org/10.1146/annurev-ento-010814-020822.

Douglas, A.E., 2019. Simple animal models for microbiome research. Nat. Rev. Microbiol. 17, 764–775. https://doi.org/10.1038/s41579-019-0242-1.

Duplouy, A., Hornett, E.A., 2018. Uncovering the hidden players in Lepidoptera biology: the heritable microbial endosymbionts. PeerJ 2018, 1–45. https://doi.org/10.7717/peerj.4629.

Dutta, B., Gitaitis, R., Barman, A., Avci, U., Marasigan, K., Srinivasan, R., 2016. Interactions between *Frankliniella fusca* and *Pantoea ananatis* in the center rot epidemic of onion (Allium cepa). Phytopathology 106, 956–962. https://doi.org/10.1094/PHYTO-12-15-0340-R.

Eigenbrode, S.D., Bosque-Pérez, N.A., Davis, T.S., 2018. Insect-borne plant pathogens and their vectors: ecology, evolution, and complex interactions. Annu. Rev. Entomol. 63, 169–191. https://doi.org/10.1146/annurev-ento-020117-043119.

Engel, P., Moran, N.A., 2013. The gut microbiota of insects—diversity in structure and function. FEMS Microbiol. Rev. 37, 699–735. https://doi.org/10.1111/1574-6976.12025.

Engel, P., Kwong, W.K., McFrederick, Q., Anderson, K.E., Barribeau, S.M., Chandler, J.A., Cornman, R.S., Dainat, J., de Miranda, J.R., Doublet, V., Emery, O., Evans, J.D., Farinelli, L., Flenniken, M.L., Granberg, F., Grasis, J.A., Gauthier, L., Hayer, J., Koch, H., Kocher, S., Martinson, V.G., Moran, N., Munoz-Torres, M., Newton, I., Paxton, R.J., Powell, E., Sadd, B.M., Schmid-Hempel, P., Schmid-Hempel, R., Song, S.J., Schwarz, R.S., vanEngelsdorp, D., Dainat, B., 2016. The bee microbiome: impact on bee health and model for evolution and ecology of host-microbe interactions. MBio 7, e02164-15. https://doi.org/10.1128/mBio.02164-15.

Erb, M., Reymond, P., 2019. Molecular interactions between plants and insect herbivores. Annu. Rev. Plant Biol. 70, 527–557. https://doi.org/10.1146/annurev-arplant-050718-095910.

Erb, M., Meldau, S., Howe, G.A., 2012. Role of phytohormones in insect-specific plant reactions. Trends Plant Sci. 17, 250–259. https://doi.org/10.1016/j.tplants.2012.01.003.

Feldhaar, H., Straka, J., Krischke, M., Berthold, K., Stoll, S., Mueller, M.J., Gross, R., 2007. Nutritional upgrading for omnivorous carpenter ants by the endosymbiont *Blochmannia*. BMC Biol. 5, 48. https://doi.org/10.1186/1741-7007-5-48.

Fishilevich, E., Vélez, A.M., Storer, N.P., Li, H., Bowling, A.J., Rangasamy, M., Worden, S.E., Narva, K.E., Siegfried, B.D., 2016. RNAi as a management tool for the western corn rootworm, *Diabrotica virgifera virgifera*. Pest Manag. Sci. 72, 1652–1663.

Flórez, L.V., Biedermann, P.H.W., Engl, T., Kaltenpoth, M., 2015. Defensive symbioses of animals with prokaryotic and eukaryotic microorganisms. Nat. Prod. Rep. 32, 879–1156. https://doi.org/10.1039/C5NP00010F.

Flórez, L.V., Scherlach, K., Gaube, P., Ross, C., Sitte, E., Hermes, C., Rodrigues, A., Hertweck, C., Kaltenpoth, M., 2017. Antibiotic-producing symbionts dynamically transition between plant pathogenicity and insect-defensive mutualism. Nat. Commun. 8, 15172. https://doi.org/10.1038/ncomms15172.

Frago, E., Dicke, M., Godfray, H.C.J., 2012. Insect symbionts as hidden players in insect-plant interactions. Trends Ecol. Evol. 27, 705–711. https://doi.org/10.1016/j.tree.2012.08.013.

Frago, E., Mala, M., Weldegergis, B.T., Yang, C., McLean, A., Godfray, H.C.J., Gols, R., Dicke, M., 2017. Symbionts protect aphids from parasitic wasps by attenuating herbivore-induced plant volatiles. Nat. Commun. 8, 1860. https://doi.org/10.1038/s41467-017-01935-0.

Fukumori, K., Koga, R., Nikoh, N., Fukatsu, T., 2017. Symbiotic bacteria associated with gut symbiotic organs and female genital accessory organs of the leaf beetle *Bromius obscurus* (Coleoptera: Chrysomelidae). Appl. Entomol. Zool. 52, 589–598. https://doi.org/10.1007/s13355-017-0513-0.

Gawande, S.J., Anandhan, S., Ingle, A., Roylawar, P., Khandagale, K., Gawai, T., Jacobson, A., Asokan, R., Singh, M., 2019. Microbiome profiling of the onion thrips, *Thrips tabaci* Lindeman (Thysanoptera: Thripidae). PLoS One 14, e0223281. https://doi.org/10.1371/journal.pone.0223281.

Gedling, C.R., Smith, C.M., LeMoine, C.M.R., Cassone, B.J., 2018. The Mexican bean beetle (*Epilachna varivestis*) regurgitome and insights into beetle-borne virus specificity. PLoS One 13, e0192003. https://doi.org/10.1371/journal.pone.0192003.

Gil, R., Silva, F.J., Zientz, E., Delmotte, F., González-Candelas, F., Latorre, A., Rausell, C., Kamerbeek, J., Gadau, J., Hölldobler, B., Van Ham, R.C.H.J., Gross, R., Moya, A., 2003. The genome sequence of *Blochmannia floridanus*: comparative analysis of reduced genomes. Proc. Natl. Acad. Sci. U. S. A. 100, 9388–9393. https://doi.org/10.1073/pnas.1533499100.

Giron, D., Frago, E., Glevarec, G., Pieterse, C.M.J., Dicke, M., 2013. Cytokinins as key regulators in plant-microbe-insect interactions: connecting plant growth and defence. Funct. Ecol. 27, 599–609. https://doi.org/10.1111/1365-2435.12042.

Giron, D., Dedeine, F., Dubreuil, G., Huguet, E., Mouton, L., Outreman, Y., Vavre, F., Simon, J.C., 2017. Influence of microbial symbionts on plant–insect interactions. Adv. Bot. Res. 81, 225–257. Elsevier Ltd. https://doi.org/10.1016/bs.abr.2016.09.007.

Graham, R.I., Zahner, V., Lucarotti, C.J., 2008. An intracellular symbiont and other microbiota associated with field-collected populations of sawflies (Hymenoptera: Symphyta). Can. J. Microbiol. 54, 758–768. https://doi.org/10.1139/W08-067.

Guo, H., Qu, Y., Liu, X., Zhong, W., Fang, J., 2014. Female-biased symbionts and tomato yellow leaf curl virus infections in *Bemisia tabaci*. PLoS One 9, e84538. https://doi.org/10.1371/journal.pone.0084538.

Guo, J., Hatt, S., He, K., Chen, J., Francis, F., Wang, Z., 2017. Nine facultative endosymbionts in aphids. A review. J. Asia Pac. Entomol. 20, 794–801.

Gutzwiller, F., Dedeine, F., Kaiser, W., Giron, D., Lopez-Vaamonde, C., 2015. Correlation between the green-island phenotype and Wolbachia infections during the evolutionary diversification of Gracillariidae leaf-mining moths. Ecol. Evol. 5, 4049–4062. https://doi.org/10.1002/ece3.1580.

Hallmann, C.A., Sorg, M., Jongejans, E., Siepel, H., Hofland, N., Schwan, H., Stenmans, W., Müller, A., Sumser, H., Hörren, T., Goulson, D., De Kroon, H., 2017. More than 75 percent decline over 27 years in total flying insect biomass in protected areas. PLoS One 12, e0185809. https://doi.org/10.1371/journal.pone.0185809.

Hammer, T.J., Bowers, M.D., 2015. Gut microbes may facilitate insect herbivory of chemically defended plants. Oecologia 179, 1–14. https://doi.org/10.1007/s00442-015-3327-1.

Hammer, T.J., Moran, N.A., 2019. Links between metamorphosis and symbiosis in holometabolous insects. Philos. Trans. R. Soc. B Biol. Sci. 374, 20190068. https://doi.org/10.1098/rstb.2019.0068.

Hammer, T.J., Janzen, D.H., Hallwachs, W., Jaffe, S.P., Fierer, N., 2017. Caterpillars lack a resident gut microbiome. Proc. Natl. Acad. Sci. U. S. A. 114, 9641–9646. https://doi.org/10.1073/pnas.1707186114.

Hammer, T.J., Dickerson, J.C., Mcmillan, W.O., Fierer, N., 2019a. Pollen-feeding Heliconius butterflies host distinctive adult-stage microbiomes. bioRxiv. https://doi.org/10.1101/820829.

Hammer, T.J., Sanders, J.G., Fierer, N., 2019b. Not all animals need a microbiome. FEMS Microbiol. Lett. 366, 1–11. https://doi.org/10.1093/femsle/fnz117.

Hammerbacher, A., Schmidt, A., Wadke, N., Wright, L.P., Schneider, B., Bohlmann, J., Brand, W.A., Fenning, T.M., Gershenzon, J., Paetz, C., 2013. A common fungal associate of the spruce bark beetle metabolizes the stilbene defenses of Norway spruce. Plant Physiol. 162, 1324–1336. https://doi.org/10.1104/pp.113.218610.

Hannula, S.E., Zhu, F., Heinen, R., Bezemer, T.M., 2019. Foliar-feeding insects acquire microbiomes from the soil rather than the host plant. Nat. Commun. 10, 1–9. https://doi.org/10.1038/s41467-019-09284-w.

Hashidoko, Y., Itoh, E., Yokota, K., Yoshida, T., Tahara, S., 2002. Characterization of five phyllosphere bacteria isolated from rosa rugosa leaves, and their phenotypic and metabolic properties. Biosci. Biotechnol. Biochem. 66, 2474–2478. https://doi.org/10.1271/bbb.66.2474.

Hegde, S., Nilyanimit, P., Kozlova, E., Anderson, E.R., Narra, H.P., Sahni, S.K., Heinz, E., Hughes, G.L., 2019. CRISPR/Cas9-mediated gene deletion of the ompA gene in symbiotic *Cedecea neteri* impairs biofilm formation and reduces gut colonization of Aedes aegypti mosquitoes. PLoS Negl. Trop. Dis. 13, e0007883.

Henry, L.M., Peccoud, J., Simon, J.-C., Hadfield, J.D., Maiden, M.J.C., Ferrari, J., Godfray, H.C.J., 2013. Horizontally transmitted symbionts and host colonization of ecological niches. Curr. Biol. 23, 1713–1717. https://doi.org/10.1016/j.cub.2013.07.029.

Hilker, M., Fatouros, N.E., 2015. Plant responses to insect egg deposition. Annu. Rev. Entomol. 60, 493–515. https://doi.org/10.1146/annurev-ento-010814-020620.

Hilker, M., Schwachtje, J., Baier, M., Balazadeh, S., Baurle, I., Geiselhardt, S., Hincha, D.K., Kunze, R., Mueller-Roeber, B., Rillig, M.C., Rolff, J., Romeis, T., Schmulling, T., Steppuhn, A., van Dongen, J., Whitcomb, S.J., Wurst, S., Zuther, E., Kopka, J., 2015. Priming and memory of stress responses in organisms lacking a nervous system. Biol. Rev. 91, 1118–1133. https://doi.org/10.1111/brv.12215.

Hodge, S., Powell, G., 2010. Conditional facilitation of an aphid vector, *Acyrthosiphon pisum*, by the plant pathogen, *pea enation mosaic virus*. J. Insect Sci. 10, 155.

Hong, Y.Y., Yi, T.Y., Tan, X.L., Su, J.W., Ge, F., 2017. Microbes affected the TYLCCNV transmission rate by the Q biotype whitefly under high O-3. Sci. Rep. 7, 14412. https://doi.org/10.1038/s41598-017-14023-6.

Howe, M., Keefover-Ring, K., Raffa, K.F., 2018. Pine engravers carry bacterial communities whose members reduce concentrations of host monoterpenes with variable degrees of redundancy, specificity, and capability. Environ. Entomol. 47, 638–645. https://doi.org/10.1093/ee/nvy032.

Hrček, J., McLean, A.H.C., Godfray, H.C.J., 2016. Symbionts modify interactions between insects and natural enemies in the field. J. Anim. Ecol. 85, 1605–1612. https://doi.org/10.1111/1365-2656.12586.

Hu, Y., Sanders, J.G., Łukasik, P., D'Amelio, C.L., Millar, J.S., Vann, D.R., Lan, Y., Newton, J.A., Schotanus, M., Kronauer, D.J.C., Pierce, N.E., Moreau, C.S., Wertz, J.T., Engel, P., Russell, J.A., 2018. Herbivorous turtle ants obtain essential nutrients from a conserved nitrogen-recycling gut microbiome. Nat. Commun. 9, 964. https://doi.org/10.1038/s41467-018-04935-w.

Hulcr, J., Stelinski, L.L., 2017. The ambrosia symbiosis: from evolutionary ecology to practical management. Annu. Rev. Entomol. 62, 285–303. https://doi.org/10.1146/annurev-ento-031616-035105.

Husnik, F., McCutcheon, J.P., 2018. Functional horizontal gene transfer from bacteria to eukaryotes. Nat. Rev. Microbiol. 16, 67–79. https://doi.org/10.1038/nrmicro.2017.137.

Idowu, A.B., Edema, M.O., Oyedepo, M.T., 2009. Extracellular enzyme production by microflora from the gut region of the variegated grasshopper *Zonocerus variegatus* (Orthoptera: Pyrgomorphidae). Int. J. Trop. Insect Sci. 29, 229–235. https://doi.org/10.1017/S1742758409990312.

Ingwell, L.L., Eigenbrode, S.D., Bosque-Pérez, N.A., 2012. Plant viruses alter insect behavior to enhance their spread. Sci. Rep. 2, 578.

Ioriatti, C., Guzzon, R., Anfora, G., Ghidoni, F., Mazzoni, V., Villegas, T.R., Dalton, D.T., Walton, V.M., 2018. *Drosophila suzukii* (Diptera: Drosophilidae) contributes to the development of sour rot in grape. J. Econ. Entomol. 111, 283–292. https://doi.org/10.1093/jee/tox292.

IPCC, 2019: Climate Change and Land: an IPCC special report on climate change, desertification, land degradation, sustainable land management, food security, and greenhouse gas fluxes in terrestrial ecosystems. [P.R. Shukla, J. Skea, E. Calvo Buendia, V. Masson-Delmotte, H.-O. Pörtner, D. C. Roberts, P. Zhai, R. Slade, S. Connors, R. van Diemen, M. Ferrat, E. Haughey, S. Luz, S. Neogi, M. Pathak, J. Petzold, J. Portugal Pereira, P. Vyas, E. Huntley, K. Kissick, M. Belkacemi, J. Malley, (Eds.)], in press.

Jia, D., Mao, Q., Chen, Y., Liu, Y., Chen, Q., Wu, W., Zhang, X., Chen, H., Li, Y., Wei, T., 2017. Insect symbiotic bacteria harbour viral pathogens for transovarial transmission. Nat. Microbiol. 2, 17025. https://doi.org/10.1038/nmicrobiol.2017.25.

Jia, D., Chen, Q., Mao, Q., Zhang, X., Wu, W., Chen, H., Yu, X., Wang, Z., Wei, T., 2018. Vector mediated transmission of persistently transmitted plant viruses. Curr. Opin. Virol. 28, 127–132. https://doi.org/10.1016/j.coviro.2017.12.004.

Jones, A.G., Mason, C.J., Felton, G.W., Hoover, K., 2019. Host plant and population source drive diversity of microbial gut communities in two polyphagous insects. Sci. Rep. 9, 1–11. https://doi.org/10.1038/s41598-019-39163-9.

Joy, J.B., 2013. Symbiosis catalyses niche expansion and diversification. Proc. R. Soc. B Biol. Sci. 280, 20122820. https://doi.org/10.1098/rspb.2012.2820.

Kaltenpoth, M., Engl, T., 2014. Defensive microbial symbionts in Hymenoptera. Funct. Ecol. 28, 315–327. https://doi.org/10.1111/1365-2435.12089.

Kaltenpoth, M., Flórez, L.V., 2020. Versatile and dynamic symbioses between insects and *Burkholderia* bacteria. Annu. Rev. Entomol. 65, 1–26. https://doi.org/10.1146/annurev-ento-011019-025025.

Kanakala, S., Ghanim, M., 2019. Global genetic diversity and geographical distribution of *Bemisia tabaci* and its bacterial endosymbionts. PLoS One 14, 21. https://doi.org/10.1371/journal.pone.0213946.

Kikuchi, Y., Hayatsu, M., Hosokawa, T., Nagayama, A., Tago, K., Fukatsu, T., 2012. Symbiont-mediated insecticide resistance. Proc. Natl. Acad. Sci. U. S. A. 109, 8618–8622. https://doi.org/10.1073/pnas.1200231109.

Kim, D.R., Cho, G., Jeon, C.W., Weller, D.M., Thomashow, L.S., Paulitz, T.C., Kwak, Y.S., 2019. A mutualistic interaction between *Streptomyces* bacteria, strawberry plants and pollinating bees. Nat. Commun. 10, 4802. https://doi.org/10.1038/s41467-019-12785-3.

Kirsch, R., Gramzow, L., Theissen, G., Siegfried, B.D., Ffrench-Constant, R.H., Heckel, D.G., Pauchet, Y., 2014. Horizontal gene transfer and functional diversification of plant cell wall degrading polygalacturonases: key events in the evolution of herbivory in beetles. Insect Biochem. Mol. Biol. 52, 33–50. https://doi.org/10.1016/j.ibmb.2014.06.008.

Koch, H., Schmid-Hempel, P., 2011. Socially transmitted gut microbiota protect bumble bees against an intestinal parasite. Proc. Natl. Acad. Sci. U. S. A. 108, 19288–19292. https://doi.org/10.1073/pnas.1110474108.

Kölsch, G., Synefiaridou, D., 2012. Shared ancestry of symbionts? Sagrinae and Donaciinae (Coleoptera, Chrysomelidae) harbor similar bacteria. Insects 3, 473–491. https://doi.org/10.3390/insects3020473.

Lamelas, A., Gosalbes, M.J., Manzano-Marin, A., Pereto, J., Moya, A., Latorre, A., 2011. *Serratia symbiotica* from the aphid Cinara cedri: a missing link from facultative to obligate insect endosymbiont. PLoS Genet. 7, e1002357. https://doi.org/10.1371/journal.pgen.1002357.

Lauzon, C.R., 2003. Symbiotic relationships of tephritids. In: Insect Symbiosis. CRC Press, Taylor & Francis Group. https://doi.org/10.1201/9780203009918.

Lazebnik, J., Frago, E., Dicke, M., van Loon, J.J.A., 2014. Phytohormone mediation of interactions between herbivores and plant pathogens. J. Chem. Ecol. 40, 730–741. https://doi.org/10.1007/s10886-014-0480-7.

Leonardo, T.E., 2004. Removal of a specialization-associated symbiont does not affect aphid fitness. Ecol. Lett. 7, 461–468. https://doi.org/10.1111/j.1461-0248.2004.00602.x.

Leroy, P.D., Sabri, A., Heuskin, S., Thonart, P., Lognay, G., Verheggen, F.J., Francis, F., Brostaux, Y., Felton, G.W., Haubruge, E., 2011. Microorganisms from aphid honeydew attract and enhance the efficacy of natural enemies. Nat. Commun. 2, 348. https://doi.org/10.1038/ncomms1347.

Leybourne, D.J., Valentine, T.A., Bos, J.I.B., Karley, A.J., 2020. A fitness cost resulting from *Hamiltonella defensa* infection is associated with altered probing and feeding behaviour in *Rhopalosiphum padi*. J. Exp. Biol. 223, 207936. https://doi.org/10.1242/jeb.207936.

Li, Q., Fan, J., Sun, J., Zhang, Y., Hou, M., Chen, J., 2019. Anti-plant defense response strategies mediated by the secondary symbiont *Hamiltonella defensa* in the wheat aphid *Sitobion miscanthi*. Front. Microbiol. 10, 1–12. https://doi.org/10.3389/fmicb.2019.02419.

Lombardero, M.J., Ayres, M.P., Lorio, P.L., Ruel, J.J., 2000. Environmental effects on constitutive and inducible resin defences of Pinus taeda. Ecol. Lett. 3, 329–339. https://doi.org/10.1046/j.1461-0248.2000.00163.x.

Lucas-Barbosa, D., van Loon, J.J.A., Gols, R., van Beek, T.A., Dicke, M., 2013. Reproductive escape: annual plant responds to butterfly eggs by accelerating seed production. Funct. Ecol. 27, 245–254. https://doi.org/10.1111/1365-2435.12004.

Ma, D., Leulier, F., 2018. The importance of being persistent: the first true resident gut symbiont in *Drosophila*. PLoS Biol. 16, e2006945. https://doi.org/10.1371/journal.pbio.2006945.

Manzano-Marin, A., d'acier, A.C., Clamens, A.L., Orvain, C., Cruaud, C., Barbe, V., Jousselin, E., 2020. Serial horizontal transfer of vitamin-biosynthetic genes enables the establishment of new nutritional symbionts in aphids' di-symbiotic systems. ISME J. 14, 259–273. https://doi.org/10.1038/s41396-019-0533-6.

Martinson, V.G., Carpinteyro-Ponce, J., Moran, N.A., Markow, T.A., 2017a. A distinctive and host-restricted gut microbiota in populations of a cactophilic *Drosophila* species. Appl. Environ. Microbiol. 83, 1–10. https://doi.org/10.1128/AEM.01551-17.

Martinson, V.G., Douglas, A.E., Jaenike, J., 2017b. Community structure of the gut microbiota in sympatric species of wild *Drosophila*. Ecol. Lett. 20, 629–639. https://doi.org/10.1111/ele.12761.

Mason, C.J., Couture, J.J., Raffa, K.F., 2014. Plant-associated bacteria degrade defense chemicals and reduce their adverse effects on an insect defoliator. Oecologia 175, 901–910. https://doi.org/10.1007/s00442-014-2950-6.

Mason, C.J., Jones, A.G., Felton, G.W., 2019a. Co-option of microbial associates by insects and their impact on plant–folivore interactions. Plant Cell Environ. 42, 1078–1086. https://doi.org/10.1111/pce.13430.

Mason, C.J., Ray, S., Shikano, I., Peiffer, M., Jones, A.G., Luthe, D.S., Hoover, K., Felton, G.W., 2019b. Plant defenses interact with insect enteric bacteria by initiating a leaky gut syndrome. Proc. Natl. Acad. Sci. U. S. A. 116, 15991–15996. https://doi.org/10.1073/pnas.1908748116.

McCutcheon, J.P., Boyd, B.M., Dale, C., 2019. The life of an insect endosymbiont from the cradle to the grave. Curr. Biol. 29, R485–R495. https://doi.org/10.1016/j.cub.2019.03.032.

McFrederick, Q.S., Thomas, J.M., Neff, J.L., Vuong, H.Q., Russell, K.A., Hale, A.R., Mueller, U.G., 2017. Flowers and wild megachilid bees share microbes. Microb. Ecol. 73, 188–200. https://doi.org/10.1007/s00248-016-0838-1.

McKenna, D.D., Shin, S., Ahrens, D., Balke, M., Beza-Beza, C., Clarke, D.J., Donath, A., Escalona, H.E., Friedrich, F., Letsch, H., Liu, S., Maddison, D., Mayer, C., Misof, B., Murin, P.J., Niehuis, O., Peters, R.S., Podsiadlowski, L., Pohl, H., Scully, E.D., Yan, E.V., Zhou, X., Ślipiński, A., Beutel, R.G., 2019. The evolution and genomic basis of beetle diversity. Proc. Natl. Acad. Sci. U. S. A. 116, 24729–24737. https://doi.org/10.1073/pnas.1909655116.

McLean, A.H.C., Parker, B.J., Hrček, J., Henry, L.M., Godfray, H.C.J., 2016. Insect symbionts in food webs. Philos. Trans. R. Soc. Lond. Ser. B Biol. Sci. 371, 45. https://doi.org/10.1098/rstb.2015.0325.

Misof, B., Liu, S., Meusemann, K., Peters, R.S., Donath, A., Mayer, C., Frandsen, P.B., Ware, J., Flouri, T., Beutel, R.G., 2014. Phylogenomics resolves timing and pattern of insect evolution. Science 346, 763–767. https://doi.org/10.5061/dryad.3c0f1.

Monticelli, L.S., Outreman, Y., Frago, E., Desneux, N., 2019. Impact of host endosymbionts on parasitoid host range—from mechanisms to communities. Curr. Opin. Insect Sci. 32, 77–82. https://doi.org/10.1016/j.cois.2018.11.005.

Moran, N.A., Tran, P., Gerardo, N.M., 2005. Symbiosis and insect diversification: an ancient symbiont of sap-feeding insects from the bacterial phylum Bacteroidetes. Appl. Environ. Microbiol. 71, 8802–8810. https://doi.org/10.1128/AEM.71.12.8802-8810.2005.

Moran, N.A., McCutcheon, J.P., Nakabachi, A., 2008. Genomics and evolution of heritable bacterial symbionts. Annu. Rev. Genet. 42, 165–190. https://doi.org/10.1146/annurev.genet.41.110306.130119.

Moran, N.A., Ochman, H., Hammer, T.J., 2019. Evolutionary and ecological consequences of gut microbial communities. Annu. Rev. Ecol. Evol. Syst. 50, 451–475. https://doi.org/10.1146/annurev-ecolsys-110617-062453.

Moreau, C.S., 2020. Symbioses among ants and microbes. Curr. Opin. Insect Sci. 39, 1–5. https://doi.org/10.1016/j.cois.2020.01.002.

Niu, J., Taning, C.N.T., Christiaens, O., Smagghe, G., Wang, J.-J., 2018. Rethink RNAi in insect pest control: challenges and perspectives. Adv. Insect Physiol. 55, 1–17.

Oliver, K.M., Martinez, A.J., 2014. How resident microbes modulate ecologically-important traits of insects. Curr. Opin. Insect Sci. 4, 1–7. https://doi.org/10.1016/j.cois.2014.08.001.

Oliver, K.M., Degnan, P.H., Burke, G.R., Moran, N.A., 2010. Facultative symbionts in aphids and the horizontal transfer of ecologically important traits. Annu. Rev. Entomol. 55, 247–266. https://doi.org/10.1146/annurev-ento-112408-085305.

Oliver, K.M., Smith, A.H., Russell, J.a., 2014. Defensive symbiosis in the real world—advancing ecological studies of heritable, protective bacteria in aphids and beyond. Funct. Ecol. 28, 341–355. https://doi.org/10.1111/1365-2435.12133.

Paniagua-Voirol, L.R., Frago, E., Kaltenpoth, M., Hilker, M., Fatouros, N.E., 2018. Bacterial symbionts in lepidoptera: their diversity, transmission, and impact on the host. Front. Microbiol. 9, 556. https://doi.org/10.3389/fmicb.2018.00556.

Paniagua-Voirol, L.R., Valsamakis, G., Lortzing, V., Weinhold, A., Johnston, P.R., Fatouros, N., Kunze, R., Hilker, M., 2020. Plant responses to insect eggs are not induced by egg-associated microbes, but by a secretion attached to the eggs. Plant Cell Environ. in press.

Peccoud, J., Simon, J.-C., 2010. The pea aphid complex as a model of ecological speciation. Ecol. Entomol. 35, 119–130. https://doi.org/10.1111/j.1365-2311.2009.01147.x.

Peccoud, J., Ollivier, A., Plantegenest, M., Simon, J.C., 2009. A continuum of genetic divergence from sympatric host races to species in the pea aphid complex. Proc. Natl. Acad. Sci. U. S. A. 106, 7495–7500. https://doi.org/10.1073/pnas.0811117106.

Phalnikar, K., Kunte, K., Agashe, D., 2019. Disrupting butterfly microbiomes does not affect host survival and development. bioRxiv. http://dx.https://doi.org/10.1101/574871.

Pieterse, C.M.J., Poelman, E.H., Van Wees, S.C.M., Dicke, M., 2013. Induced plant responses to microbes and insects. Front. Plant Sci. 4, 475. https://doi.org/10.3389/fpls.2013.00475.

Pilon, F.M., Visôtto, L.E., Guedes, R.N.C., Oliveira, M.G.A., 2013. Proteolytic activity of gut bacteria isolated from the velvet bean caterpillar *Anticarsia gemmatalis*. J. Comp. Physiol. B Biochem. Syst. Environ. Physiol. 183, 735–747. https://doi.org/10.1007/s00360-013-0744-5.

Pineda, A., Kaplan, I., Bezemer, T.M., 2017. Steering soil microbiomes to suppress aboveground insect pests. Trends Plant Sci. 22, 770–778. https://doi.org/10.1016/j.tplants.2017.07.002.

Pinheiro, P.V., Kliot, A., Ghanim, M., Cilia, M., 2015. Is there a role for symbiotic bacteria in plant virus transmission by insects? Curr. Opin. Insect Sci. 8, 69–78. https://doi.org/10.1016/j.cois.2015.01.010.

Popa, V., Déziel, E., Lavallée, R., Bauce, E., Guertin, C., 2012. The complex symbiotic relationships of bark beetles with microorganisms: a potential practical approach for biological control in forestry. Pest Manag. Sci. 68, 963–975.

Poulsen, M., Bot, A.N.M., Boomsma, J.J., 2003. The effect of metapleural gland secretion on the growth of a mutualistic bacterium on the cuticle of leaf-cutting ants. Naturwissenschaften 90, 406–409. https://doi.org/10.1007/s00114-003-0450-3.

Qi, L.D., Sun, J.T., Hong, X.Y., Li, Y.X., 2019. Diversity and phylogenetic analyses reveal horizontal transmission of endosymbionts between whiteflies and their parasitoids. J. Econ. Entomol. 112, 894–905. https://doi.org/10.1093/jee/toy367.

Ramya, S.L., Venkatesan, T., Murthy, K.S., Jalali, S.K., Varghese, A., 2016. Degradation of acephate by *Enterobacter asburiae, Bacillus cereus* and *Pantoea agglomerans* isolated from diamondback moth *Plutella xylostella* (L), a pest of cruciferous crops. J. Environ. Biol. 37, 611.

Rao, Q., Rollat-Farnier, P.A., Zhu, D.T., Santos-Garcia, D., Silva, F.J., Moya, A., Latorre, A., Klein, C.C., Vavre, F., Sagot, M.F., Liu, S.S., Mouton, L., Wang, X.W., 2015. Genome reduction and potential metabolic complementation of the dual endosymbionts in the whitefly *Bemisia tabaci*. BMC Genomics 16, 226. https://doi.org/10.1186/s12864-015-1379-6.

Ras, E., Beukeboom, L.W., Cáceres, C., Bourtzis, K., 2017. Review of the role of gut microbiota in mass rearing of the olive fruit fly, *Bactrocera oleae*, and its parasitoids. Entomol. Exp. Appl. 164. https://doi.org/10.1111/eea.12609.

Ravenscraft, A., Berry, M., Hammer, T., Peay, K., Boggs, C., 2019. Structure and function of the bacterial and fungal gut microbiota of Neotropical butterflies. Ecol. Monogr. 89, e01346. https://doi.org/10.1002/ecm.1346.

Renoz, F., Pons, I., Hance, T., 2019. Evolutionary responses of mutualistic insect–bacterial symbioses in a world of fluctuating temperatures. Curr. Opin. Insect Sci. 35, 20–26. https://doi.org/10.1016/j.cois.2019.06.006.

Robert, C.A.M., Frank, D.L., Leach, K.A., Turlings, T.C.J., Hibbard, B.E., Erb, M., 2013. Direct and indirect plant defenses are not suppressed by endosymbionts of a specialist root herbivore. J. Chem. Ecol. 39, 507–515. https://doi.org/10.1007/s10886-013-0264-5.

Rock, D.I., Smith, A.H., Joffe, J., Albertus, A., Wong, N., O'Connor, M., Oliver, K.M., Russell, J.A., 2018. Context-dependent vertical transmission shapes strong endosymbiont community structure in the pea aphid, *Acyrthosiphon pisum*. Mol. Ecol. 27, 2039–2056. https://doi.org/10.1111/mec.14449.

Rohfritsch, O., 2008. Plants, gall midges, and fungi: a three-component system. Entomol. Exp. Appl. 128, 208–216. https://doi.org/10.1111/j.1570-7458.2008.00726.x.

Rohrscheib, C.E., Brownlie, J.C., 2013. Microorganisms that manipulate complex animal behaviours by affecting the host's nervous system. Springer Sci. Rev. 1, 133–140. https://doi.org/10.1007/s40362-013-0013-8.

Rombaut, A., Guilhot, R., Xuéreb, A., Benoit, L., Chapuis, M.P., Gibert, P., Fellous, S., 2017. Invasive *Drosophila suzukii* facilitates *Drosophila melanogaster* infestation and sour rot outbreaks in the vineyards. R. Soc. Open Sci. 4, 170117. https://doi.org/10.1098/rsos.170117.

Ronquist, F., Nieves-Aldrey, J.-L., Buffington, M.L., Liu, Z., Liljeblad, J., Nylander, J.A.A., 2015. Phylogeny, evolution and classification of gall wasps: the plot thickens. PLoS One 10, e0123301. https://doi.org/10.1371/journal.pone.0123301.

Russell, J.A., Moreau, C.S., Goldman-Huertas, B., Fujiwara, M., Lohman, D.J., Pierce, N.E., 2009. Bacterial gut symbionts are tightly linked with the evolution of herbivory in ants. Proc. Natl. Acad. Sci. U. S. A. 106, 21236–21241. https://doi.org/10.1073/pnas.0907926106.

Sabri, A., Vandermoten, S., Leroy, P.D., Haubruge, E., Hance, T., Thonart, P., De Pauw, E., Francis, F., 2013. Proteomic investigation of aphid honeydew reveals an unexpected diversity of proteins. PLoS One 8, e74656. https://doi.org/10.1371/journal.pone.0074656.

Sacchetti, P., Granchietti, A., Landini, S., Viti, C., Giovannetti, L., Belcari, A., 2008. Relationships between the olive fly and bacteria. J. Appl. Entomol. 132, 682–689. https://doi.org/10.1111/j.1439-0418.2008.01334.x.

Salem, H., Florez, L., Gerardo, N., Kaltenpoth, M., 2015. An out-of-body experience: the extracellular dimension for the transmission of mutualistic bacteria in insects. Proc. Biol. Sci. 282, 20142957. https://doi.org/10.1098/rspb.2014.2957.

Salem, H., Bauer, E., Kirsch, R., Berasategui, A., Cripps, M., Weiss, B., Koga, R., Fukumori, K., Vogel, H., Fukatsu, T., Kaltenpoth, M., 2017. Drastic genome reduction in an herbivore's pectinolytic symbiont. Cell 171, 1520–1531.e13. https://doi.org/10.1016/j.cell.2017.10.029.

Salzman, S., Whitaker, M., Pierce, N.E., 2018. Cycad-feeding insects share a core gut microbiome. Biol. J. Linn. Soc. 123, 728–738. https://doi.org/10.1093/biolinnean/bly017.

Samways, M.J., McGeoch, M.A., New, T.R., 2010. Insect Conservation: A Handbook of Approaches and Methods. Oxford University Press.

Santos-Garcia, D., Juravel, K., Freilich, S., Zchori-Fein, E., Latorre, A., Moya, A., Morin, S., Silva, F.J., 2018. To B or Not to B: comparative genomics suggests *Arsenophonus* as a source of B vitamins in whiteflies. Front. Microbiol. 9, 16. https://doi.org/10.3389/fmicb.2018.02254.

Sapountzis, P., Nash, D.R., Schiott, M., Boomsma, J.J., 2019. The evolution of abdominal microbiomes in fungus-growing ants. Mol. Ecol. 28, 879–899. https://doi.org/10.1111/mec.14931.

Schausberger, P., 2018. Herbivore-associated bacteria as potential mediators and modifiers of induced plant defense against spider mites and thrips. Front. Plant Sci. 9, 1107. https://doi.org/10.3389/fpls.2018.01107.

Schoonhoven, L.M., van Loon, J.J.A., Dicke, M., 2005. Insect-Plant Biology. Oxford University Press.

Schwartzberg, E.G., Tumlinson, J.H., 2014. Aphid honeydew alters plant defence responses. Funct. Ecol. 28, 386–394. https://doi.org/10.1111/1365-2435.12182.

Shelomi, M., Lo, W.-S., Kimsey, L.S., Kuo, C.-H., 2013. Analysis of the gut microbiota of walking sticks (Phasmatodea). BMC. Res. Notes 6, 368. https://doi.org/10.1186/1756-0500-6-368.

Shelomi, M., Sitepu, I.R., Boundy-Mills, K.L., Kimsey, L.S., 2015. Review of the gross anatomy and microbiology of the Phasmatodea digestive tract. J. Orthop. Res. 24, 29–40. https://doi.org/10.1665/034.024.0105.

Shelomi, M., Danchin, E.G., Heckel, D., Wipfler, B., Bradler, S., Zhou, X., Pauchet, Y., 2016. Horizontal gene transfer of pectinases from bacteria preceded the diversification of stick and leaf insects. Sci. Rep. 6, 26388. https://doi.org/10.1038/srep26388.

Shikano, I., Rosa, C., Tan, C.-W., Felton, G.W., 2017. Tritrophic interactions: microbe-mediated plant effects on insect herbivores. Annu. Rev. Phytopathol. 55, 313–331. https://doi.org/10.1146/annurev-phyto-080516-035319.

Sochard, C., Leclair, M., Simon, J.-C., Outreman, Y., 2019. Host plant effects on the outcomes of defensive symbioses in the pea aphid complex. Evol. Ecol. 33, 651–669. https://doi.org/10.1007/s10682-019-10005-4.

Stam, J.M., Kroes, A., Li, Y., Gols, R., van Loon, J.J.A., Poelman, E.H., Dicke, M., 2013. Plant interactions with multiple insect herbivores: from community to genes. Annu. Rev. Plant Biol. 65, 6.1–6.25. https://doi.org/10.1146/annurev-arplant-050213-035937.

Starmer, W.T., Fogleman, J.C., 1986. Coadaptation of Drosophila and yeasts in their natural habitat. J. Chem. Ecol. 12, 1037–1055. https://doi.org/10.1007/BF01638995.

Staudacher, H., Kaltenpoth, M., Breeuwer, J.A., Menken, S.B., Heckel, D.G., Groot, A.T., 2016. Variability of bacterial communities in the moth *Heliothis virescens* indicates transient association with the host. PLoS One 11, e0154514. https://doi.org/10.1371/journal.pone.0154514.

Stoll, S., Gadau, J., Gross, R., Feldhaar, H., 2007. Bacterial microbiota associated with ants of the genus *Tetraponera*. Biol. J. Linn. Soc. 90, 399–412. https://doi.org/10.1111/j.1095-8312.2006.00730.x.

Stoops, J., Crauwels, S., Waud, M., Claes, J., Lievens, B., Van Campenhout, L., 2016. Microbial community assessment of mealworm larvae (*Tenebrio molitor*) and grasshoppers (*Locusta migratoria migratorioides*) sold for human consumption. Food Microbiol. 53, 122–127. https://doi.org/10.1016/j.fm.2015.09.010.

Stork, N.E., 2018. How many species of insects and other terrestrial arthropods are there on Earth? Annu. Rev. Entomol. 63, 31–45. https://doi.org/10.1146/annurev-ento-020117-043348.

Su, Q., Pan, H., Liu, B., Chu, D., Xie, W., Wu, Q., Wang, S., Xu, B., Zhang, Y., 2013a. Insect symbiont facilitates vector acquisition, retention, and transmission of plant virus. Sci. Rep. 3, 1367. https://doi.org/10.1038/srep01367.

Su, Q., Zhou, X., Zhang, Y., 2013b. Symbiont-mediated functions in insect hosts. Commun. Integr. Biol. 6, e23804. https://doi.org/10.4161/cib.23804.

Su, Q., Oliver, K.M., Xie, W., Wu, Q., Wang, S., Zhang, Y., 2015. The whitefly-associated facultative symbiont *Hamiltonella defensa* suppresses induced plant defences in tomato. Funct. Ecol. 29, 1007–1018. https://doi.org/10.1111/1365-2435.12405.

Sudakaran, S., Kost, C., Kaltenpoth, M., 2017. Symbiont acquisition and replacement as a source of ecological innovation. Trends Microbiol. 25, 375–390. https://doi.org/10.1016/j.tim.2017.02.014.

Sugio, A., Dubreuil, G., Giron, D., Simon, J.C., 2015. Plant-insect interactions under bacterial influence: ecological implications and underlying mechanisms. J. Exp. Bot. 66, 467–478. https://doi.org/10.1093/jxb/eru435.

Sun, B.F., Xiao, J.H., He, S.M., Liu, L., Murphy, R.W., Huang, D.W., 2013. Multiple ancient horizontal gene transfers and duplications in lepidopteran species. Insect Mol. Biol. 22, 72–87. https://doi.org/10.1111/imb.12004.

Szenteczki, M.A., Pitteloud, C., Casacci, L.P., Kešnerová, L., Whitaker, M.R.L., Engel, P., Vila, R., Alvarez, N., 2019. Bacterial communities within *Phengaris* (Maculinea) alcon caterpillars are shifted following transition from solitary living to social parasitism of *Myrmica* ant colonies. Ecol. Evol. 9, 4452–4464. https://doi.org/10.1002/ece3.5010.

Teseo, S., van Zweden, J.S., Pontieri, L., Kooij, P.W., Sørensen, S.J., Wenseleers, T., Poulsen, M., Boomsma, J.J., Sapountzis, P., 2019. The scent of symbiosis: gut bacteria may affect social interactions in leaf-cutting ants. Anim. Behav. 150, 239–254. https://doi.org/10.1016/j.anbehav.2018.12.017.

Thierry, M., Hrček, J., Lewis, O.T., 2019. Mechanisms structuring host–parasitoid networks in a global warming context: a review. Ecol. Entomol. 44, 581–592. https://doi.org/10.1111/een.12750.

Toby Kiers, E., Palmer, T.M., Ives, A.R., Bruno, J.F., Bronstein, J.L., 2010. Mutualisms in a changing world: an evolutionary perspective. Ecol. Lett. 13, 1459–1474. https://doi.org/10.1111/j.1461-0248.2010.01538.x.

Tsuchida, T., Koga, R., Shibao, H., Matsumoto, T., Fukatsu, T., 2002. Diversity and geographic distribution of secondary endosymbiotic bacteria in natural populations of the pea aphid, *Acyrthosiphon pisum*. Mol. Ecol. 11, 2123–2135. https://doi.org/10.1046/j.1365-294X.2002.01606.x.

Tsuchida, T., Koga, R., Fukatsu, T., 2004. Host plant specialization governed by facultative symbiont. Science 303, 1989. https://doi.org/10.1126/science.1094611.

Turlings, T.C.J., Erb, M., 2018. Tritrophic interactions mediated by herbivore-induced plant volatiles: mechanisms, ecological relevance, and application potential. Annu. Rev. Entomol. 63, 433–452. https://doi.org/10.1146/annurev-ento-020117-043507.

van den Bosch, T.J.M., Welte, C.U., 2017. Detoxifying symbionts in agriculturally important pest insects. Microb. Biotechnol. 10, 531–540. https://doi.org/10.1111/1751-7915.12483.

Van den Heuvel, J.F.J.M., Verbeek, M., Van der Wilk, F., 1994. Endosymbiotic bacteria associated with circulative transmission of potato leafroll virus by *Myzus persicae*. J. Gen. Virol. 75, 2559–2565. https://doi.org/10.1099/0022-1317-75-10-2559.

Vandermoten, S., Harmel, N., Mazzucchelli, G., De Pauw, E., Haubruge, E., Francis, F., 2014. Comparative analyses of salivary proteins from three aphid species. Insect Mol. Biol. 23, 67–77. https://doi.org/10.1111/imb.12061.

Via, S., Bouck, A.C., Skillman, S., 2000. Reproductive isolation between divergent races of pea aphids on two hosts. II. Selection against migrants and hybrids in the parental environments. Evolution (N. Y.) 54, 1626–1637. https://doi.org/10.1111/j.0014-3820.2000.tb00707.x.

Vigneron, A., Masson, F., Vallier, A., Balmand, S., Rey, M., Vincent-Monégat, C., Aksoy, E., Aubailly-Giraud, E., Zaidman-Rémy, A., Heddi, A., 2014. Insects recycle endosymbionts when the benefit is over. Curr. Biol. 24, 2267–2273. https://doi.org/10.1016/j.cub.2014.07.065.

Visôtto, L.E., Oliveira, M.G.A., Guedes, R.N.C., Ribon, A.O.B., Good-God, P.I.V., 2009. Contribution of gut bacteria to digestion and development of the velvetbean caterpillar, *Anticarsia gemmatalis*. J. Insect Physiol. 55, 185–191. https://doi.org/10.1016/j.jinsphys.2008.10.017.

Vorburger, C., 2018. Symbiont-conferred resistance to parasitoids in aphids—challenges for biological control. Biol. Control 116, 17–26. https://doi.org/10.1016/j.biocontrol.2017.02.004.

Voulgari-Kokota, A., McFrederick, Q.S., Steffan-Dewenter, I., Keller, A., 2019. Drivers, diversity, and functions of the solitary-bee microbiota. Trends Microbiol. 27, 1034–1044. https://doi.org/10.1016/j.tim.2019.07.011.

Wang, J., Chung, S.H., Peiffer, M., Rosa, C., Hoover, K., Zeng, R., Felton, G.W., 2016. Herbivore oral secreted bacteria trigger distinct defense responses in preferred and non-preferred host plants. J. Chem. Ecol. 42, 463–474. https://doi.org/10.1007/s10886-016-0712-0.

Wang, J., Peiffer, M., Hoover, K., Rosa, C., Zeng, R., Felton, G.W., 2017. *Helicoverpa zea* gut-associated bacteria indirectly induce defenses in tomato by triggering a salivary elicitor(s). New Phytol. 214, 1294–1306. https://doi.org/10.1111/nph.14429.

Wari, D., Kabir, M.A., Mujiono, K., Hojo, Y., Shinya, T., Tani, A., Nakatani, H., Galis, I., 2019. Honeydew-associated microbes elicit defense responses against brown planthopper in rice. J. Exp. Bot. 70, 1683–1696. https://doi.org/10.1093/jxb/erz041.

Wernegreen, J.J., 2004. Endosymbiosis: lessons in conflict resolution. PLoS Biol. 2, e68. https://doi.org/10.1371/journal.pbio.0020068.

Wernegreen, J.J., 2012. Mutualism meltdown in insects: bacteria constrain thermal adaptation. Curr. Opin. Microbiol. 15, 255–262. https://doi.org/10.1016/j.mib.2012.02.001.

Whitaker, M.R.L., Salzman, S., Sanders, J., Kaltenpoth, M., Pierce, N.E., 2016. Microbial communities of lycaenid butterflies do not correlate with larval diet. Front. Microbiol. 7, 1920. https://doi.org/10.3389/fmicb.2016.01920.

Wiens, J.J., Lapoint, R.T., Whiteman, N.K., 2015. Herbivory increases diversification across insect clades. Nat. Commun. 6, 8370. https://doi.org/10.1038/ncomms9370.

Wong, A.C.N., Luo, Y., Jing, X., Franzenburg, S., Bost, A., Douglas, A.E., 2015. The host as the driver of the microbiota in the gut and external environment of *Drosophila melanogaster*. Appl. Environ. Microbiol. 81, 6232–6240. https://doi.org/10.1128/AEM.01442-15.

Wu, W., Huang, L., Mao, Q., Wei, J., Li, J., Zhao, Y., Zhang, Q., Jia, D., Wei, T., 2019. Interaction of viral pathogen with porin channels on the outer membrane of insect bacterial symbionts mediates their joint transovarial transmission. Philos. Trans. R. Soc. B 374, 20180320.

Wybouw, N., Dermauw, W., Tirry, L., Stevens, C., Grbic, M., Feyereisen, R., Van Leeuwen, T., 2014. A gene horizontally transferred from bacteria protects arthropods from host plant cyanide poisoning. elife 2014, 1–17. https://doi.org/10.7554/eLife.02365.

Wybouw, N., Pauchet, Y., Heckel, D.G., Van Leeuwen, T., 2016. Horizontal gene transfer contributes to the evolution of arthropod herbivory. Genome Biol. Evol. 8, 1785–1801. https://doi.org/10.1093/gbe/evw119.

Xiang, H., Wei, G.-F., Jia, S., Huang, J., Miao, X.-X., Zhou, Z., Zhao, L.-P., Huang, Y.-P., 2006. Microbial communities in the larval midgut of laboratory and field populations of cotton bollworm (*Helicoverpa armigera*). Can. J. Microbiol. 52, 1085–1092. https://doi.org/10.1139/w06-064.

Yeoman, C.J., Brutscher, L.M., Esen, Ö.C., Ibaoglu, F., Fowler, C., Eren, A.M., Wanner, K., Weaver, D.K., 2019. Genome-resolved insights into a novel *Spiroplasma* symbiont of the Wheat Stem Sawfly (*Cephus cinctus*). PeerJ 7, e7548. https://doi.org/10.7717/peerj.7548.

Zheng, H., Steele, M.I., Leonard, S.P., Motta, E.V.S., Moran, N.A., 2018. Honey bees as models for gut microbiota research. Lab Anim. (NY) 47, 317–325. https://doi.org/10.1038/s41684-018-0173-x.

Zytynska, S.E., Meyer, S.T., 2019. Effects of biodiversity in agricultural landscapes on the protective microbiome of insects—a review. Entomol. Exp. Appl. 167, 2–13. https://doi.org/10.1111/eea.12751.

Zytynska, S.E., Weisser, W.W., 2016. The natural occurrence of secondary bacterial symbionts in aphids. Ecol. Entomol. 41, 13–26. https://doi.org/10.1111/een.12281.

Zytynska, S.E., Tighiouart, K., Frago, E., 2019. A meta-analysis on the benefits and costs of hosting secondary endosymbionts in sap-sucking insects. bioRxiv. https://doi.org/10.1101/563031.

CHAPTER FIVE

Symbiotic solutions to nitrogen limitation and amino acid imbalance in insect diets

Allison K. Hansen[a,*], Daniel Pers[a], Jacob A. Russell[b]
[a]Department of Entomology, University of California, Riverside, CA, United States
[b]Department of Biology, Drexel University, Philadelphia, PA, United States
*Corresponding author: e-mail address: ahans003@ucr.edu

Contents

Advances in Insect Physiology, Volume 58
ISSN 0065-2806
https://doi.org/10.1016/bs.aiip.2020.03.001

Abstract

Although comprising ~80% of the Earth's air, nitrogen (N), in a usable form, is a limited resource for herbivorous insects. Additionally, much of the consumed dietary N is lost through-excretion, further challenging such insects, while limiting their growth and reproduction. To meet this challenge, many insect groups have evolved symbiotic relationships with microbes, enabling success within otherwise inhospitable niches. These mutualistic microbes contribute to their hosts' N-economies through nitrogen recycling, fixation and/or upgrading, using divergent metabolic strategies with varying precursors and physiological requirements. In this chapter we highlight each strategy, organizing our discussion by insect taxonomy, while enumerating the various microbes and innovations that have converged upon N-centric, nutritive functions. Our overview places a large emphasis on N-recycling, due in part to a lack of recent reviews on this topic. In reviewing N-fixation, we take a termite-centred approach, capitalizing upon an extensive bank of research performed across several decades. We also emphasize essential amino acid (EAA) and precursor biosynthesis, discussing non-fixing, non-recycling mutualisms in which symbionts make up for dietary shortcomings and missing pathways for nutrient biosynthesis in their hosts. In these explorations, we discuss the specificity and evolutionary histories between herbivorous insect hosts and their often ancient N-metabolizing symbionts. We focus, further, on correlations between dietary evolution and altered symbioses, which cast light on the causes and consequences of these nutritional relationships. We, finally, describe the evidence supporting prior arguments for N-centric mutualisms, emphasizing how a pairing of genomics and experimentation can uncover mechanism, while pinpointing just how symbiont metabolism shapes the fitness and N-budgets of diverse, herbivorous hexapods.

1. Introduction

Nutrient-limitation, especially nitrogen (N), is central to the ecology of many organisms. Comprising 78% of dry air, this element is environmentally plentiful, but not in forms usable by plants and animals. Needed in large quantities to build cells, tissues, and their building blocks, N is primarily found in the form of proteins, and their constituent amino acids (AAs) when measured inside living organisms (Behar et al., 2005; Janssen et al., 2017; Nardi et al., 2002). Other N-containing molecules consist of nucleotides, hormones, vitamins, and—for many arthropods—the chitin molecules that are used to build the cuticle (Elser et al., 1996; Finke, 2007, 2015). In the face of high demand, a scarcity of usable N can be a limiting factor for growth, reproduction and survival (McNeil and Southwood, 1978).

While potentially limiting under some conditions for predators (Denno and Fagan, 2003), N-limitation is paramount for many insect herbivores, whose plant-based diets lack sufficient overall quantities, or sufficient quantities of essential N-metabolites that animals cannot make themselves (Mattson, 1980). In response, insects, like other animals have evolved a range of mechanisms for efficient acquisition and conservation of this element (summarized in Mattson, 1980; McNeil and Southwood, 1978; for a few examples see Fig. 1). These include prolonged development, high rates of feeding (Kheirallah, 1978; Slansky and Feeny, 1977), as well as manipulations of food plants (e.g. galls) that engender higher nutritive content (Sandström et al., 2000). Some insects use seasonal shifts in consumed plant varieties, or in the tissues of preferred food plants, to coincide with peak periods of N-availability (McNeil and Southwood, 1978). Others increase dietary N through diet mixing, including coprophagy (faecal consumption),

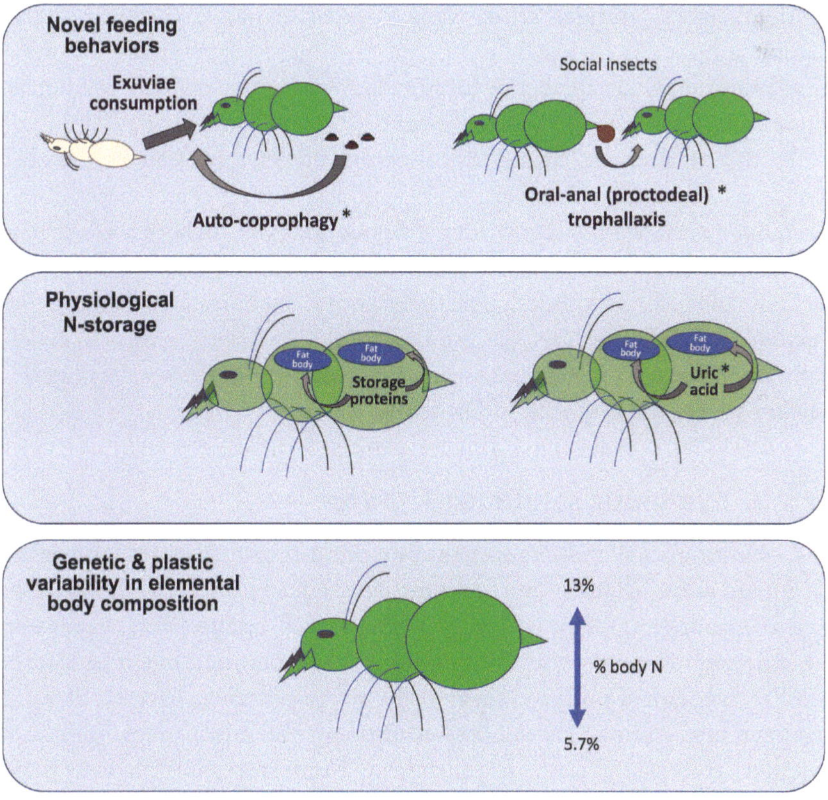

Fig. 1 Insect-evolved strategies to improve N-economies. Asterisks indicate strategies that evolved in the context of interactions with microbes, often internally housed symbionts. Several other host-evolved N-conserving and N-acquiring mechanisms are discussed in the text and are reviewed elsewhere (e.g. Mattson, 1980).

proctodeal trophallaxis (consumption of liquids from the anus), supplementary predation or cannibalism, and consumption of shed cuticles (Denno and Fagan, 2003; Machida et al., 2001; Mira, 2000; but see Clay et al., 2017). Physiological mechanisms aiding this endeavour include the storage of proteins or uric acid in the fat body (Burmester, 1999; Martinez and Wheeler, 1994; Mullins, 2015) and modestly flexible N content of the insect body plan (Fagan et al., 2002).

While endogenously controlled, the efficacy of some aforementioned mechanisms relies on exogenous parties, namely microbial symbionts. Housed either within the insect gut, hemolymph, or specialized host cells, and often involving eubacteria or fungi, such symbionts draw from biochemical pathways that are common in microbes but absent from insects. These metabolisms allow symbionts to concentrate, conserve, and improve the forms of N-containing nutrients to the direct benefit of insects (Douglas, 2009). Tightly integrated, occasionally ancient, and highly specialized nutritional mutualisms applying these mechanisms are posited as major evolutionary innovations for several herbivorous insects (Bennett and Moran, 2015).

In this chapter we discuss the diversity of ways by which microbial symbionts acquire and metabolize N, and how their insect hosts obtain symbiont-synthesized N-containing compounds. We highlight the dietary niches that are correlated with the maintenance of such symbioses, along with the physiological, anatomical, and behavioural modifications that support them. We also discuss how varying methodologies have shaped our understanding of established or hypothesized N-provisioning symbioses, highlighting key knowledge gaps in need of further study.

2. Symbiotic solutions: A primer

Nutritional symbionts are common in insects with N-imbalanced or N-limited diets, whether they consume live or dead plant tissue. The mechanisms enabling symbiont contributions to host N-budgets vary, necessitating different morphological and physiological adaptations, but at least three general strategies are utilized (Fig. 2). In *nitrogen recycling*, bacteria or fungi conserve nitrogenous wastes obtained through the insect's metabolism, or the diet, converting waste products into forms accessible to insects. In another strategy, *nitrogen fixation*, bacteria derive forms of N accessible to insects though energy-intensive fixation of atmospheric dinitrogen (N_2). In the third, *amino acid (and precursor) provisioning*, symbionts synthesize essential amino acids (EAAs) or precursors for these molecules, which their hosts

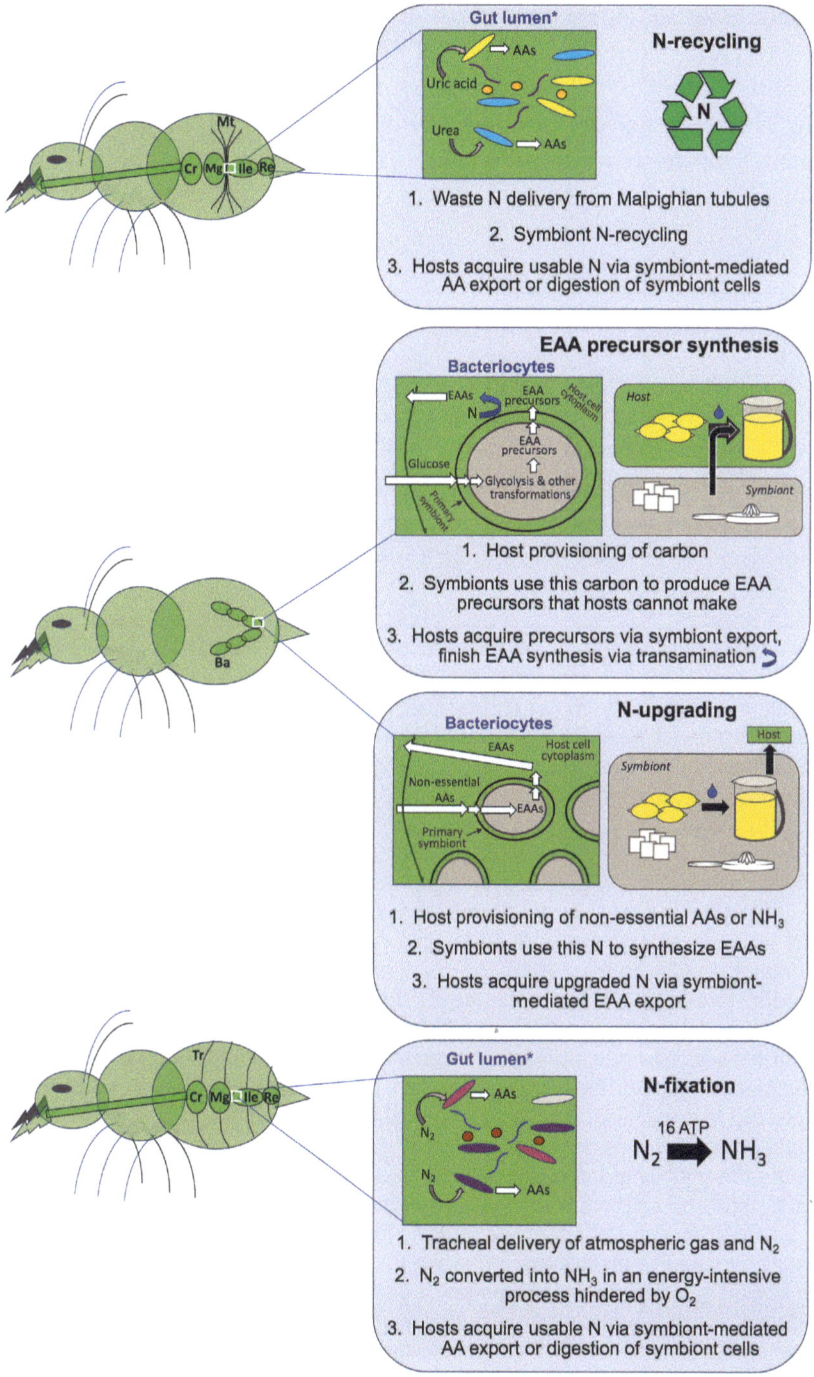

Fig. 2 See legend on next page.

cannot make, occasionally through use of a process termed N-upgrading. N-fixers and N-recyclers often fulfil this latter function, after assimilating inorganic nitrogen, but intracellular symbionts from numerous hemipteran insects make major contributions through EAA and precursor biosynthesis without apparent fixation or recycling of N. In doing so, they address amino acid imbalance but potentially not the overall shortage of N in host diets.

3. Nitrogen recycling symbioses

One major way by which microbes improve host N-economies is though conservation. Towards this end, some symbionts plug into their hosts' waste management systems to acquire, catabolize, assimilate, and upgrade compounds otherwise destined for excretion (Fig. 3). Others utilize typical insect N-wastes obtained from the diet to scavenge N from molecules that would otherwise pass through undigested. Like other animals, insects

Fig. 2 Symbiont contributions to host N-budgets and amino acid pools require varying means of physiological integration. *Top panel*—Some bacteria or fungi colonizing the gut lumen (*or hemolymph or fat body) recycle N-waste products obtained from the diet or host metabolism. Such activity may be mediated by delivery of wastes through Malpighian tubules (mt) into the hindgut or ileum (Ile). But for fat body bacteriocyte-colonizing symbionts (e.g. *Blattabacterium* of cockroaches), host derivation of urea from fat body stores of uric acid may be key. Delivery mechanisms for another N-recycler, midgut (mg) bacteriocyte-colonizing *Blochmannia*, remain unclear. *Middle panel*—The first example of AA/precursor provisioning. Bacteriocyte (Ba) colonizing microbes receive metabolites from hosts, including glucose, that are transformed (e.g. via glycolysis) into essential amino acid (EAA) precursors using metabolic pathways absent from animals. After hand-offs back to hosts, insect-encoded enzymes add N to complete the terminal steps in EAA synthesis. We use the analogy of two chefs, making lemonade together and providing varying ingredients/materials. *Third panel*—The second example of AA/precursor provisioning. Hosts provide trans-aminating N-donor molecules to symbionts, including non-EAAs (e.g. glutamate or glutamine), through bacteriocyte-expressed transporters. Symbionts use these to make EAAs in a process termed N-upgrading. This is common for symbioses with intracellular bacteria in hemipterans. We use the analogy of 1 chef (the symbiont) making lemonade out of lemons (with the aid of sugar and a juicer). *Bottom panel*—Bacteria in the gut lumen (*and possibly the hemolymph) fix atmospheric N, provisioning fixed N to hosts as AAs. Innervation of gut tissues by tracheae (tr) delivers atmospheric gas containing N, but also nitrogenase-inhibiting O_2, suggesting the importance of O_2 consumption by gut wall-associated microbes. For all extracellular gut symbioses, regardless of N-metabolic mechanisms, hosts can obtain nutrients by gut absorption after symbiont export. But some hosts obtain nutrients by digesting symbiont cells.

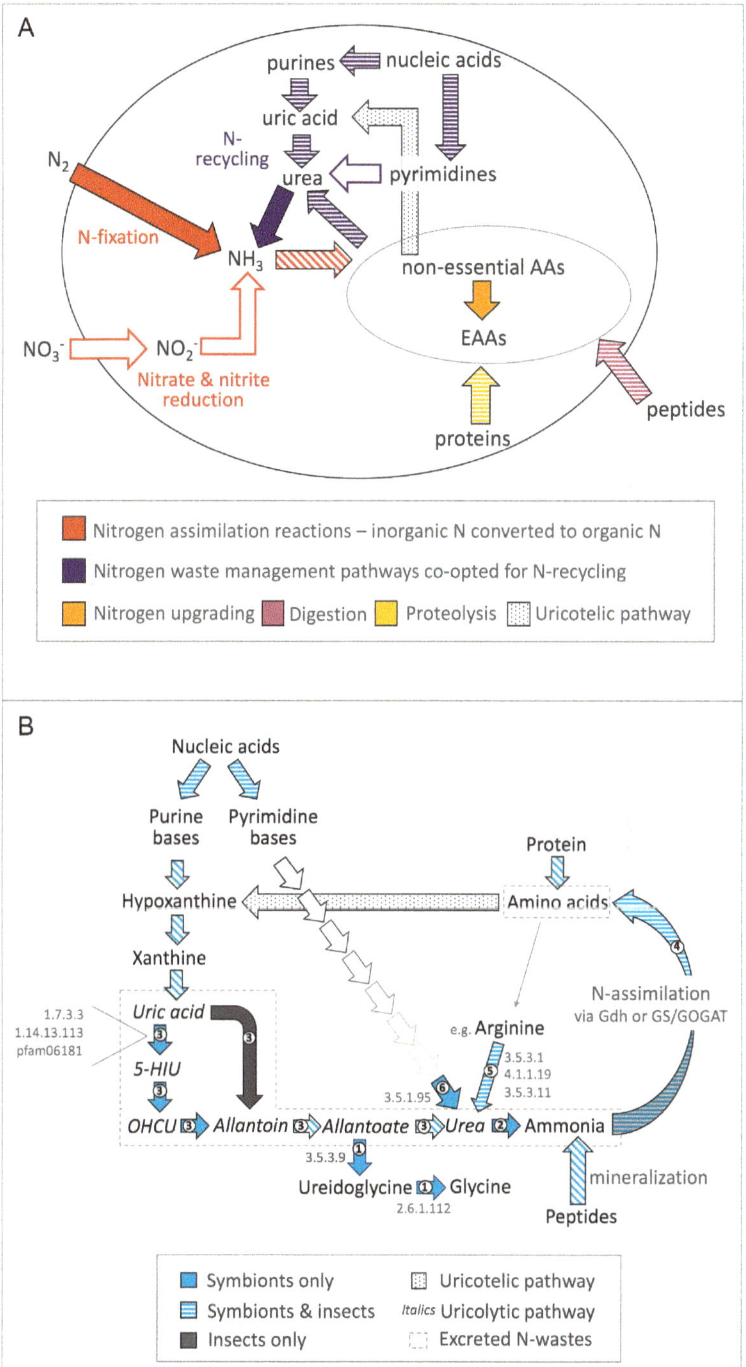

Fig. 3 See legend on next page.

produce a range of metabolites enabling removal of excess N (Bursell, 1967). Uric acid is one of the most common forms of waste, but some insects produce allantoin or allantoate, transporting these metabolites to the hindgut through Malpighian tubules, for eventual removal in the excreta (O'Donnell, 2008). A subset of insects can also derive urea through uricolytic pathways (Scaraffia et al., 2008) or through arginine catabolism, as part of the urea cycle (Fig. 3; Panfilio et al., 2019). Although uric acid may serve as an antioxidant in some species (Souza et al., 1997), the aforementioned waste products are of limited nutritional use to insects due to their inability to reacquire the constituent N. Microbial metabolism, in contrast, can clear such hurdles, due to urease capacities to derive ammonia and to other parts of the purine catabolism pathway assisting with the re-assimilation of waste-product N (Fig. 3). Collectively, these activities suggest several avenues for symbiont-mediated N-recycling. The range of possibilities expands with the knowledge that ammonia can also serve as the primary form of insect N-waste.

Fig. 3 Key aspects of insect and symbiont N-metabolism. (A) Terminology used in the manuscript, showing symbiont-contributed functions (coloured arrows) using inorganic (red arrows—assimilation), or organic (yellow, pink, orange, and purple) forms of N. For several metabolic transformations, insects lack the necessary enzymes (solid arrows), and for others, at least a subset can participate (hatched arrows—e.g. a subset of insects can generate urea from uric acid). Symbiont-mediated N-fixation, N-recycling, and N-upgrading are diagrammed and are of demonstrated importance to insects. Contributions through nitrate and nitrite reduction and peptide-enabled digestion, in contrast, are not yet known to be significant. (B) Aspects of N-waste/N-recycling pathways controlled by symbionts and/or hosts. The use of colour and hatched arrows generally follows the use in (A). Shown is the uricolytic pathway to derive urea from ammonia (*italicized font*) and the uricotelic pathway (polka-dot arrow), a mechanism by which amino acids and various C-metabolites are combined to make hypoxanthine. Rather than nucleic acid breakdown, this is posited to be the main way by which insects derive N-wastes including uric acid and its derivatives (Bursell, 1967). Dashed box highlights compounds excreted in some insects as waste products. Numbered circles highlight: (1) Glycine derivation from a uricolytic product, allantoate, as possibly fulfilled by symbionts of olive flies. (2) Urease activity, not encoded by insects, but fulfilled by some symbionts. (3) The uricolytic pathway encoded by a subset of symbionts and, partially, by some insects. (4) The assimilation of ammonia by GS-GOGAT or glutamate dehydrogenase, resulting in glutamine and/or glutamate for direct use or transamination (i.e. N-upgrading). (5) Arginase or arginine decarboxylase/agmatinase activity to derive urea. (6) A fairly under-studied enzyme, ureidomalonase, found in Actinomycetales bacteria, that makes urea from through pyrimidine catabolism. For these examples EC numbers are provided for key steps, when available (Pfam IDs provided, when not).

3.1 Nitrogen-recycling in termites

Among the best-studied and most fascinating groups of insects engaging in N-centric symbioses are those from the Blattodea, commonly known as the cockroaches and termites. The diets of most termites consist of dead wood or grass, but those in the "higher" termites (family Termitidae) have expanded to include cultivated fungi, detritus, and soil organic matter (Donovan et al., 2001). Arguably less is known about the diets of cockroaches. Presently, most are thought to be detritivores (Bell et al., 2007), despite common beliefs that most are omnivorous. With the exception of soil-feeding termites, and their peptide-rich diets, many cockroaches and termites require adaptations for a balanced N-economy. Conspicuous concentrations of microbes in both groups were recognized decades ago (Buchner, 1965; Leidy, 1881), and have long been hypothesized to fulfil nutritional services (e.g. Brooks and Richards, 1955).

One of the first demonstrations of nutritive symbiont function in the Blattodea was shown for the lower, wood-feeding termite, *Reticulitermes flavipes*. This insect was known to store large quantities of uric acid within the fat body, the primary site of uric acid synthesis; and uric acid stores showed impressive capacities to accumulate over time (Potrikus and Breznak, 1980a). Surveys of other "lower" termites, from the Kalotermitidae and Rhinotermitidae, showed that uric acid comprised ~1–14% of dry body mass and ~4–38% of total termite N (Potrikus and Breznak, 1980a). Since uric acid was extremely rare (0.04% of dry faecal mass) in termite excreta (Potrikus and Breznak, 1980a), and uricase activity was not found in microbe-free termite tissues of *R. spearatus* (Potrikus and Breznak, 1981), gut microbes became prime suspects in a hypothesized N-recycling symbiosis.

In a seminal test of this hypothesis, Potrikus and Breznak (1981) detected high levels of uricase activity in hindgut luminal contents of *R. flavipes*, and suppression of this activity with bacteria-targeting antibiotics. The authors also cultivated uric acid-degrading bacteria from termite hindguts. And they, further, detected labelled N—derived from dietary uric acid—within termite tissues, via isotope ratio mass spectrometry. These findings illustrate that N-recycling is occurring in termites, with the authors positing that hosts mobilize and deliver their own N-wastes to gut symbionts, by way of hemolymph and Malpighian tubules. This delivery route, however, has gone unconfirmed, leaving some to hypothesize that cannibalism of uric acid-rich siblings is the key means of N-waste delivery (Slaytor and Chappell, 1994).

This above work cemented a symbiont-mediated role in the re-use of insect nitrogenous waste (Bursell, 1967). But nearly 40 years since this discovery, the generality of this phenomenon across termites remains largely unknown. In one exception, uric acid-degrading bacteria were isolated from the guts of eight additional termite species (Thong-on et al., 2012). These termites spanned four of the six families, including the higher termite grouping Termitidae, expanding the range of hosts in which microbe-mediated N-recycling may be important. Lacking still, however, have been efforts to quantify recycling in vivo or the abundance of N-recycling bacteria within the termite gut. Also observed is that cultivated uricolytic bacteria from termite guts show high relatedness to free-living bacteria (Potrikus and Breznak, 1981; Thong-on et al., 2012). So, despite evidence for ancient specialization of many gut-associated termite symbionts (Hongoh et al., 2005), and evolved trophallaxis behaviours promoting partner fidelity (Nalepa et al., 2001), it is unclear whether N-recycling is carried out by coevolved, termite-confined microbes. Additional questions remain, as well. For example, based on NCBI searches, the genome of *Coptotermes secundus* (Kalotermitidae) encodes a uricase enzyme (accession #s: XP_023702356.1; XP_023702357.1). No uricase genes were reported in the genomes of two other termite species, *Nasutitermes exitiosus* (Termitidae) and *Zootermopsis nevadensis* (Termopsidae), and our efforts to identify these through BLAST searches, using the aforementioned termite uricases, have accordingly come up negative. This suggests the potential for a variety of entry, or hand-off points, for N-recycling gut symbionts utilizing the uricolytic pathway, in vivo, within these eusocial insects (Fig. 3B).

3.2 Nitrogen-recycling in cockroaches

The N-recycling symbioses of cockroaches show several deviations from those seen in termites. To begin, the focal contributors are intracellular *Blattabacterium* symbionts, hailing from the order Flavobacteriales (phylum Bacteroidetes). Confined to cockroaches and the basal termite lineage Mastotermitidae, phylogenetic analyses have shown that these specialized bacteria have largely cospeciated with their hosts (Lo et al., 2003). After the origin of termites from within the cockroach lineage, and retention by early-branching termites, *Blattabacterium* was lost from the crown group of these eusocial insects (Bandi et al., 1995). The symbiont has, contrastingly, been retained by the majority of cockroach lineages, including asocial and subsocial species (Evangelista et al., 2019).

Blattabacterium cells colonize fat body-associated bacteriocytes, and are passaged transovarially (reviewed in Mullins, 2015), supporting partner fidelity across this ancient, >200 million-year-old symbiosis

(Evangelista et al., 2019). Discovered in the late 1800s by Blochmann (1887), early research hinted at N-recycling, including the observation that *Blattabacterium*'s bacteriocyte domiciles were found next to fat body cells containing urate (uric acid) crystals (Park et al., 2013). These N reserves were depleted in cockroaches fed N-poor diets (Cochran, 1985; Mullins and Cochran, 1974), suggesting their use to sustain host N-requirements. Additional observations showed uric acid accumulates in large quantities in cockroaches deprived of *Blattabacterium*, further implicating these microbes in the breakdown of nitrogenous waste (Valovage and Brooks, 1979).

So, do *Blattabacterium* recycle N-waste to make derived products that are useful for the host? Support for this role was obtained upon sequencing the *Blattabacterium* genome (Sabree et al., 2009). While lacking genes for the breakdown of uric acid, the bacterium encodes a urease enzyme that catalyses the hydrolysis of urea into ammonia and carbon dioxide (Fig. 3B). This transformation represents the only task within the uricolytic pathway that insects are universally unable to fulfil. In this same study, the authors reported that cockroaches encode enzymes to convert uric acid into urea. Since this time, research has documented the expression of all required host-encoded N-recycling genes—uricase, allantoinase, and allantoicase (Fig. 3B)—in fat body tissues, observing their upregulation in cockroaches consuming low N diets (Patiño-Navarrete et al., 2014). Enzymatic urease activity has also been reported in the symbiont-housing cockroach fat body (López-Sánchez et al., 2009), consistent with symbiont conversion of unusable urea into usable ammonia.

To summarize, the body of work to date shows that *Blattabacterium*: (1) can generate ammonia through the metabolism of host-derived urea, (2) that it can likely assimilate this ammonia into glutamate (via glutamate dehydrogenase, *gdhA*), and (3) that it can then use glutamate to synthesize up to 10 EAAs, in part through use of this glutamate-incorporated, recycled N (Fig. 3; Henry, 1962; Sabree et al., 2009). While this detailed understanding has transformed the cockroach-*Blattabacterium* interaction into an exemplar N-centric symbiosis, important work remains. For example, while isotope-labelled N in dietary uric acid is detectable in amino acids found in cockroach oothecae (Mullins et al., 1992), it is not clear if these amino acids were confined to cells of *Blattabacterium* queued up for transovarial transfer, or if they were instead confined to cockroach tissue—indicating host acquisition. Experiments more directly testing for such acquisition, the identities of acquired nutrients that are synthesized from recycled N, and the conditions favouring such provisioning, will be useful areas for future study.

3.3 Nitrogen-recycling in ants: Blochmannia

The above examples in termites and cockroaches illustrate two distinct mechanisms by which symbionts recycle forms of N commonly derived from host waste metabolism. The symbionts of other insects may recycle similar waste products, or based on known variability in N-waste metabolism (Bursell, 1967), utilize other host-delivered metabolites. Ants and their symbionts help to illustrate this. In their ~150 million-year evolutionary history (Moreau and Bell, 2013), these insects (Hymenoptera: Formicidae) have evolved a variety of lifestyles during their rise towards dominance of terrestrial ecosystems (e.g. Nelsen et al., 2018). While frequently viewed as predators or omnivores, behavioural observations and stable isotope analyses suggest that many arboreal ants feed on a range of N-poor or N-inaccessible diets (Blüthgen et al., 2003; Davidson et al., 2003). Such diets include sugar-rich extrafloral nectar and hemipteran honeydew, which show either a paucity or imbalance of amino acids (Auclair, 1963; Baker et al., 1978; Fischer et al., 2002). In addition, some ants are attracted to vertebrate excreta, including N-waste-rich bird droppings and urine (Powell, 2008). This led to the hypothesis that such "herbivorous" ants may rely on nutritional activities of symbiotic bacteria (Davidson et al., 2003), and the separate origins of these diets raised the prospect of functional convergence in symbiont roles (Russell et al., 2017).

The first symbiosis used to test this hypothesis involved ants in the tribe Camponotini (i.e. carpenter ants and relatives) and their *Blochmannia* symbionts (Gammaproteobacteria: Enterobacteriales). Colonizing bacteriocytes associated with the midgut cells of their hosts, these transovarially transmitted bacteria (Sauer et al., 2002; Schröder et al., 1996) have cospeciated with camponotines for an estimated 40 million years (Degnan et al., 2004; Sauer et al., 2000; Wernegreen et al., 2009). Sequencing of the first *Blochmannia* genome revealed a capacity to synthesize several EAAs and to convert urea to ammonia via urease activity (Gil et al., 2003). Full gene sets for functional urease enzymes are encoded in all five *Blochmannia* genomes that have been subsequently sequenced (Williams and Wernegreen, 2015). In an important demonstration of this symbiont's contributions, Feldhaar et al. (2007) showed that *Blochmannia*-infected *Camponotus floridanus* metabolize ^{15}N-labelled urea, and that the urea-derived N is used to make EAAs and non-essential amino acids (non-EAAs) found circulating in ant hemolymph.

Capacities for amino acid biosynthesis are supported by genome sequencing across five *Blochmannia* species bracketing the diversity within the host Camponotini (Gil et al., 2003; Williams and Wernegreen, 2015). For instance, all strains can synthesize 9 of the 10 EAAs (all but arginine), with the exception of one lacking the *ilvA* gene for isoleucine synthesis. Curiously, while most can process ammonia derived from urease activity, glutamine synthetase was lost independently in two separate *Blochmannia*. The absence of glutamate dehydrogenase and glutamine oxoglutarate aminotransferase (GOGAT) suggest a potential inability to assimilate the ammonia derived from urea catabolism (Fig. 3B)—making these symbionts highly unusual in comparison to other bacteria (Williams and Wernegreen, 2010). In examining other ammonia-adding steps from the KEGG pathways that could possibly make-up for this absence, we saw only one AA biosynthesis gene that could fulfil this capacity in *Blochmannia*—cysteine conjugate beta lyase. To our knowledge, cysteine is not a major amino group donor in the synthesis of other amino acids, unlike glutamate and glutamine, suggesting a novel means of ammonia processing in some camponotine-*Blochmannia* symbioses. Host involvement may be key, based on the precedent from the pea aphid-*Buchnera* system in which symbionts receive host-derived glutamate and glutamine for use in N-upgrading and EAA synthesis (Hansen and Moran, 2011).

Additional studies in the camponotine system give insight into aspects of host N-demand making *Blochmannia* useful. To begin, *Blochmannia* show peak abundance during pupation and immediately after eclosion (Wolschin et al., 2004). Patterns of gene expression reveal the symbiont's investment in N-recycling during and around metamorphosis. Importantly, urease genes, agmatinase (involved in deriving urea from arginine; Fig. 3B), and glutamine synthetase genes show peak expression late in larval development through early adulthood (Stoll et al., 2009; Zientz et al., 2006). Since aromatic amino acid biosynthesis genes show peak expression in mid-to-late pupation, and symbiont numbers decline in mature adults, *Blochmannia*'s N-metabolic activities have been argued to contribute towards construction of the camponotine cuticle.

As a likely N-recycling symbiosis, questions arise as to how hosts deliver N-wastes and in what form. The host *Camponotus floridanus* genome (Bonasio et al., 2010) encodes uricase, allantoinase, and allantoicase enzymes (NCBI accession #s: XP_011255385.1, XP_019883740.2, XP_025268351.1) that would collectively enable ants to generate symbiont-usable urea (Fig. 3B; Scaraffia et al., 2008). While dietary urea can reach these symbionts (Feldhaar et al., 2007), *Blochmannia*'s intracellular midgut localization, and absence from the hindgut and fat body, raise the question of whether urea produced from host waste metabolism is the main ingredient for recycling, due to unresolved routes for urea delivery to bacteriocytes. Future research on these questions is warranted, as is work on the importance of host- and symbiont-encoded enzymes that derive urea from arginine (or its derivatives), including arginase (accession #: XP_011264771.1) and agmatinase (Stoll et al., 2009).

3.4 Nitrogen-recycling in ants: Beyond Blochmannia and the camponotines

Two recent studies have further supported the hypothesis that transitions to low N diets in arboreal ants have been facilitated by convergent N-metabolic activities of symbiotic bacteria (Bisch et al., 2018; Hu et al., 2018). In the first, researchers focused on *Cephalotes* turtle ants (Hu et al., 2018), which have diets resembling those of camponotines (Davidson et al., 2003), including foods with limited N (e.g. plant wound secretions, extrafloral nectar) and inaccessible forms of N (urine, bird droppings—e.g. de Andrade and Baroni Urbani, 1999). Using stable isotope labelling in manipulative dietary experiments on *C. varians*, Hu and colleagues found that N from ^{15}N-labelled dietary urea was used to make at least seven EAAs and eight non-EAAs that were eventually acquired by workers. Suppression of symbiotic gut bacteria with antibiotics erased these recycling, upgrading, and acquisition capabilities. The authors found no evidence for symbiont contributions though N-fixation, and detected only subtle degrees of symbiont N-upgrading via diet-derived glutamate.

Extracellular symbionts from *Cephalotes* ants colonize the gut environment, including the hindgut (a.k.a. ileum; Roche and Wheeler, 1997), where metabolic N-wastes from the hemocoel are delivered by Malpighian tubules. Phylogenetic analyses have illustrated that these bacteria are generally host clade specific (Hu et al., 2014), with symbiont community similarity dendrograms showing partial, but significant matching with the host phylogeny (Sanders et al., 2014). This pattern is consistent with host-microbe cospeciation and is perhaps mediated through social transmission-enabled partner fidelity (Lanan et al., 2016). Shotgun metagenomic analysis of the specialized bacteria from 17 *Cephalotes* species (Hu et al., 2018) revealed a strong degree of conservation, both in the bacterial lineages colonizing these ants and the presence of key N-recycling genes encoded in specific symbiont lineages. Bacteria from the ubiquitous symbiont genus *Cephaloticoccus*, for instance, universally encoded urease genes, allowing those tested to derive ammonia, from urea, in in vitro assays. Metagenomics also demonstrated that at least 13 microbiomes harboured unnamed members of the

Burkholderiales (Betaproteobacteria) encoding uricase genes. Several encoded full uricolytic pathways enabling the synthesis of urea; and cultured *Burkholderia* isolates from multiple *Cephalotes* could synthesize urea from the uric acid derivative allantoin.

In the second study of relevance beyond the Camponotini-*Blochmannia* symbiosis (Bisch et al., 2018), researchers focused on ants in the genus *Dolichoderus*; a group known for their N-limited and N-imbalanced diets of sap-feeding hemipteran honeydew (Cook and Davidson, 2006). Hailing from a different subfamily (Dolichoderinae) relative to cephalotines (subfamily Myrmicinae) and camponotines (subfamily Formicinae), these arboreal ants harbour highly conserved microbiomes largely dominated by a monophyletic group of genus-specific bacteria in the Rhizobiales (Alphaproteobacteria) showing relatedness to *Bartonella*. Metagenomes from four *Dolichoderus* species characterized these midgut-colonizing Rhizobiales (provisionally assigned to the genus *Tokpelaia*) as urease-encoding N-recyclers. Their maintenance and apparent, long-term cospeciation are arguably due to proctodeal trophallactic transfer. This mechanism mirrors those used for passaging hindgut bacteria of termites and cephalotine ants. But the localization of *Dolichoderus* symbionts raises questions about the mechanisms of waste N-delivery, like those raised for *Blochmannia* and camponotines, as their midgut domicile lies anterior to the site of Malpighian tubule-driven N-waste delivery (Fig. 2).

Experimentation to demonstrate host acquisition of recycled N in *Dolichoderus* clearly awaits, as does work to understand the importance of urease genes in related bacteria found in predatory ants (Neuvonen et al., 2016). Also, of interest will be studies on roles of hosts and symbionts in scavenging N from large urate stores in *Dolichoderus* fat body organs (Cook and Davidson, 2006). N-recycling is a posited function of symbionts in other ants, as well (van Borm et al., 2002), including some hosting Rhizobiales symbionts with relatedness to those in *Dolichoderus* (Russell et al., 2009; Stoll et al., 2007). Interestingly, a subset of Rhizobiales found in cephalotine ants encode ureases (Hu et al., 2018). So concentrations of related Rhizobiales in at least four ant herbivore groups (also *Tetraponera* and *Cataulacus*) suggests an intriguing potential for functional convergence due to N-recycling symbioses (Russell et al., 2017).

3.5 Nitrogen-recycling in Coleoptera and Diptera

Beetles (Coleoptera) have evolved a range of diverse symbioses with microbes, and a subset have clear or hypothesized impacts on beetle N-economies. Unlike ants and cockroaches, there is not yet evidence for ancient, N-recycling symbioses of importance to beetle biology. But a collection of findings has begun to hint at such importance. Perhaps the best illustration of bacterial N-recycling in beetles was obtained for the Asian long-horned beetle, *Anoplophora glabripennis* (Cerambycidae), a major invasive tree pest in North America. In this study (Ayayee et al., 2014), the authors used ^{15}N-urea-labelled diets, finding the heavy N signal in degutted beetle tissues and, thus, demonstrating that beetles had acquired symbiont-processed N from their gut-confined microbes. Stable isotope ratio mass spectrometry identified that this recycled N was used to synthesize several EAAs and non-EAAs. The authors also used a combination of PCR and reverse transcriptase PCR, discovering the presence and expression of urease genes (i.e. *ureC*) in eggs and larvae. In this same study the authors provided evidence for symbiont-mediated N-fixation, suggesting a multi-faceted set of mechanisms that can improve host N-economies. Since this publication, researchers have sequenced nucleic acids from gut tissue-extracts for this same beetle species (Scully et al., 2014). Amplicon sequencing of bacterial 16S rRNA genes and fungal ITS loci identified a suite of recurring bacteria and yeasts comprising the microbiome. And shotgun-metatranscriptomic analyses supported the prior finding of bacterial-encoded urease genes, while implicating symbiotic yeasts in both uricolysis and urea degradation. It was later shown that defecation into oviposition sites can transmit symbionts vertically (Mason et al., 2019), opening the potential for coevolved N-metabolic mutualisms in this invasive insect system.

On a different branch of the coleopteran phylogeny, a root- and leaf-feeding scarab beetle, *Melolontha hippocastani* (family Scarabaeidae) harbours several varieties of gut bacteria, including members of the Betaproteobacteria in the order Burkholderiales (Alonso-Pernas et al., 2017). Density gradient centrifugation coupled with 16S rRNA sequencing revealed that these bacteria incorporate ^{15}N from host-consumed dietary urea into their own DNA. Through further in vivo experiments, this labelled N was found in symbiont-free insect tissue (e.g. muscle), using isotope ratio mass spectrometry. It is seemingly not yet understood whether these are highly specialized symbioses with evolved mechanisms for transmission, or whether hosts, instead, acquire beneficial microbes from the environment to navigate their N-poor diets.

Beyond these examples, there is modest evidence for more widespread distribution of N-recycling in beetles. One potential case involves weevil species in the Scolytinae, known commonly as bark beetles. A subset of these insects harbour vertically transmitted fungi (Six and Klepzig, 2004) that concentrate N in host tree phloem sap through unknown mechanisms. Fungal mutualist-harbouring beetles also feed on fungal tissue, further alleviating dietary N-shortages (Six, 2012). Beetles with more phloem-heavy diets are expected to face greater shortfalls in dietary N and increased phloem consumption is one strategy to address this (Ayres et al., 2000). Another may involve nutritional symbiotic bacteria. In a cultivation-based study, researchers grew N-fixing and uricolytic bacteria from *Dendroctonus* bark beetles (Morales-Jiménez et al., 2013). The candidate N-recyclers were closely related to free-living microbes from three orders in the Gammaproteobacteria and made up 1.7% of the total culturable bacterial community—approximately sixfold lower than population size estimates for uricolytic bacteria from termites (i.e. *R. flavipes*; Potrikus and Breznak, 1980b). Like termite N-recycling symbionts, modest quantities of uric acid were found in the tissues and eggs of the studied bark beetles (Morales-Jiménez et al., 2013), raising the potential that these reserves could be mobilized for N-scavenging by bacteria to benefit beetle nutrition.

Diptera and Lepidoptera round out the remaining orders of holometabolous insects; and in the latter group there is no present evidence for N-recycling (or N-fixing) symbioses, to the best of our knowledge. But at least one instance of N-recycling has been reported in the Diptera, involving true fruit flies from the family Tephritidae. Larvae of the olive fruit fly, *Bactrocera oleae* feed on unripened fruit, while adults consume diets such as plant exudates, honeydew and bird excreta, which are low in EAAs or available N. Like so many of the aforementioned examples, the limiting nature of these food items suggests potential benefits through N-provisioning symbioses, and research into this possibility has focused on gut bacteria. Through a series of manipulative in vivo experiments, *B. oleae* fitness was shown to be reduced upon antibiotic suppression of its gut microbiome (Ben-Yosef et al., 2014) on diets of urea or bird droppings, but not on sucrose-only diets or diets containing all amino acids. The authors suggested that gut bacteria—including the nearly ubiquitous symbiont, *Erwinia dacicola*—play a N-recycling role, enabling hosts' use of waste N from these dietary sources. *E. dacicola* symbionts switch from intracellular stages in larvae to extracellular gut habitats in adults, and it is argued that intracellularity may allow them to survive metamorphosis (Estes et al., 2009; Hammer and Moran, 2019). They are, further, found associated with female ovipositors and, hence, arguably transmitted from mother to offspring through egg smearing. Such partner fidelity may explain why *B. oleae* harbour a monophyletic lineage of these bacteria, which is divergent from other *Erwinia* (Estes et al., 2009).

A recent study on *E. dacicola* partially supported its hypothesized N-recycling role, with two urease accessory protein-encoding genes being reported in its draft genome (Estes et al., 2018). Follow-up examination of this genome (NCBI BioProject accession #: PRJNA288714) through our own efforts identified a seemingly full set of urease genes on a single scaffold (accession #: LJAM02000124), including the core enzyme components (*ureA, ureB,* and *ureC*), the initially reported accessory genes (*ureF* and *ureG*), as well as two other genes playing likely accessory/activation roles (*ureD* and *ureJ*). Beyond their likely urease capacities, Estes et al. (2018) invoked additional roles in N-waste metabolism for *E. dacicola* through their discovery of two possible allantoate transporter genes. After locating the scaffold containing these genes (accession #: LJAM02000068.1), our BLAST-based study of adjacent genes suggested the existence of two others of likely importance. The first encodes an enzyme that can convert allantoate into ureidoglycine (i.e. allantoate amidohydrolase, protein accession #: RAP72056.1). The second encodes an enzyme that could upgrade ureidoglycine (i.e. ureidoglycine-glyoxylate aminotransferase; accession #: RAP72061.1), combining this molecule with glyoxylate to synthesize glycine (Fig. 3). KEGG pathway maps illustrate that glycine can be directly used to synthesize threonine or serine, also serving as a more distant precursor for amino acids like lysine, cysteine, and methionine. Its carbon skeleton can also be used in leucine, valine, and isoleucine synthesis; in these latter processes, recycled N—originally incorporated into glycine—would be freed as ammonia and made available for assimilation into other amino acids. An inspection of 51 genomes from *Erwinia* species in the IMG (Integrated Microbial Genomes) database reveals that nearly all strains encode the ability to make glycine from allantoate and to then transform it into the above-described amino acids. Most also have the capacity to make allantoate from uric acid, suggesting a potential loss of this function from *E. dacicola.* In addition to their work on this symbiont, Estes et al. (2018) reported on a symbiont from the genus *Enterobacter* that is potentially uricolytic. This symbiont is common in some populations of *B. oleae,* suggesting that a multipartite relationship may occur in at least some populations to facilitate N-recycling.

Beyond olive fruit flies, there are few clearly documented instances of N-recycling in the Diptera. For instance, a candidate yeast symbiont (genus *Meyerozyma*) from the digestive systems of blood-feeding sand flies encodes a uricolytic pathway and can use uric acid as a sole N-source (Martin et al., 2018). But low population prevalence—i.e. it is found in ~9% of surveyed *Phlebotomus perniciosus* sand fly hosts—raises questions about this symbiont's importance to the host N-economy, as does a lack of demonstrated uricolysis in vivo.

4. Nitrogen-fixing symbioses

Nitrogen fixation is a bacterially driven, energy-intensive process that generates ammonia for eventual assimilation and upgrading (Figs 2 and 3). Fixation of just a single molecule of N_2 requires $\geq$16 ATP molecules, indicating that the use of a plentiful, but otherwise inaccessible form of this element might be expected for symbiont-housing animals with N-poor, energy-rich diets. Several groups of wood-feeding termites appear to meet these requirements, given the prevalence of symbiont-driven cellulolytic activity and reductive acetogenesis (Brune and Dietrich, 2015; Hongoh, 2011). And indeed, the first demonstration of nitrogenase activity in insects was obtained for termites, through use of acetylene reduction assays (Benemann, 1973; Breznak, 2000; Breznak et al., 1973). This tool has been widely applied across free-living and symbiotic systems, as active nitrogenase enzymes convert acetylene into ethylene—molecules that are easily detectable through techniques including flame ionization gas chromatography.

4.1 Experimental verification of N-fixing and its importance to termites

As pointed out by Tayasu et al. (1994), woody tissues contain 0.03–0.7% N and their C/N ratios range from 70 to 500. Wood-feeding termites, in contrast, have body tissues comprised of 10% N, with C/N ratios ranging from 4 to 12. With needs met in part by N-recycling in at least some termites, support for N-fixation has been a further revelation for this group. The first direct demonstration that atmospheric N_2 was converted to ammonia for eventual incorporation into host tissues was obtained through use of stable isotope-tracing experiments. In particular, Bentley (1984) experimented on *Nasutitermes corniger*, a higher termite (Termitidae), examining fates of newly fixed N_2 across tissues of workers and soldiers with isotope ratio mass spectrometry. Within 24 h of incubation, fixed N was detectable in head and body tissues. Greater quantities were found in body tissues of workers compared to soldiers, and soldiers received greater quantities of fixed N through trophallactic exchanges with workers. This study was important in being the first to directly show fixed N-assimilation, but also because it elucidated the social component of this N-metabolic symbiosis, suggesting that trophallactically acquired nutrients or microbial cells are necessary for termites' assimilation of fixed N.

Since this early study, other authors have estimated symbiont-mediated N-fixation contributions using similar analytical chemistry, stable isotope profiling, and comparisons of naturally occurring isotope ratios in food vs. hosts. Fujita and Abe (2006), for instance, quantified the degree to which symbionts of a lower termite, *Reticulitermes spearatus* (Rhinotermitidae), enabled assimilation of fixed atmospheric N_2. The authors reared these insects in chambers with controlled quantities of ^{15}N-labelled dinitrogen for up to 28 days, subsequently performing gas chromatography-coupled mass spectrometry on various tissues to detect assimilated N. Their findings confirmed Bentley's (1984) discovery that fixed N is incorporated into symbiont-free termite body tissues. Through a detection sequence ranging from first observations of fixed N in the hindgut, subsequent detection in the crop/midgut, and later ^{15}N arrival in head plus gut-free body tissues, the authors also supported earlier arguments that fixed N is obtained by termites through trophallaxis. Separate research has added to our current understanding on the mechanisms behind this nutrient acquisition, revealing hosts obtain symbiont-derived nutrition from post-trophallactic digestion of symbiont cells in the midgut (Brune, 2014; Fujita and Abe, 2002; Fujita et al., 2001).

In other research, isotope measures in food and termite tissues were used to estimate the fraction of termite N that is derived from atmospheric N-fixation (Täyasu et al., 1994). The authors detected lower ratios of heavy (^{15}N) vs. light (^{14}N) stable N isotopes in *Neotermes* termite tissues (Kalotermitidae) than those detected in their food. Recognizing that insects are typically enriched for heavy N when compared to their diets (due to fractionation), and also recognizing the low ratios of $^{15}N/^{14}N$ for the atmosphere, it was concluded that substantial fractions of the N in termite tissue were derived from symbiont-enacted atmospheric N-fixation. Quantitative estimates suggested that these proportions ranged from, and perhaps exceeded, 30–50%. Since this work was done on field-collected termites, it is of heightened importance due to tendencies of lab-rearing to under-estimate N-fixation (Breznak, 2000).

Follow-up work (Täyasu et al., 1997) shows wood-feeding termites exhibit N-isotope ratios closely matched to diets, while those of soil feeders—in strong contrast—exceed those of their diets, in a more typical fractionation pattern. These results support the prediction that consumption of peptide-rich, humified foods by soil-feeding termites (e.g. Brune and Ohkuma, 2010; Mattson, 1980) should lessen the need for N-fixation. Instead, host and symbionts of soil-feeding termites appear devoted to N-mineralization activities (Ji and Brune, 2006), hinting at drastically different roles for symbioses across ecologically divergent termites (e.g. Eggleton and Täyasu, 2001).

4.2 Genetic evidence for N-fixation in the termite system

Molecular approaches have been applied to further knowledge on the termite gut microbe symbiosis. In one early example (Noda et al., 1999), researchers leveraged the predicted reliance on substantial N-fixation by the aforementioned *Neotermes* species (Täyasu et al., 1994) to develop a priori expectations for high rates of in vivo nitrogenase gene expression. Through a focus on the *nifH* (dinitrogenase reductase) family of homologous genes, they performed reverse transcriptase PCR, cloning, and sequencing, and T-RFLP analysis on nucleic acids isolated from termite guts. Estimated numbers of *nifH* transcripts across diets correlated with acetylene reduction levels for these *Neotermes* specimens; and consumption of ammonium in the diet drove expression down to undetectable levels. These mRNA-derived findings contrasted with findings of no change in *nifH* gene copy number using DNA extracted from termites under the same conditions. This suggested transcriptional plasticity by bacteria as a response to changing N levels, rather than changes in population size. It was also discovered that a majority of expressed *nifH* genes belonged to the *anfH* family of molybdenum- and vanadium-independent, "iron-only" nitrogenases. This finding was bolstered by T-RFLP analysis, which showed enrichment of *anfH* vs. other nitrogenases in the mRNA, but not the DNA fraction. Hence, while multiple N-fixers may colonize gut tissues, not all are similarly active.

Explorations for *nifH* genes have been carried out in other studies, including an early metagenomic analysis (Warnecke et al., 2007) that reported 12 nearly full-length *nifH* gene sequences in lumen-extracted DNA of wood-feeding *Nasutitermes* (Termitidae). Shortly after this report, genome sequencing of one Bacteroidetes symbiont provided evidence for a N-fixing role. After lysing a single protist cell from the gut of a wood-feeding *Coptotermes* termite (Rhinotermitidae), researchers obtained endosymbiotic bacteria, which they subjected to whole genome amplification and sequencing (Hongoh et al., 2008). Annotation of the streamlined 1.1 Mbp genome revealed genes encoding all required components for Mo-Fe type N-fixation, including the nitrogenase (NifHDK), molybdenum-iron cofactor synthesis proteins, a regulator of the *nif*-operon (NifA), and a molybdenum transporter. This microbe also encoded a urease, so its additionally encoded glutamine synthetase might assimilate NH_3 produced from both N-fixation and N-recycling. Evidence that this microbe can, further, synthesize 19 amino acids, suggests that this symbiosis may be important in N-upgrading.

Like this prior example, researchers used similar methodologies to sequence genomes of a Spirochaete bacterium (genus *Treponema*) from protists in the guts of wood-feeding *Hodotermopsis* (Termopsidae; Ohkuma et al., 2015). Isolating and sequencing individual bacterial cells, they found similar capacities for N-fixation, N-recycling, NH_3 assimilation through glutamine synthetase, and the synthesis of most amino acids. Having isolated these bacteria from protists associated with high rates of N_2-fixation, the authors argue that reductive acetogenesis, further encoded by this symbiont, couples an energy-generating process with the ATP-intensive reduction of N_2 molecules. Confinement of these symbionts to cellulose-digesting protists may further concentrate the materials sustaining these energetic requirements. It is noteworthy that this system shows strong parallels, in these regards, to the aforementioned Bacteroidetes-protist-*Coptotermes* association.

4.3 Macroevolutionary patterns inferred for termites and their nutritional gut bacteria

In a broad attempt to characterize *nifH* gene diversity across a variety of Blattodea, and to relate this to host ecology and phylogeny, Yamada et al. (2007) performed PCR with conserved primers, cloning, and Sanger sequencing. Their sampling design involved 5 termite families, including three subfamilies within the Termitidae as well as the wood-feeding cockroach *Cryptocercus*, the closest relative of termites, in which N-fixation activity was previously detected (Breznak et al., 1974). Using sequence similarity, phylogenetics, and ordination analyses, the authors found a trend whereby termites with similar diets and lifestyles harbour related N-fixing bacteria, or at least bacteria with related *nifH* genes.

The first group united by similar *nifH* gene content included termites in the Rhinotermitidae and Mastotermitidae. Enriched within four out of the five species in this paraphyletic grouping were *nifH* alleles from the III-3 phylogenetic lineage, which have thus far been predominantly isolated from termites. It was also noted that the termites in this group were all wood-feeders of the "intermediate life type," whereby termite colonies consume the wood that they colonize, but change their joint nesting/feeding sites throughout their lifespan. Another paraphyletic grouping, united through their sharing of related *nifH* alleles, included termites from the Kalotermitidae and Termopsidae, in addition to the wood-feeding *Cryptocercus* cockroach. Five of the six studied species from these collective taxa were enriched for clade III-3 *nifH* genes, and all were enriched for *nifH* alleles in the group II clade, which included homologues from free-living

Clostridium and Spirochaetes. Termites in this group were wood-feeding, "single-piece nesters," which colonize and consume one piece of wood for the duration of each colony's existence. The final grouping of termites united by *nifH* similarity included seven of the nine species of sampled Termitidae. Seeming to lack, completely, *nifH* allele copies from the II and III-3 lineages, these termites were instead enriched for bacteria encoding III-2 clade *nifH* alleles, showing relatedness to those from various *Clostridium*, including one from the bovine rumen. Termitidae termites harbouring bacteria with such *nifH* were united by a lifestyle of "separate-piece nesting," meaning that they colonize separate pieces of wood from those they eat.

It is intriguing that patterns here seem to transcend host relatedness, being better predicted by termite life type. Prior authors have hypothesized important differences for symbioses across these habits (Eggleton and Täyasu, 2001; Higashi et al., 1992). Differences among wood-feeders were hypothesized due to diverging interests in preserving wood biomass, a strategy presumably favoured by single-piece nesters. For such insects, conservative use of C to derive energy to fuel N_2-fixation was posited as the optimal strategy for overcoming N-limitation. Separate piece foragers were, contrastingly, hypothesized to have divergent interests, investing more in metabolisms to "burn off" carbon from rampantly consumed non-nest wood, through use of fungi or prokaryotic methanogens, Through this strategy, they might uniquely solve the challenge of N-shortages through restoration of a more ideal C:N balance.

In the final ecological pattern from Yamada et al. (2007), the authors observed that soil- and fungus-feeding termite species (*Pericapritermes nitobei* and *Odontotermes formosanus*) hosted a low relative abundance of bacteria with functional *nifH*. Among this paraphyletic, two-species grouping, most sequenced *nifH* clones grouped into cluster IV, which is presumed to have some other function aside from N-fixation. While limited by small sample size, this discovery fits previous observations that high N-fixation activity is most pronounced in wood-feeding termites, and less so in termites consuming fungi, fungus-conditioned food, and soil (e.g. Breznak, 2000; but see Sapountzis et al., 2016). Also of importance is just how often higher termites invest in N-fixation. Altered expectations arise in part due derived tendencies towards separate-piece foraging, but also because of their loss of energy-generating cellulolytic protists.

In conjunction with these questions are broader queries on the fidelity between N-fixing symbionts and their hosts. Despite similarity among relatives (Hongoh et al., 2005), termite gut communities have undergone changes in composition across evolutionary timescales. Based on amplicon sequencing of 16S rRNA, it was posited that a subset of termite gut microbiota was harboured in the common ancestor shared with cockroaches (Dietrich et al., 2014). Using this same dataset, researchers found additional ecological and evolutionary correlates of microbiome change. Interpreting beta diversity in light of the host phylogeny, it was argued that bacterial gut communities were altered after the termite/*Cryptocercus* wood roach lineage diverged from the remaining Blattodea (i.e. cockroaches), when they acquired protists and wood-based diets. The subsequent loss of protists and shifts beyond wood-feeding in higher termites appear correlated with another major reorganization in gut bacterial composition (Dietrich et al., 2014). Within the higher termites (Termitidae), amplicon 16S rRNA sequencing has also shown a link between dietary convergence and convergence in bacterial gut microbiomes; and it has further detailed that bouts of cospeciation are ephemeral and punctuated by occasional host-switching (Mikaelyan et al., 2015). But in the Rhinotermitidae (lower termites), cospeciation has been documented among hosts, protists of the genus *Pseudotrichonympha*, and their endosymbiotic bacteria from the Bacteroidetes (*Azobacteroides pseudotrichonymphae*; Noda et al., 2007). The apparent N-fixing activity of the Bacteroidetes bacteria (Hongoh et al., 2008) hints that symbionts fulfilling such functions can be tightly coevolved with both termites and protist hosts, and this is likely facilitated by social, trophallactic symbiont transfer. Phylogenetic clustering of *nifH* genes from bacteria of related termites suggests similar specialization, to some degree, even though the encoding bacteria may sometimes move between distant termites with similar life types (Yamada et al., 2007).

4.4 N-fixation beyond termites

Outside the Blattodea, N-fixing symbioses have been demonstrated or hypothesized in a number of other insect systems, ranging from wood wasps, cerambycid beetles, bark beetles, cochineal scale insects, and tephritid fruit flies. Such discoveries have been recently reviewed in Ulyshen (2015) and Bar-Shmuel et al. (2019). We thus, refrain from elaboration here. Patterns emerging from these studies suggest enrichment within xylophagous insects, but there appear to be exceptions, with detection in some phloem-feeders, as well. Future studies on whether energy-rich, N-limited diets exclusively favour convergence in N-fixing symbioses will be of considerable interest to the field. Of additional interest will be concerted efforts to test whether preliminary findings (e.g. *nifH* detection, cultivation of N-fixing bacteria) frequently translate to in vivo nitrogenase activity, with definitive impacts on host N-budgets. Of further importance will be efforts to understand whether N-fixers commonly coevolve with their hosts, or whether hosts instead acquire their diazotrophs with regularity from the environment.

5. Hemipteran symbioses: Synthesis of amino acids and their precursors as a central service

Symbioses without N_2-fixation or N-recycling can still be of benefit to host amino acid budgets, as commonly seen across sap-feeding hemipteran insects. Diets of sap-feeders include phloem and xylem. Phloem is a sugar-rich, N-poor liquid, and its limited N pools are primarily comprised of free amino acids. Non-EAAs are most predominant, with a high representation of asparagine, glutamine, or glutamate (Andersen et al., 1992; Brodbeck et al., 1990; Douglas, 2006; Sandström and Moran, 1999). Xylem sap, in comparison, has less sugar and even lower N content, with a similar amino acid imbalance (Brodbeck et al., 1990; Mattson, 1980).

These facets pose a strong challenge to sap-feeding hemipterans, and are compounded by the shortage of EAAs. In phloem and xylem, EAAs comprise an average of 20–25% of the total free amino acid concentration (Andersen et al., 1989; Sandström and Moran, 1999). But in the proteins of sap-feeding insects EAAs make up 50% of the total (Sandström and Moran, 1999). Challenges posed by animals' irreversible loss of EAA biosynthesis genes, and by these insufficient levels of EAAs must, therefore, be solved by nutritional mutualists (e.g. Bernays and Klein, 2002; Gündüz and Douglas, 2008), frequently termed primary, or obligate, symbionts. Such microbes cospeciate with hemipteran hosts (Moran et al., 2008), and are required for their growth, development, and reproduction (e.g. Fisher et al., 2017).

5.1 Tight integration of hemipteran/EAA(precursor)-synthesizing bacterial symbioses

A widespread theme for the majority of symbioses in the sap-feeding hemipteran suborders, Auchenorrhyncha, Sternorrhyncha, and the lesser-studied Coleorrhyncha (e.g. Santos-Garcia et al., 2014), is the intracellular localization of primary symbionts. These reside within host-membrane derived vesicles, inside of the cytoplasm of specialized, polyploid cells called bacteriocytes, which often aggregate to form an organ-like structure called the bacteriome (Fig. 2; Baumann, 2005; Braendle et al., 2003; Douglas, 1997; Moran et al., 2008; Munson et al., 1991). Most bacteriomes are located adjacent to the ovarioles in the abdominal haemocoel (Braendle et al., 2003; Douglas, 1997), and these nutritional endosymbionts are transmitted to the next generation maternally and internally, mainly by transovarial infection (Salem et al., 2015). In such cases, entire bacteriocytes (whiteflies, psyllids) or exocytosed symbionts (aphids, mealybugs, planthoppers) migrate from the bacteriome to the nearby egg or ovarioles, respectively, where they are endocytosed into the egg or developing embryo (Braendle et al., 2003; Douglas, 1997; Hongoh and Ishikawa, 1997; Koga et al., 2012; Michalik et al., 2018; Salem et al., 2015; Sasaki et al., 1996; Thao and Baumann, 2004; von Dohlen et al., 2001; Wernegreen, 2002). For such symbionts, not only have hosts evolved mechanisms for faithful vertical transmission, several have also evolved regulatory mechanisms that integrate their own N-metabolism with that of their symbiont (Hansen and Moran, 2013).

Before discussing such integration, we note first that an impressive body of research has shown benefits from obligate symbionts in sap-feeding hemipterans that stem largely from EAA, or EAA-precursor provisioning (Fig. 2). Most impressive has been work performed in the aphid-*Buchnera* system though use of antibiotic treatments, artificial diets, isotope-labelling, quantification of amino acid concentrations in sap and aphid proteins, and genome sequencing (Douglas, 2006; Douglas and Prosser, 1992; Douglas et al., 2001; Febvay et al., 1995, 1999; Gündüz and Douglas, 2008; Hansen and Moran, 2013). For example, artificial diets containing labelled amino acids or sucrose have been fed to either symbiotic or symbiont-cured aphids. Inferences from these studies show incorporation of dietary N and carbon into EAAs, but only in the presence of symbiotic *Buchnera* (Douglas et al., 2001; Febvay et al., 1999). Genomes of both aphids and *Buchnera* (Gammaproteobacteria: Enterobacteriacae) help validate that most of the leg work in the EAA biosynthesis is carried out by the symbionts, with a few key exceptions discussed below (Baumann et al., 1995; Charles and Ishikawa, 1999; IAGC, 2010; Moran and Baumann, 2000; Shigenobu et al., 2000; but see Hansen and Moran, 2013).

While the aphid-*Buchnera* system has proven adept for experimentation, many other hemipteran systems lack such tractability. Genomic data, hence, remain the central line of evidence for roles of symbionts in EAA provisioning (McCutcheon et al., 2009). Because these symbionts are irreversibly specialized, with highly eroded genomes (Moran and Bennett, 2014), retention of genes implies that they are of high functional relevance.

At minimum, they enable symbionts to sustain their own existence. But almost universally accepted is the idea that retention signifies direct benefits to hosts, given prior precedent (e.g. aphids) and known host dietary N-limitations. The importance of EAA pathways for these symbioses are further reinforced by trends of metabolic complementarity—obligate symbionts of sap-feeding insects tend to keep N-metabolic genes that hosts do not have (i.e. those for EAA synthesis), while losing genes that are redundant with those in hosts (i.e. those for non-EAA synthesis; Wilson and Duncan, 2015). When multiple obligate bacteria are harboured within the same host, metabolic complementary also exists between these "co-primary" symbionts. In these cases, a subset of one or more EAA pathways is encoded by one symbiont, while the remaining genes are encoded by its ancient co-inhabitant, in a jigsaw puzzle-like pattern (McCutcheon et al., 2009).

Another approach for analysing EAA synthesis in these endosymbiont systems involves regulatory analyses on bacteriocytes. Such methodology has been key to identifying an impressive degree of host-microbe integration in the process of EAA synthesis. In the best-studied example, the primary symbionts of many sap-feeding hemipterans are missing the terminal enzymes for several EAA pathways. It is predicted that the insect host expresses homologues of these missing enzymes to complete EAA production. This was confirmed through studies of *Buchnera* and pea aphids (*Acyrthosiphon pisum*), a model symbiotic system with published host and symbiont genomes. Through use of bacteriocyte-targeted RNAseq and proteomics, key aphid-encoded homologues for amino acid biosynthesis were shown to be up-regulated in bacteriocytes (Hansen and Moran, 2011; Poliakov et al., 2011). Two key genes of particular interest were branched chain-amino acid transaminase (EC number 2.6.1.42) and aspartate transaminase (EC number 2.6.1.10), which, respectively, encode the terminal steps for fabricating the EAAs isoleucine, valine, and leucine, and phenylalanine, respectively. Transcription of these genes was 4.73- and 3.29-fold higher in bacteriocytes relative to other body tissues (Hansen and Moran, 2011). When coupled with the absence of these genes in the *Buchnera* genome, these findings helped confirm host-participation in completion of EAA biosynthesis (Figs 2 and 4), raising the likelihood that the aphids can regulate this metabolism. Similar results have been found with RNAseq in subsequent studies within mealybugs, psyllids, whiteflies, and leafhopper bacteriocytes, suggesting that the integration of insect-encoded enzymes in the production of EAAs may be widespread in Hemipteran-microbe symbioses (Husnik et al., 2013; Luan et al., 2015; Mao et al., 2018; Sloan et al., 2014).

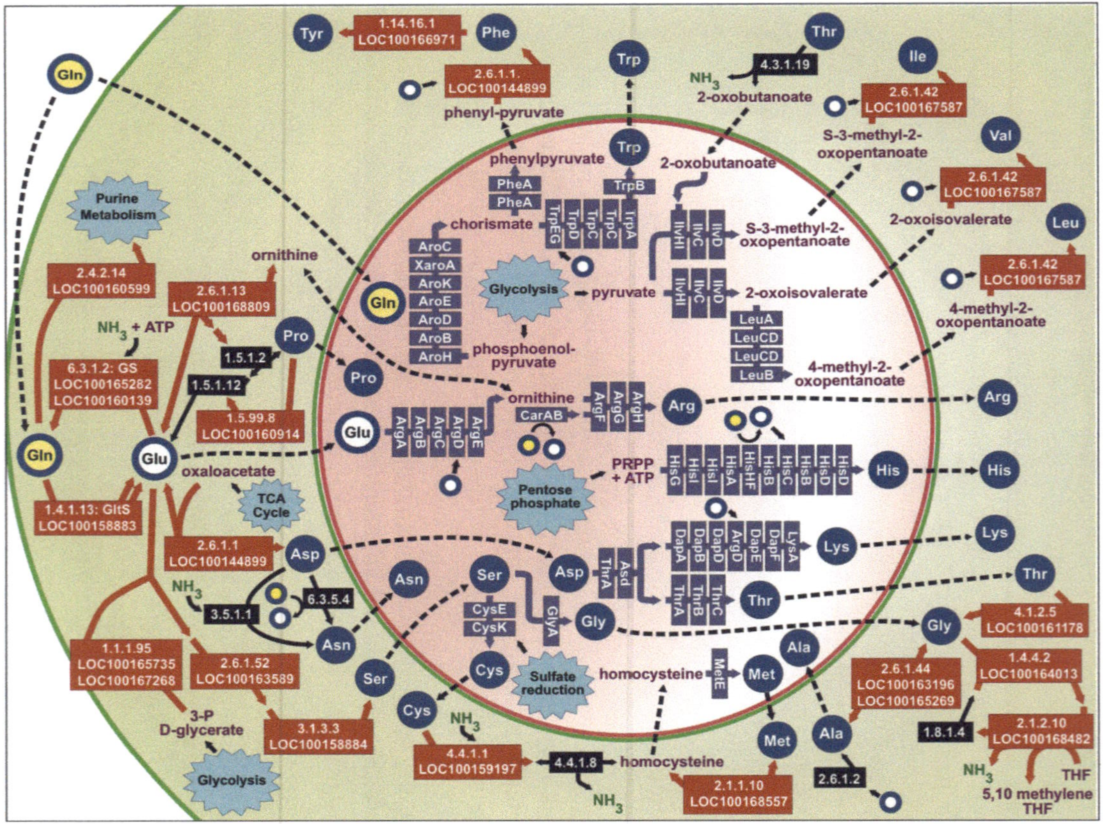

Fig. 4 A cartoon image of a specialized aphid host cell that harbours its nutritional symbiont, *Buchnera* (bacteriocyte); green = cytosol of aphid, pink = cytosol of *Buchnera*. RNAseq data reveal aphid host-encoded enzymes for amino acid pathways that are up-regulated in bacteriocytes compared to body cells (red rectangles). These are hypothesized to collaborate with *Buchnera*-encoded enzymes for essential amino acid biosynthesis (purple boxes). Aphid enzymes are labelled with Enzyme Commission and locus numbers from NCBI annotations. Brown aphid enzymes indicate that the gene is not up-regulated in bacteriocytes. *Reprinted figure obtained with permission from Hansen, A.K., Moran, N.A., 2011. Aphid genome expression reveals host-symbiont cooperation in the production of amino acids. Proc. Nat. Acad. Sci.U S A 108, 2849–2854.*

An extensive review on the presence and absence of these EAA pathways across a diversity of sap-feeding symbioses, and the complementary host-encoded enzymes, is provided in Hansen and Moran (2013).

5.2 Symbionts help sap-feeding hemipterans to overcome amino acid imbalance, with the aid of derived host bacteriocyte metabolism. But have hosts evolved additional features to improve their N-economies?

Measures of symbiont contributions to host amino acid budgets have been key to showing how EAA-provisioning symbionts can benefit hosts. For instance, Gündüz and Douglas (2008) showed that pea aphids ingested eight times their body N-requirements from diets of fava bean phloem. But ingested fluid was not sufficient to meet demands for seven of the nine examined EAAs on at least one of eight replicate plants. Using artificial diets, measures of aphid growth, and total protein content, the authors showed that aphid performance was suppressed in the absence of *Buchnera*, even more so when any of the nine studied EAAs were omitted from the diet. Variable protein growth-suppression on each of these omission diets, for aposymbiotic aphids, revealed that *Buchnera* provisions all nine studied EAAs in varying quantities, with especially large devotion to leucine. Taking a different approach, Bernays and Klein (2002) studied a different aphid, *Uroleucon ambrosiae*. They quantified phloem consumption, honeydew excretion, growth and reproductive rates, and the amino acid content of both body tissues and ingested food. Through a series of calculations, they then estimated the percentage of aphid requirements that must be met by *Buchnera*. Non EAA concentrations were considerably higher than those of EAAs in phloem, with notable excesses of asparagine, alanine, glutamine, and serine. Among the EAAs, quantities of histidine, methionine, and phenylalanine in the phloem came close to meeting aphid requirements. But the remainder were in scarce supply, with estimates suggesting *Buchnera* meets 15–80% of *U. ambrosiae* body requirements for seven EAAs, most drastically tryptophan.

In overcoming dietary amino acid imbalance, by covering for the hosts' inabilities to produce EAAs, bacteriocyte-colonizing primary symbionts of hemipterans are clearly useful. Symbiont-provided services align with prodigious sap-consumption rates, helping hemipterans succeed despite specialization on insufficient food. Several hemipterans also manipulate plant sap diets to their benefit, receiving increased concentrations of key EAAs (Sandström et al., 2000). Beyond these strategies it is unclear whether hemipterans, or their ancient bacteriocyte-dwelling symbionts, have evolved adaptations for elevated conservation of overall N content.

While it was thought that *Buchnera* and other obligate symbionts recycle ammonia derived from host waste metabolism (e.g. Prosser and Douglas, 1991), *Buchnera* genome sequencing revealed a limited capacity for NH_3-assimilation (Shigenobu et al., 2000). Subsequent work using transcriptome data concluded that host-derived glutamine synthetase assimilates host-derived ammonia in bacteriocytes, and that glutamine can be further converted to glutamate through hosts' glutamine synthetase (GS) and glutamine oxoglutarate aminotransferase (GOGAT) cycle (Hansen and Moran, 2011; Fig. 4). This hypothesis was supported by modelling work and ammonia concentration measurements (MacDonald et al., 2012), suggesting that N from ammonia-derived amino acids is primarily used in host-directed transamination reactions to make phenylalanine, isoleucine, valine, and threonine.

The GS-GOGAT cycle enzymes are also widespread within other Hemipteran insects (Hansen and Moran, 2011; Husnik et al., 2013; Luan et al., 2015; Sloan et al., 2014). For example, transcriptomic data show the GS-GOGAT cycle is up-regulated in bacteriocytes compared to other body tissues in aphids, mealybugs, psyllids, whiteflies, and leafhoppers (Hansen and Moran, 2011; Husnik et al., 2013; Luan et al., 2015; Mao et al., 2018; Sloan et al., 2014), and is hypothesized to play an important role in ammonia assimilation to fuel N-upgrading in sap-feeding insects (Hansen and Moran, 2011). This collectively suggests that the insect hosts, and not the symbionts, conduct N-recycling in these symbioses (Fig. 4). The efficiency with which they shunt useful forms of N in their bodies to bacteriocytes for these purposes remains unclear, as does the exact nature of hemipteran N-waste management. These questions lie at the heart of whether aphids, and other sap-feeding hemipterans have evolved thrifty N-economies, or whether excessive feeding, host plant manipulation, and regulatory changes to bacteriocyte-localized N-metabolism comprise the bulk of host-derived solutions to N-imbalance and N-limitation.

5.3 Supplanting symbionts spur sporadic instances of N-recycling and N-fixation across the Hemiptera

In addition to findings of host involvement in N-waste management, initial discoveries that symbionts like *Buchnera* did not fix- or recycle N did not rule out the potential for symbiont-driven N-concentration across all sap-feeding hemipterans. But the accumulation of genomes across this group has revealed that N-fixing and N-recycling enzymes are generally absent from obligate symbiont genomes. To date, the only primary symbiont encoding an enzyme with plausible N-recycling function is *Sulcia*-PSPU, which is harboured in the spittlebug, *Philaenus spumarius* (Koga and Moran, 2014). This endosymbiont encodes glutamate dehydrogenase (*gdhA*) (Koga and Moran, 2014), which may recycle ammonia and α-ketoglutarate into glutamate, an important amino donor for EAA biosynthesis. This *Sulcia*-PSPU enzyme was lost from other *Sulcia* lineages and, notably, its most closely related homologue is encoded by related *Blattabacterium* symbionts of cockroaches (Koga and Moran, 2014). Beyond this example, there is scant evidence for other N-recycling genes across hemipteran primary symbionts. The absence of urease genes is notable, but perhaps expected due to the lack of the urea cycle in many hemipterans (KEGG website: Kanehisa and Goto, 2000; Kanehisa et al., 2019; also see Panfilio et al., 2019).

In looking more broadly for N-fixation and N-recycling across insects normally dominated by EAA-provisioning symbioses, a pattern emerges whereby hemipterans—having lost their long-standing obligate symbionts—have sometimes acquired new symbionts with one or both capacities. Demonstration of N-recycling was first fulfilled for the brown planthopper *Nilaparvata lugens* (family Delphacidae), and a filamentous ascomycete fungus related to *Ophiocordyceps* (Matsuura et al., 2018; Suh et al., 2001). Previous research indicated that *N. lugens* produced substantial quantities of uric acid, which were conspicuously stored in its tissues (e.g. Hongoh and Ishikawa, 1997; Sasaki et al., 1996). The "yeast-like symbiont," which replaced ancestral primary bacterial symbionts, expressed uricase activity, but virtually none was detected in host tissues or in aposymbiotic planthoppers (Sasaki et al., 1996). Quantities of stored uric acid increased with increasing amounts of dietary amino acids. When insects were switched from rich diets to amino acid-poor diets, uric acid stores were drastically depleted, though only in the presence of symbionts (Hongoh and Ishikawa, 1997). Colonizing fat body-localized bacteriocytes (Noda, 1977), these fungi sustain host fitness in the face of N-poor diets, perhaps due to their N-metabolic functions (Hongoh and Ishikawa, 1997).

PCR and Sanger sequencing supported this symbiont's N-recycling role through detection of symbiont-encoded uricase genes (Hongoh and Ishikawa, 2000), findings verified by subsequent genome sequencing (Xue et al., 2014). In addition to uricase, the yeast-like symbiont was also reported to encode urease genes and several intervening genes in the uricolytic pathway. Inferences from the host genome (Xue et al., 2014) reveal that planthoppers encode uricase, and at least one step in the uricolytic pathway, but that they lack allantoicase enzymes for the synthesis of urea. Further genomic assessment revealed a glutamate dehydrogenase gene, and genes for the GS-GOGAT cycle, indicating that the symbiont can directly assimilate ammonia derived from uricolysis. The fungal symbiont also retains genes to synthesize the 10 EAAs, arguing that it can upgrade the recycled and assimilated N, perhaps through use of transaminating N-donors including glutamate (Xue et al., 2014).

A related N-recycling fungus has similarly supplanted the ancient aphid EAA-provisioning symbiont *Buchnera*. In gall-forming aphids of the tribe Cerataphidini (Sternorrhyncha: Aphididae), *Buchnera*-replacing yeast-like symbionts colonize hemolymph and, occasionally, fat body tissues (Buchner, 1965). Earlier work had detected uricase genes within these fungi (Hongoh and Ishikawa, 2000) and genome sequencing for one species revealed an intact uricase gene (Vogel and Moran, 2013). Our own efforts to detect urease enzymes, through a tBLASTn search, revealed a match with 87.5% amino acid identity to the urease enzyme of a related fungus (*Ophiocordyceps sinensis* accession #: EQK99008.1—tBLASTn vs. assembly accession #: GCA_000372705.1). This match spanned just 24% of the query, however, meaning that while the finding is consistent with symbiont-encoded urease function, this capacity—and the broader role of the symbiont in N-recycling—requires further investigation. Interestingly, the aphids' fungal symbiont does encode a full set of genes for the synthesis of all amino acids, including all 10 EAAs. It can, further, reduce nitrate and nitrite to ammonia, and it encodes a complete GS-GOGAT cycle. Given the symbionts' coevolved history with their aphid hosts, these discoveries reveal potentially impactful mechanisms for the assimilation of symbiont-derived ammonia, and subsequent upgrading for the hypothesized benefit of hosts.

Across the Sternorrhyncha in the scale insect subfamily Coccoidea, sap-feeding cochineals from the genus *Dactylopius* harbour symbionts with suspected roles in both N-recycling and N-fixation (Vera-Ponce de Leon et al., 2017). Chief among these is a betaproteobacterium, *Candidatus* Dactylopiibacterium carminicum. Colonizing hemolymph and ovaries, this symbiont was found to encode genes for the molybdenum-iron nitrogenase (*nifD-nifK*) and dinitrogenase reductase (*nifH*) enzymes. Also detected in its genome were genes encoding various N-regulatory proteins and transporters for nitrogenase cofactors, including *nifQpp-nifA* and 4Fe-4S ferredoxin. Expression of *nifH* was found in hemolymph, ovaries, and eggs; and acetylene reduction assays detected likely N-fixation in the former two tissues. The authors also reported genes in the uricolytic pathway, including urease, revealing a second means by which the bacterial symbiont could concentrate N for its cactus-phloem-feeding host. The *D. carminicum* symbiont also encodes biosynthesis of all amino acids except arginine, further supporting an important N-provisioning role.

In a related study on this system, researchers focused on fungal symbionts, including those colonizing the cochineal gut and Malpighian tubules (Vera-Ponce de Leon et al., 2016). Evidence from experiments involving gene expression, metagenomics, uric acid quantification, symbiont suppression, and microscopy, suggests a role for this fungus in the recycling of host N-waste. And evidence for the symbiont's vertical transmission combines to suggest that a second mutualist with N-concentrating capacities colonizes cocchineals alongside *D. carminicum*, bacterial symbionts.

The common ancestor of scale insects was likely symbiotic with EAA-provisioning bacteria (Sudakaran et al., 2017), suggesting the above example in cochineals represents an additional instance of symbiont replacement with N-recycling fungi. Bacterial symbionts have been similarly replaced in the Coccidae, or "soft scales." In this case, amplicon sequencing of 16S rRNA identified no consistent bacterial associates across soft scale species. Instead, a mixture of sequence-, phylogenetic-, and microscopy-based assessments argued for consistent presence of vertically transferred *Ophiocordyceps*-like fungi (Gomez-Polo et al., 2017). More work remains to ascertain this symbiont's N-metabolic capacities.

Other examples of fungi replacing long-standing bacterial N-provisioning mutualists have been reported recently in cicadas (Cicadoidea), distant relatives of the delphacid *N. lugens* planthoppers within the suborder Auchenorrhyncha (Matsuura et al., 2018). In particular, several Japanese cicada lineages no longer harbour *Hodgkinia*, one of their two ancient bacterial mutualists. These cicadas, instead, contribute to the emerging theme, whereby ancient bacterial mutualists have been replaced by specialized *Ophiocordyceps*-like symbionts. These fungi are somewhat related to the yeast-like symbionts of aphids and planthoppers; and in cicadas, they colonize fat body tissue surrounding bacteriocytes, along with sheath cells enveloping the bacteriome. Microscopic observations suggest transovarial vertical transmission. And genome sequencing of one cultivated *Ophiocordyceps* symbiont revealed complete pathways for the biosynthesis of all amino acids. Importantly, the genomes were argued to encode "all synthesis pathway genes for…nitrogen recycling." Using the uricase protein from the symbiont of *N. lugens* planthoppers (accession #: BAA89004.1), we performed a tBLASTn search against the genome assembly in NCBI (assembly accession #: GCA_002939055.1). A putative protein with

73.2% identity (191/261 identical amino acids) to the uricase query was identified. In similar tBLASTn searches, a urease gene from a *Metarhizium* fungus (accession #: XP_007823647.2) and another *Ophiocordyceps* (accession #: EQK909008.1) yielded hits with 69.3% (251/319 amino acids) and 82.4% (74/91, 26/29, and 97/119 amino acids) identity to a putative protein from the cicada symbiont, across most of their lengths. The presence of likely N-recycling genes, and the localization of symbionts to the fat body where uric acid is often synthesized, reveal the potential for a N-recycling mutualism.

Outside of sap-feeding hemipterans, the shield bug *Parastrachia japonensis* (suborder Heteroptera; Parastrachiidae) feeds on drupes (i.e. fleshy fruits), which are available for a very short time each year. Between bouts of food availability, the shield bug enters reproductive diapause, ingesting scarce amounts of food for many months. During this time, viability appears to depend on a bacterial N-recycling symbiont (Kashima et al., 2006), identified as *Benitsuchiphilus tojoi* (Hosokawa et al., 2010). This gammaproteobacterium is transmitted vertically through egg smearing and colonizes midgut crypts of its shield bug hosts. Symbiont removal with antibiotics increases mortality during this diapause, while also suppressing quantities of free amino acids in the hemolymph. But when symbionts are undisturbed, total N levels appear remarkably unchanged in these shield bugs when supplied with only water for 8 month rearing periods. Midgut extract from diapausing females revealed uricase, allantoinase, and allantoicase activity, and a re-test of uricase activity showed suppression upon antibiotic treatment (Kashima et al., 2006). These data, combined, suggest beneficial actions of yet another N-recycling symbiont. Contributions to hemipterans feeding on foods beyond phloem sap are an exception to the current rule. This raises the interesting possibility that some nutritional symbionts evolved their current host relationships due to their capacities to help hosts weather periods of dietary stress.

6. Conclusions and future directions

Nutritional symbionts contribute a range of diet-limited compounds to their hosts, and N is often at the center of such interactions. Thanks to improving technologies and, often, research into pest insects, a growing body of literature has illuminated our understanding of how symbioses improve the N-budgets of insects consuming N-limited and N-inaccessible diets. But numerous questions remain. For instance, do insects' symbionts commonly incorporate N through nitrate or nitrite reduction (Ngugi et al., 2011;

Vogel and Moran, 2013), as seen in some aquatic systems (Girguis et al., 2000)? Do symbionts make major improvements to hosts' N-budgets through assisted digestion, perhaps through protease secretion (Chu et al., 2013; Sethi et al., 2011)? How commonly do hosts invest in the strategy of "burning off" excess C to improve C:N ratios (Higashi et al., 1992)? Did fungus-cultivating mutualisms evolve for this purpose (Mueller et al., 2005), and can we hence view symbionts with particular forms of carbon metabolism as contributors to insect N-economies? Also of interest is the question of just how much our methods have limited our knowledge, whether through impacts of the lab environment on activities like termite N-fixation (Breznak, 2000), spurious impacts of artificial diets on sap-feeding insects and their symbionts (Pers and Hansen, 2019), or the over-reliance on genomics and existing gene annotations to predict function without experimental validation. Future research in these areas, and further applications of cutting-edge technologies (e.g. Volland et al., 2018), will be part of an exciting era for symbiosis research across insects and beyond.

While much exciting work awaits, we should not minimize how far we have come in recent decades. As far back as the mid-1900s, many of the patterns seen today for insect-symbiont associations had been at least roughly outlined, with one set of authors noting, "Most of the literature on endosymbiosis is of a descriptive rather than an experimental nature, and the theories on the role of the symbiotes are usually based on circumstantial evidence. Practically all symbiote-bearing insects feed on diets which are incomplete or inadequate in certain substances known to be required by non-symbiotic insects or by vertebrates," (Brooks and Richards, 1955). Clearly, the inseparable nature of many host symbiont relationships was a serious obstacle at that time. But technological advances involving genomics and analytical chemistry have allowed researchers to circumnavigate roadblocks deriving from mutual dependence. Through the resulting hard-won discoveries on symbiont roles we have uncovered a diversity of ways that microbes have aided in the exploitation of diets rich in C, but depauperate in N. Although insects have evolved many stand-alone mechanisms for the use of such plant-based diets (Berner et al., 2005; Karban and Agrawal, 2002; Lavoie and Oberhauser, 2004), many have opted for a collaborative approach. Their microbial menageries bring diverse C- and N-metabolic repertoires, helping to exploit incredibly abundant, nutritionally challenging plant resources. In doing so, symbionts have helped to revolutionize terrestrial ecosystems, supporting lifestyles of a broad swath of the numerically dominant and hyper diverse herbivorous insects (Grimaldi and Engel, 2005).

References

Alonso-Pernas, P., et al., 2017. In vivo isotopic labeling of symbiotic bacteria involved in cellulose degradation and nitrogen recycling within the gut of the forest Cockchafer (*Melolontha hippocastani*). Front. Microbiol. 8, 1970. https://doi.org/10.3389/fmicb.2017.01970.

Andersen, P.C., Brodbeck, B.V., Mizell, R.F., 1989. Metabolism of amino acids, organic acids, and sugars extracted from the xylem fluid of four host plants by adult *Homalodisca coagulata*. Entomol. Exp. Appl. 50, 149–160.

Andersen, P.C., Brodbeck, B.V., Mizell, R.F., 1992. Feeding by the leafhopper, *Homalodisca coagulata*, in relation to xylem fluid chemistry and tension. J. Insect Physiol. 38, 611–622.

Auclair, J.L., 1963. Aphid feeding and nutrition. Annu. Rev. Entomol. 8, 439–490.

Ayayee, P., et al., 2014. Gut microbes contribute to nitrogen provisioning in a wood-feeding cerambycid. Environ. Entomol. 43 (4), 903–912. https://doi.org/10.1603/en14045.

Ayres, M.P., et al., 2000. Nitrogen budgets of phloem-feeding bark beetles with and without symbiotic fungi (Coleoptera: Scolytidae). Ecology 81, 2198–2210.

Baker, H.G., Opler, P.A., Baker, I., 1978. A comparison of the amino acid complements of floral and extrafloral nectars. Bot. Gaz. 139, 322–332.

Bandi, C., et al., 1995. The establishment of intracellular symbiosis in an ancestor of cockroaches and termites. Proc. R. Soc. B 259 (1356), 293–299. https://doi.org/10.1098/rspb.1995.0043.

Bar-Shmuel, N., Behar, A., Segoli, M., 2019. What do we know about biological nitrogen fixation in insects? Evidence and implications for the insect and the ecosystem. Insect Sci. https://doi.org/10.1111/1744-7917.12697. (Early View).

Baumann, P., 2005. Biology of bacteriocyte-associated endosymbionts of plant sap-sucking insects. Annu. Rev. Microbiol. 59, 155–189.

Baumann, P., et al., 1995. Genetics, physiology and evolutionary relationships of the genus *Buchnera*: intracellular symbionts of aphids. Annu. Rev. Microbiol. 49, 55–94.

Behar, A., Yuval, B., Jurkevitch, E., 2005. Enterobacteria-mediated nitrogen fixation in natural populations of the fruit fly *Ceratitis capitata*. Mol. Ecol. 14, 2637–2643. https://doi.org/10.1111/j.1365-294X.2005.02615.x.

Bell, W.J., Roth, L.M., Nalepa, C.A., 2007. Cockroaches Ecology, Behavior, and Natural History. Johns Hopkins University Press, Baltimore.

Benemann, J.R., 1973. Nitrogen fixation in termites. Science 181 (4095), 164–165.

Bennett, G.M., Moran, N.A., 2015. Heritable symbiosis: the advantages and perils of an evolutionary rabbit hole. Proc. Natl. Acad. Sci. U. S. A. 112 (33), 10169–10176. https://doi.org/10.1073/pnas.1421388112.

Bentley, B.L., 1984. Nitrogen fixation in termites: fate of newly fixed nitrogen. J. Insect Physiol. 30, 653–655.

Ben-Yosef, M., et al., 2014. Symbiotic bacteria enable olive flies (*Bactrocera oleae*) to exploit intractable sources of nitrogen. J. Evol. Biol. 27 (12), 2695–2705. https://doi.org/10.1111/jeb.12527.

Bernays, E.A., Klein, B.A., 2002. Quantifying the symbiont contribution to essential amino acids in aphids: the importance of tryptophan for *Uroleucon ambrosiae*. Physiol. Entomol. 27 (4), 275–284. https://doi.org/10.1046/j.1365-3032.2002.00297.x.

Berner, D., Blanckenhorn, W.U., Körner, C., 2005. Grasshoppers cope with low host plant quality by compensatory feeding and food selection: N limitation challenged. Oikos 111 (3), 525–533. https://doi.org/10.1111/j.1600-0706.2005.14144.x.

Bisch, G., et al., 2018. Genome evolution of Bartonellaceae symbionts of ants at the opposite ends of the trophic scale. Genome Biol. Evol. 10 (7), 1687–1704. https://doi.org/10.1093/gbe/evy126.

Blochmann, F., 1887. Occurrence of bacteria-like particles in tissues and eggs of different insects. Biol. Zent. Bl. 7, 606–608.
Blüthgen, N., Gebauer, G., Fiedler, K., 2003. Disentangling a rainforest food web using stable isotopes: dietary diversity in a species-rich ant community. Oecologia 137, 426–435. https://doi.org/10.1007/s00442-003-1347-8.
Bonasio, R., et al., 2010. Genomic comparison of the ants *Camponotus floridanus* and *Harpegnathos saltator*. Science 329 (5995), 1068–1071. https://doi.org/10.1126/science.1192428.
Braendle, C., et al., 2003. Developmental origin and evolution of bacteriocytes in the aphid-*Buchnera* symbiosis. PLoS Biol. 1 (1), 70–76. https://doi.org/10.1371/journal.pbio.0000021.
Breznak, J.A., 2000. Ecology of prokaryotic microbes in the guts of wood- and litter-feeding termites. In: Abe, T., Bignell, D.E., Higashi, M. (Eds.), Termites: Evolution, Sociality, Symbioses, Ecology. Kluwer Academic, Dordrect, The Netherlands, pp. 209–231.
Breznak, J.A., et al., 1973. Nitrogen fixation in termites. Nature 244, 577–580.
Breznak, J., Mertins, J., Coppel, H., 1974. Nitrogen fixation and methane production in a wood-eating cockroach, *Cryptocercus punctulatus* Scudder (Orthoptera: Blattidae). University of Wisconsin Forestry Research Notes. 184. Madison, WI.
Brodbeck, B.V., et al., 1990. Amino acids as determinants of host preference for the xylem feeding leafhopper, *Homalodisca coagulata* (Homoptera: Cicadellidae). Oecologia 83, 338–345.
Brooks, M.A., Richards, A.G., 1955. Intracellular symbiosis in cockroaches: II. Mitotic division of mycetocytes. Science 122 (3162), 242. https://doi.org/10.1126/science.122.3162.242.
Brune, A., 2014. Symbiotic digestion of lignocelluloase in termite guts. Nat. Rev. Microbiol. 12, 168–180. https://doi.org/10.1038/nrmicro3182.
Brune, A., Dietrich, C., 2015. The gut microbiota of termites: digesting the diversity in the light of ecology and evolution. Annu. Rev. Microbiol. 69, 145–166. https://doi.org/10.1146/annurev-micro-092412-155715.
Brune, A., Ohkuma, M., 2010. Role of the termite gut microbiota in symbiotic digestion. In: Bignell, D., Roisin, Y., Lo, N. (Eds.), Biology of Termites: A Modern Synthesis. Springer, Dordecht, pp. 439–475. https://doi.org/10.1007/978-90-481-3977-4_16.
Buchner, P., 1965. Endosymbiosis of Animals With Plant Micro-Organisms. John Wiley and Sons, New York.
Burmester, T., 1999. Evolution and function of the insect hexamerins. Eur. J. Entomol. 96, 213–225.
Bursell, E., 1967. The excretion of nitrogen in insects. Adv. Insect Physiol. 4, 33–67.
Charles, H., Ishikawa, H., 1999. Physical and genetic map of the genome of *Buchnera*, the primary endosymbiont of the pea aphid *Acyrthosiphon pisum*. J. Mol. Evol. 48, 142–150.
Chu, C.-C., et al., 2013. Gut bacteria facilitate adaptation to crop rotation in western corn rootworm. Proc. Natl. Acad. Sci. U. S. A. 110 (29), 11917–11922.
Clay, N.A., Lehrter, R.J., Kaspari, M., 2017. Towards a geography of omnivory: omnivores increase carnivory when sodium is limiting. J. Anim. Ecol. 86 (6), 1523–1531. https://doi.org/10.1111/1365-2656.12754.
Cochran, A.G., 1985. Nitrogen-excretion in cockroaches. Annu. Rev. Entomol. 30, 29–49.
Cook, S.C., Davidson, D.W., 2006. Nutritional and functional biology of exudate-feeding ants. Entomol. Exp. Appl. 118 (1), 1–10. https://doi.org/10.1111/j.1570-7458.2006.00374.x.
Davidson, D.W., et al., 2003. Explaining the abundance of ants in lowland tropical rainforest canopies. Science 300 (5621), 969–972. https://doi.org/10.1126/science.1082074.
de Andrade, M., Baroni Urbani, C., 1999. Diversity and adaptation in the ant genus *Cephalotes* past and present. Stutt. Beitr. Naturkunde B 271, 1–889.

Degnan, P.H., et al., 2004. Host-symbiont stability and fast evolutionary rates in an ant-bacterium association: cospeciation of *Camponotus* species and their endosymbionts, *Candidatus Blochmannia*. Syst. Biol. 53 (1), 95–110. https://doi.org/10.1080/10635150490264842.

Denno, R.F., Fagan, W.F., 2003. Might nitrogen limitation promote omnivory among carnivorous arthropods? Ecology 84 (10), 2522–2531. https://doi.org/10.1890/02-0370.

Dietrich, C., Köhler, T., Brune, A., 2014. The cockroach origin of the termite gut microbiota: patterns in bacterial community structure reflect major evolutionary events. Appl. Environ. Microbiol. 80 (7), 2261–2269. https://doi.org/10.1128/AEM.04206-13.

Donovan, S.E., Eggleton, P., Bignell, D.E., 2001. Gut content analysis and a new feeding group classification of termites. Ecol. Entomol. 26 (4), 356–366. https://doi.org/10.1046/j.1365-2311.2001.00342.x.

Douglas, A.E., 1997. Parallels and contrasts between symbiotic bacteria and bacterial-derived organelles: evidence from *Buchnera*, the bacterial symbiont of aphids. FEMS Microbiol. Ecol. 24, 1–9.

Douglas, A.E., 2006. Phloem-sap feeding by animals: problems and solutions. J. Exp. Bot. 57 (4), 747–754. https://doi.org/10.1093/jxb/erj067.

Douglas, A.E., 2009. The microbial dimension in insect nutritional ecology. Funct. Ecol. 23 (1), 38–47. https://doi.org/10.1111/j.1365-2435.2008.01442.x.

Douglas, A.E., Prosser, W.A., 1992. Synthesis of the essential amino acid tryptophan in the pea aphid (*Acyrthosiphon pisum*) symbiosis. J. Insect Physiol. 38, 565–568.

Douglas, A.E., Minto, L.B., Wilkinson, T.L., 2001. Quantifying nutrient production by the microbial symbionts in an aphid. J. Exp. Biol. 204, 349–358.

Eggleton, P., Tayasu, I., 2001. Feeding groups, lifetypes and the global ecology of termites. Ecol. Res. 16 (5), 941–960. https://doi.org/10.1046/j.1440-1703.2001.00444.x.

Elser, J.J., et al., 1996. Organism size, life history, and N: P stoichiometry toward a unified view of cellular and ecosystem processes. Bioscience 46 (9), 674–684. https://doi.org/10.2307/1312897.

Estes, A.M., et al., 2009. The olive fly endosymbiont, "*Candidatus* Erwinia dacicola" switches from an intracellular existence to an extracellular existence during host insect development. Appl. Environ. Microbiol. 75 (22), 7097–7106. https://doi.org/10.1128/AEM.00778-09.

Estes, A.M., et al., 2018. Comparative genomics of the *Erwinia* and *Enterobacter* olive fly endosymbionts. Sci. Rep. 8, 15936.

Evangelista, D.A., et al., 2019. An integrative phylogenomic approach illuminates the evolutionary history of cockroaches and termites (Blattodea). Proc. R. Soc. B 286 (1895), 1–9. https://doi.org/10.1098/rspb.2018.2076.

Fagan, W.F., et al., 2002. Nitrogen in insects: implications for trophic complexity and species diversification. Am. Nat. 160 (6), 784–802. https://doi.org/10.1086/343879.

Febvay, G., et al., 1995. Analysis of energetic amino acid metabolism in *Acyrthosiphon pisum*: a multidimensional approach to amino acid metabolism in aphids. Arch. Insect Biochem. Physiol. 29, 45–69.

Febvay, G., et al., 1999. Fate of dietary sucrose and neosynthesis of amino acids in the pea aphid, *Acyrthosiphon pisum*, reared on different diets. J. Exp. Biol. 202, 2639–2652.

Feldhaar, H., et al., 2007. Nutritional upgrading for omnivorous carpenter ants by the endosymbiont *Blochmannia*. BMC Biol. 5, 48. https://doi.org/10.1186/1741-7007-5-48.

Finke, M.D., 2007. Estimate of chitin in raw whole insects. Zoo Biol. 26 (2), 105–115. https://doi.org/10.1002/zoo.20123.

Finke, M.D., 2015. Complete nutrient content of four species of commercially available feeder insects fed enhanced diets during growth. Zoo Biol. 34 (6), 554–564. https://doi.org/10.1002/zoo.21246.

Fischer, M.K., et al., 2002. Age-specific patterns in honeydew production and honeydew composition in the aphid *Metopeurum fuscoviride*: implications for ant-attendance. J. Insect Physiol. 48, 319–326.

Fisher, R.M., et al., 2017. The evolution of host-symbiont dependence. Nat. Commun. 8, 15973. https://doi.org/10.1038/ncomms15973.

Fujita, A., Abe, T., 2002. Amino acid concentration and distribution of lysozyme and protease activities in the guts of higher termites. Physiol. Entomol. 27 (1), 76–78. https://doi.org/10.1046/j.1365-3032.2001.00265.x.

Fujita, A., Abe, T., 2006. Atmospheric nitrogen assimilation by a wood-feeding termite, *Reticulitermes speratus* (Isoptera: Rhinotermitidae). Sociobiology 47, 175–188.

Fujita, A., Shimizu, I., Abe, T., 2001. Distribution of lysozyme and protease, and amino acid concentration in the guts of a wood-feeding termite, *Reticulitermes speratus* (Kolbe): possible digestion of symbiont bacteria transferred by trophallaxis. Physiol. Entomol. 26 (2), 116–123. https://doi.org/10.1046/j.1365-3032.2001.00224.x.

Gil, R., et al., 2003. The genome sequence of *Blochmannia floridanus*: comparative analysis of reduced genomes. Proc. Natl. Acad. Sci. U. S. A. 100 (16), 9388–9393. https://doi.org/10.1073/pnas.1533499100.

Girguis, P.R., et al., 2000. Fate of nitrate acquired by the tubeworm *Riftia pachyptila*. Appl. Environ. Microbiol. 66 (7), 2783–2790. https://doi.org/10.1128/AEM.66.7.2783-2790.2000.

Gomez-Polo, P., et al., 2017. An exceptional family: *Ophiocordyceps*-allied fungus dominates the microbiome of soft scale insects (Hemiptera: Sternorrhyncha: Coccidae.). Mol. Ecol. 26 (20), 5855–5868. https://doi.org/10.1111/mec.14332.

Grimaldi, D., Engel, M.S., 2005. Evolution of the Insects. Cambridge University Press, New York.

Gündüz, E.A., Douglas, A.E., 2008. Symbiotic bacteria enable insect to use a nutritionally inadequate diet. Proc. R. Soc. B 276, 987–991. https://doi.org/10.1098/rspb.2008.1476.

Hammer, T.J., Moran, N.A., 2019. Links between metamorphosis and symbiosis in holometabolous insects. Philos. Trans. R. Soc. B 374, 20190068. https://doi.org/10.1098/rstb.2019.0068.

Hansen, A.K., Moran, N.A., 2011. Aphid genome expression reveals host-symbiont cooperation in the production of amino acids. Proc. Natl. Acad. Sci. U. S. A. 108, 2849–2854.

Hansen, A.K., Moran, N.A., 2013. The impact of microbial symbionts on host plant utilization by herbivorous insects. Mol. Ecol. 23, 1473–1496.

Henry, S.M., 1962. The significance of microorganisms in the nutrition of insects. Trans. NY Acad. Sci. 24 (6.2), 676–683. https://doi.org/10.1111/j.2164-0947.1962.tb01905.x.

Higashi, M., Abe, T., Burns, T.P., 1992. Carbon-nitrogen balance and termite ecology. Proc. R. Soc. B 249 (1326), 303–308. https://doi.org/10.1098/rspb.1992.0119.

Hongoh, Y., 2011. Toward the functional analysis of uncultivable, symbiotic microorganisms in the termite gut. Cell. Mol. Life Sci. 68, 1311–1325. https://doi.org/10.1007/s00018-011-0648-z.

Hongoh, Y., Ishikawa, H., 1997. Uric acid as a nitrogen resource for the brown planthopper, *Nilaparvata lugens*: studies with synthetic diets and aposymbiotic insects. Zoolog. Sci. 14 (4), 581–586. https://doi.org/10.2108/zsj.14.581.

Hongoh, Y., Ishikawa, H., 2000. Evolutionary studies on uricases of fungal endosymbionts of aphids and planthoppers. J. Mol. Evol. 51 (3), 265–277. https://doi.org/10.1007/s002390010088.

Hongoh, Y., et al., 2005. Intra- and interspecific comparisons of bacterial diversity and community structure support coevolution of gut microbiota and termite host. Appl. Environ. Microbiol. 71 (11), 6590–6599. https://doi.org/10.1128/AEM.71.11.6590.

Hongoh, Y., et al., 2008. Genome of an endosymbiont coupling N_2 fixation to cellulolysis within protist cells in termite gut. Science 322 (5904), 1108–1109. https://doi.org/10.1126/science.1165578.

Hosokawa, T., et al., 2010. Phylogenetic position and peculiar genetic traits of a midgut bacterial symbiont of the stinkbug *Parastrachia japonensis*. Appl. Environ. Microbiol. 76 (13), 4130–4135. https://doi.org/10.1128/AEM.00616-10.

Hu, Y., et al., 2014. Correlates of gut community composition across an ant species (*Cephalotes varians*) elucidate causes and consequences of symbiotic variability. Mol. Ecol. 23, 1284–1300.

Hu, Y., et al., 2018. Herbivorous turtle ants obtain essential nutrients from a conserved nitrogen-recycling gut microbiome. Nat. Commun. 9 (1), 964. https://doi.org/10.1038/s41467-018-03357-y.

Husnik, F., et al., 2013. Horizontal gene transfer from diverse bacteria to an insect genome enables a tripartite nested mealybug symbiosis. Cell 153, 1567–1578.

International Aphid Genomics Consortium (IAGC), 2010. Genome sequence of the pea aphid *Acyrthosiphon pisum*. PLoS Biol. 8, e1000313. https://doi.org/10.1371/journal.pbio.1000313.

Janssen, R.H., et al., 2017. Nitrogen-to-protein conversion factors for three edible insects: *Tenebrio molitor*, *Alphitobius diaperinus*, and *Hermetia illucens*. J. Agric. Food Chem. 65 (11), 2275–2278. https://doi.org/10.1021/acs.jafc.7b00471.

Ji, R., Brune, A., 2006. Nitrogen mineralization, ammonia accumulation, and emission of gaseous NH_3 by soil-feeding termites. Biogeochemistry 78, 267–283.

Kanehisa, M., Goto, S., 2000. KEGG: Kyoto encyclopedia of genes and genomes. Nucleic Acids Res. 28 (1), 27–30. https://doi.org/10.1093/nar/28.1.27.

Kanehisa, M., et al., 2019. New approach for understanding genome variations in KEGG. Nucleic Acids Res. 47 (D1), D590–D595. https://doi.org/10.1093/nar/gky962.

Karban, R., Agrawal, A.A., 2002. Herbivore offense. Annu. Rev. Ecol. Syst. 33, 641–664.

Kashima, T., Nakamura, T., Tojo, S., 2006. Uric acid recycling in the shield bug, *Parastrachia japonensis* (Hemiptera: Parastrachiidae) during diapause. J. Insect Physiol. 52, 816–825.

Kheirallah, A.M., 1978. The consumption and utilization of two different species of leaf litter by a laboratory population of *Orthomorpha gracilis* (Diplopoda: Polydesmoidea). Entomol. Exp. Appl. 23 (1), 14–19. https://doi.org/10.1111/j.1570-7458.1978.tb02726.x.

Koga, R., Moran, N., 2014. Swapping symbionts in spittlebugs: evolutionary replacement of a reduced genome symbiont. ISME J. 8, 1237–1246. https://doi.org/10.1038/ismej.2013.235.

Koga, R., et al., 2012. Cellular mechanism for selective vertical transmission of an obligate insect symbiont at the bacteriocyte–embryo interface. Proc. Natl. Acad. Sci. U. S. A. 109 (20), E1230–E1237. https://doi.org/10.1073/pnas.1119212109.

Lanan, M.C., et al., 2016. A bacterial filter protects and structures the gut microbiome of an insect. ISME J. 10, 1866–1876.

Lavoie, B., Oberhauser, K.S., 2004. Compensatory feeding in *Danaus plexippus* (Lepidoptera: Nymphalidae) in response to variation in host plant quality. Environ. Entomol. 33 (4), 1062–1069. https://doi.org/10.1603/0046-225X-33.4.1062.

Leidy, J., 1881. Parasites of the termites. J. Phil. Acad. Sci. 8, 425–447. Philadelphia.

Lo, N., et al., 2003. Evidence for cocladogenesis between diverse dictyopteran lineages and their intracellular endosymbionts. Mol. Biol. Evol. 20 (6), 907–913. https://doi.org/10.1093/molbev/msg097.

López-Sánchez, M.J., et al., 2009. Evolutionary convergence and nitrogen metabolism in *Blattabacterium* strain Bge, primary endosymbiont of the cockroach *Blattella germanica*. PLoS Genet. 5 (11), e1000721. https://doi.org/10.1371/journal.pgen.1000721.

Luan, J.B., et al., 2015. Metabolic coevolution in the bacterial symbiosis of whiteflies and related plant sap-feeding insects. Genome Biol. Evol. 7, 2635–2647.

MacDonald, S.J., et al., 2012. The central role of the host cell in symbiotic nitrogen metabolism. Proc. R. Soc. B 279 (1740), 2965–2973. https://doi.org/10.1098/rspb.2012.0414.
Machida, M., et al., 2001. Nitrogen recycling through proctodeal trophallaxis in the Japanese damp-wood termite *Hodotermopsis japonica* (Isoptera, Termopsidae). Insect. Soc. 48, 52–56. https://doi.org/10.1007/PL00001745.
Mao, M., Yang, X., Bennett, G.M., 2018. Evolution of host support for two ancient bacterial symbionts with differentially degraded genomes in a leafhopper host. Proc. Natl. Acad. Sci. U. S. A. 50, E11691–E11700.
Martin, E., et al., 2018. The mycobiota of the sand fly *Phlebotomus perniciosus*: involvement of yeast symbionts in uric acid metabolism. Environ. Microbiol. 20 (3), 1064–1077. https://doi.org/10.1111/1462-2920.14044.
Martinez, T., Wheeler, D.E., 1994. Storage proteins in adult ants: roles in colony founding by queens and in larval rearing by workers. J. Insect Physiol. 40, 723–729.
Mason, C.J., et al., 2019. Bacterial and fungal midgut community dynamics and transfer between mother and brood in the Asian longhorned beetle (*Anoplophora glabripennis*), an invasive xylophage. Microb. Ecol. 77 (1), 230–242. https://doi.org/10.1007/s00248-018-1205-1.
Matsuura, Y., et al., 2018. Recurrent symbiont recruitment from fungal parasites in cicadas. Proc. Natl. Acad. Sci. U. S. A. 115 (26), E5970–E5979. https://doi.org/10.1073/pnas.1803245115.
Mattson, W.J., 1980. Herbivory in relation to plant nitrogen content. Annu. Rev. Ecol. Syst. 11 (1), 119–161. https://doi.org/10.1146/annurev.es.11.110180.001003.
McCutcheon, J.P., McDonald, B.R., Moran, N.A., 2009. Convergent evolution of metabolic roles in bacterial co-symbionts of insects. Proc. Natl. Acad. Sci. U. S. A. 106 (36), 15394–15399. https://doi.org/10.1073/pnas.0906424106.
McNeil, S., Southwood, T.R.E., 1978. The role of nitrogen in the development of insect plant relationships. In: Harborne, J.B. (Ed.), Biochemical Aspects of Plant and Animal Coevolution. Academic Press, London, pp. 77–98.
Michalik, A., et al., 2018. Symbiotic cornucopia of the monophagous planthopper *Ommatidiotus dissimilis* (Fallén, 1806) (Hemiptera: Fulgoromorpha: Caliscelidae). Protoplasma 255, 1317–1329.
Mikaelyan, A., et al., 2015. Diet is the primary determinant of bacterial community structure in the guts of higher termites. Mol. Ecol. 24 (20), 5284–5295. https://doi.org/10.1111/mec.13376.
Mira, A., 2000. Exuviae eating: a nitrogen meal? J. Insect Physiol. 46 (4), 605–610. https://doi.org/10.1016/S0022-1910(99)00146-8.
Morales-Jiménez, J., et al., 2013. Nitrogen-fixing and uricolytic bacteria associated with the gut of *Dendroctonus rhizophagus* and *Dendroctonus valens* (Curculionidae: Scolytinae). Microb. Ecol. 66 (1), 200–210. https://doi.org/10.1007/s00248-013-0206-3.
Moran, N.A., Baumann, P., 2000. Bacterial endosymbionts in animals. Curr. Opin. Microbiol. 3, 270–275.
Moran, N.A., Bennett, G.M., 2014. The tiniest tiny genomes. Annu. Rev. Microbiol. 68, 195–215.
Moran, N.A., McCutcheon, J.P., Nakabachi, A., 2008. Genomics and evolution of heritable bacterial symbionts. Annu. Rev. Genet. 42, 165–190. https://doi.org/10.1146/annurev.genet.41.110306.130119.
Moreau, C.S., Bell, C.D., 2013. Testing the museum versus cradle tropical biological diversity hypothesis: phylogeny, diversification, and ancestral biogeographical range evolution of the ants. Evolution 67 (8), 2240–2257. https://doi.org/10.1111/evo.12105.
Mueller, U.G., et al., 2005. The evolution of agriculture in insects. Annu. Rev. Ecol. Evol. Syst. 36, 563–595. https://doi.org/10.1146/annurev.ecolsys.36.102003.152626.

Mullins, D.E., 2015. Physiology of environmental adaptations and resource acquisition in cockroaches. Annu. Rev. Entomol. 60, 473–492. https://doi.org/10.1146/annurev-ento-011613-162036.

Mullins, D.E., Cochran, D.G., 1974. Nitrogen metabolism in the American cockroach: an examination of whole body and fat body regulation of cations in response to nitrogen balance. J. Exp. Biol. 61, 557–570.

Mullins, D.E., Keil, C.B., White, R.H., 1992. Maternal and paternal nitrogen investment in *Blattella germanica* (l.) (Dictyoptera; Blattellidae). J. Exp. Biol. 162, 55–72.

Munson, M.A., Baumann, P., Kinsey, M.G., 1991. *Buchnera* gen. nov. and *Buchnera aphidicola* sp. nov., a taxon consisting of the mycetocyte-associated, primary endosymbionts of aphids. Int. J. Syst. Bacteriol. 41 (4), 566–568.

Nalepa, C.A., Bignell, D.E., Bandi, C., 2001. Detritivory, coprophagy, and the evolution of digestive mutualisms in Dictyoptera. Insect. Soc. 48 (3), 194–201. https://doi.org/10.1007/PL00001767.

Nardi, J.B., Mackie, R.I., Dawson, J.O., 2002. Could microbial symbionts of arthropod guts contribute significantly to nitrogen fixation in terrestrial ecosystems? J. Insect Physiol. 48, 751–763.

Nelsen, M.P., Ree, R.H., Moreau, C.S., 2018. Ant-plant interactions evolved through increasing interdependence. Proc. Natl. Acad. Sci. U. S. A. 115 (48), 12253–12258. doi.org/10.1073/pnas.1719794115.

Neuvonen, M.M., et al., 2016. The genome of Rhizobiales bacteria in predatory ants reveals urease gene functions but no genes for nitrogen fixation. Sci. Rep. 6, 1–11. https://doi.org/10.1038/srep39197.

Ngugi, D.K., Ji, R., Brune, A., 2011. Nitrogen mineralization, denitrification, and nitrate ammonification by soil-feeding termites: a ^{15}N-based approach. Biogeochemistry 103, 355–369. https://doi.org/10.1007/s10533-010-9478-6.

Noda, H., 1977. Histological and histochemical observation of intracellular yeast-like symbiotes in the fat body of the smaller brown planthopper, *Laodelphax striatellus* (Homoptera: Delphacidae). Appl. Entomol. Zool. 12, 134–141.

Noda, S., et al., 1999. Culture-independent characterization of a gene responsible for nitrogen fixation in the symbiotic microbial community in the gut of the termite *Neotermes koshunensis*. Appl. Environ. Microbiol. 65 (11), 4935–4942.

Noda, S., et al., 2007. Cospeciation in the triplex symbiosis of termite gut protists (*Pseudotrichonympha* spp.), their hosts, and their bacterial endosymbionts. Mol. Ecol. 16 (6), 1257–1266. https://doi.org/10.1111/j.1365-294X.2006.03219.x.

O'Donnell, M., 2008. Insect excretory mechanisms. Adv. Insect Physiol. 35, 1–122. https://doi.org/10.1016/S0065-2806(08)00001-5.

Ohkuma, M., et al., 2015. Acetogenesis from H_2 plus CO_2 and nitrogen fixation by an endosymbiotic spirochete of a termite-gut cellulolytic protist. Proc. Natl. Acad. Sci. U. S. A. 112 (33), 10224–10230. https://doi.org/10.1073/pnas.1423979112.

Panfilio, K.A., et al., 2019. Molecular evolutionary trends and feeding ecology diversification in the Hemiptera, anchored by the milkweed bug genome. Genome Biol. 20 (1), 64. https://doi.org/10.1186/s13059-019-1660-0.

Park, M.S., Park, P., Takeda, M., 2013. Roles of fat body trophocytes, mycetocytes and urocytes in the American cockroach, *Periplaneta americana* under starvation conditions: an ultrastructural study. Arthropod Struct. Dev. 42 (4), 287–295. https://doi.org/10.1016/j.asd.2013.03.004.

Patiño-Navarrete, R., et al., 2014. The cockroach *Blattella germanica* obtains nitrogen from uric acid through a metabolic pathway shared with its bacterial endosymbiont. Biol. Lett. 10 (7), 7–10. https://doi.org/10.1098/rsbl.2014.0407.

Pers, D., Hansen, A.K., 2019. The effects of different diets and transgenerational stress on *Acyrthosiphon pisum* development. Insects 10 (9), 260. https://doi.org/10.3390/insects10090260.

Poliakov, A., et al., 2011. Large-scale label-free quantitative proteomics of the pea aphid-*Buchnera* symbiosis. Mol. Cell. Proteomics 10. M110.007039.

Potrikus, C.J., Breznak, J.A., 1980a. Uric acid in wood-eating termites. Insect Biochem. 10, 19–27. https://doi.org/10.1016/0020-1790(80)90034-7.

Potrikus, C.J., Breznak, J.A., 1980b. Uric acid-degrading bacteria in guts of termites [*Reticulitermes flavipes* (Kollar)]. Appl. Environ. Microbiol. 40, 117–124.

Potrikus, C.J., Breznak, J.A., 1981. Gut bacteria recycle uric acid nitrogen in termites: a strategy for nutrient conservation. Proc. Natl. Acad. Sci. U. S. A. 78 (7), 4601–4605. https://doi.org/10.1073/pnas.78.7.4601.

Powell, S., 2008. Ecological specialization and the evolution of a specialized caste in *Cephalotes* ants. Funct. Ecol. 22 (5), 902–911.

Prosser, W.A., Douglas, A.E., 1991. The aposymbiotic aphid: an analysis of chloretracycline-treated pea aphid, *Acyrthosiphon pisum*. J. Insect Physiol. 10, 713–719. https://doi.org/10.1016/0022-1910(91)90104-8.

Roche, R.K., Wheeler, D.E., 1997. Morphological specializations of the digestive tract of *Zacryptocerus rohweri* (Hymenoptera: Formicidae). J. Morphol. 234 (3), 253–262. https://doi.org/10.1002/(SICI)1097-4687(199712)234:3<253::AID-JMOR4>3.0.CO;2-A.

Russell, J.A., et al., 2009. Bacterial gut symbionts are tightly linked with the evolution of herbivory in ants. Proc. Natl. Acad. Sci. U. S. A. 106 (50), 21236–21241. https://doi.org/10.1073/pnas.0907926106.

Russell, J.A., Sanders, J.G., Moreau, C.S., 2017. Hotspots for symbiosis: function, evolution, and specificity of ant-microbe associations from trunk to tips of the ant phylogeny (Hymenoptera: Formicidae). Myrmecol. News 24, 43–69. https://doi.org/10.25849/myrmecol.news_024:043.

Sabree, Z.L., Kambhampati, S., Moran, N.A., 2009. Nitrogen recycling and nutritional provisioning by Blattabacterium, the cockroach endosymbiont. Proc. Natl. Acad. Sci. U. S. A. 106 (46), 19521–19526.

Salem, H., et al., 2015. An out-of-body experience: the extracellular dimension for the transmission of mutualistic bacteria in insects. Proc. R. Soc. B 282 (1804), 20142957.

Sanders, J.G., et al., 2014. Stability and phylogenetic correlation in gut microbiota: lessons from ants and apes. Mol. Ecol. 23, 1268–1283.

Sandström, J., Moran, N.A., 1999. How nutritionally imbalanced is phloem sap for aphids? Entomol. Exp. Appl. 91, 203–210.

Sandström, J., Telang, A., Moran, N.A., 2000. Nutritional enhancement of host plants by aphids - a comparison of three aphid species on grasses. J. Insect Physiol. 46 (1), 33–40. https://doi.org/10.1016/S0022-1910(99)00098-0.

Santos-Garcia, D., et al., 2014. Small but powerful, the primary endosymbiont of moss bugs, *Candidatus* Evansia muelleri, holds a reduced genome with large biosynthetic capabilities. Genome Biol. Evol. 6 (7), 1875–1893. https://doi.org/10.1093/gbe/evu149.

Sapountzis, P., et al., 2016. Potential for nitrogen fixation in the fungus-growing termite symbiosis. Front. Microbiol. 7, 1993. https://doi.org/10.3389/fmicb.2016.01993.

Sasaki, T., Kawamura, M., Ishikawa, H., 1996. Nitrogen recycling in the brown planthopper, *Nilaparvata lugens*: involvement of yeast-like endosymbionts in uric acid metabolism. J. Insect Physiol. 42 (2), 125–129. https://doi.org/10.1016/0022-1910(95)00086-0.

Sauer, C., et al., 2000. Systematic relationships and cospeciation of bacterial endosymbionts and their carpenter ant host species: proposal of the new taxon *Candidatus* Blochmannia gen. Nov. Int. J. Syst. Evol. Microbiol. 50 (5), 1877–1886. https://doi.org/10.1099/00207713-50-5-1877.

Sauer, C., et al., 2002. Tissue localization of the endosymbiotic bacterium "*Candidatus* Blochmannia floridanus" in adults and larvae of the carpenter ant *Camponotus floridanus*. Appl. Environ. Microbiol. 68 (9), 4187–4193. https://doi.org/10.1128/AEM.68.9.4187.

Scaraffia, P.Y., et al., 2008. Discovery of an alternate metabolic pathway for urea synthesis in adult *Ades aegypti* mosquitos. Proc. Natl. Acad. Sci. U. S. A. 105 (2), 518–523. https://doi.org/10.1073/pnas.0708098105.

Schröder, D., et al., 1996. Intracellular endosymbiotic bacteria of *Camponotus* species (carpenter ants): systematics, evolution and ultrastructural characterization. Mol. Microbiol. 21 (3), 479–489. https://doi.org/10.1111/j.1365-2958.1996.tb02557.x.

Scully, E.D., et al., 2014. Functional genomics and microbiome profiling of the Asian longhorned beetle (*Anoplophora glabripennis*) reveal insights into the digestive physiology and nutritional ecology of wood feeding beetles. BMC Genomics 15 (1), 1096. https://doi.org/10.1186/1471-2164-15-1096.

Sethi, A., et al., 2011. Dual origin of gut proteases in Formosan subterranean termites (*Coptotermes formosanus* Shiraki) (Isoptera: Rhinotermitidae). Comp. Biochem. Physiol. A Mol. Integr. Physiol. 159 (3), 261–267. https://doi.org/10.1016/j.cbpa.2011.03.012.

Shigenobu, S., et al., 2000. Genome sequence of the endocellular bacterial symbiont of aphids *Buchnera* sp. APS. Nature 407, 81–86.

Six, D.L., 2012. Ecological and evolutionary determinants of bark beetle—fungus symbioses. Insects 3 (1), 339–366. https://doi.org/10.3390/insects3010339.

Six, D.L., Klepzig, K.D., 2004. *Dendroctonus* bark beetles as a model system for studies on symbiosis. Symbiosis 37 (1), 207–232.

Slansky, F., Feeny, P., 1977. Stabilization of the rate of nitrogen accumulation by larvae of the cabbage butterfly on wild and cultivated food plants. Ecol. Monogr. 47 (2), 209–228. https://doi.org/10.2307/1942617.

Slaytor, M., Chappell, D.J., 1994. Nitrogen metabolism in termites. Comp. Biochem. Physiol. B 107 (1), 1–10. https://doi.org/10.1016/0305-0491(94)90218-6.

Sloan, D.B., et al., 2014. Parallel histories of horizontal gene transfer facilitated extreme reduction of endosymbiont genomes in sap-feeding insects. Mol. Biol. Evol. 31, 857–871.

Souza, A.V.G., et al., 1997. Urate protects a blood-sucking insect against hemin-induced oxidative stress. Free Radic. Biol. Med. 22 (1–2), 209–214. https://doi.org/10.1016/S0891-5849(96)00293-6.

Stoll, S., et al., 2007. Bacterial microbiota associated with ants of the genus *Tetraponera*. Biol. J. Linn. Soc. 90 (3), 399–412. https://doi.org/10.1111/j.1095-8312.2006.00730.x.

Stoll, S., Feldhaar, H., Gross, R., 2009. Transcriptional profiling of the endosymbiont *Blochmannia floridanus* during different developmental stages of its holometabolous ant host. Environ. Microbiol. 11 (4), 877–888. https://doi.org/10.1111/j.1462-2920.2008.01808.x.

Sudakaran, S., Kost, C., Kaltenpoth, M., 2017. Symbiont acquisition and replacement as a source of ecological innovation. Trends Microbiol. 25 (5), 375–390. https://doi.org/10.1016/j.tim.2017.02.014.

Suh, S.-O., Noda, H., Blackwell, M., 2001. Insect symbiosis: derivation of yeast-like endosymbionts within an entomopathogenic filamentous lineage. Mol. Biol. Evol. 18 (6), 995–1000. https://doi.org/10.1093/oxfordjournals.molbev.a003901.

Täyasu, I., et al., 1994. Xylophagous termites depending on atmospheric nitrogen. Naturwissenschaften 81, 229–231. https://doi.org/10.1007/BF01138550.

Täyasu, I., et al., 1997. Nitrogen and carbon isotope ratios in termites: an indicator of trophic habit along the gradient from wood-feeding to soil-feeding. Ecol. Entomol. 22 (3), 343–351. https://doi.org/10.1046/j.1365-2311.1997.00070.x.

Thao, M.L., Baumann, P., 2004. Evolutionary relationships of primary prokaryotic endosymbionts of whiteflies and their hosts. Appl. Environ. Microbiol. 70 (6), 3401–3406. https://doi.org/10.1128/AEM.70.6.3401.

Thong-on, A., et al., 2012. Isolation and characterization of anaerobic bacteria for symbiotic recycling of uric acid nitrogen in the gut of various termites. Microbes Environ. 27 (2), 186–192. https://doi.org/10.1264/jsme2.ME11325.

Ulyshen, M.D., 2015. Insect-mediated nitrogen dynamics in decomposing wood. Ecol. Entomol. 40 (Suppl. 1), 97–112. https://doi.org/10.1111/een.12176.

Valovage, W.D., Brooks, M.A., 1979. Uric acid quantities in the fat body of normal and aposymbiotic German cockroaches, *Blattella germanica*. Ann. Entomol. Soc. Am. 72 (5), 687–689. https://doi.org/10.1093/aesa/72.5.687.

van Borm, S., et al., 2002. *Tetraponera* ants have gut symbionts related to nitrogen-fixing root-nodule bacteria. Proc. R. Soc. B 269, 2023–2027. https://doi.org/10.1098/rspb.2002.2101.

Vera-Ponce de Leon, A., et al., 2016. Fungal community associated with *Dactylopius* (Hemiptera: Coccoidea: Dactylopiidae) and its role in uric acid metabolism. Front. Microbiol. 7, 954. https://doi.org/10.3389/fmicb.2016.00954.

Vera-Ponce de Leon, A., et al., 2017. *Candidatus* Dactylopiibacterium carminicum, a nitrogen-fixing symbiont of *Dactylopius* cochineal insects (Hemiptera: Coccoidea: Dactylopiidae). Genome Biol. Evol. 9 (9), 2237–2250. https://doi.org/10.1093/gbe/evx156.

Vogel, K.J., Moran, N.A., 2013. Functional and evolutionary analysis of the genome of an obligate fungal symbiont. Genome Biol. Evol. 5 (5), 891–904. https://doi.org/10.1093/gbe/evt054.

Volland, J.M., et al., 2018. NanoSIMS and tissue autoradiography reveals symbiont carbon fixation and organic carbon transfer to giant ciliate host. ISME J. 12, 714–727.

von Dohlen, C.D., et al., 2001. Mealybug β-proteobacterial endosymbionts contain γ-proteobacterial symbionts. Nature 412, 433–436.

Warnecke, F., et al., 2007. Metagenomic and functional analysis of hindgut microbiota of a wood-feeding higher termite. Nature 450, 560–565.

Wernegreen, J.J., 2002. Genome evolution in bacterial endosymbionts of insects. Nat. Rev. Genet. 3, 850–861. https://doi.org/10.1038/nrg931.

Wernegreen, J.J., et al., 2009. One nutritional symbiosis begat another: phylogenetic evidence that the ant tribe Camponotini acquired *Blochmannia* by tending sap-feeding insects. BMC Evol. Biol. 9 (1), 292. https://doi.org/10.1186/1471-2148-9-292.

Williams, L.E., Wernegreen, J.J., 2010. Unprecedented loss of ammonia assimilation capability in a urease-encoding bacterial mutualist. BMC Genomics 11 (1), 687. https://doi.org/10.1186/1471-2164-11-687.

Williams, L.E., Wernegreen, J.J., 2015. Genome evolution in an ancient bacteria-ant symbiosis: parallel gene loss among *Blochmannia* spanning the origin of the ant tribe Camponotini. PeerJ 3, e881. https://doi.org/10.7717/peerj.881.

Wilson, A.C.C., Duncan, R.P., 2015. Signatures of host/symbiont genome coevolution in insect nutritional endosymbiosis. Proc. Natl. Acad. Sci. U. S. A. 112 (33), 10255–10261. https://doi.org/10.1073/pnas.1423305112.

Wolschin, F., Gross, R., Zientz, E., 2004. Replication of the endosymbiotic bacterium *Blochmannia floridanus* is correlated with the developmental and reproductive stages of its ant host. Appl. Environ. Microbiol. 70 (7), 4096–4102. https://doi.org/10.1128/AEM.70.7.4096.

Xue, J., et al., 2014. Genomes of the rice pest brown planthopper and its endosymbionts reveal complex complementary contributions for host adaptation. Genome Biol. 15 (12), 521. https://doi.org/10.1186/s13059-014-0521-0.

Yamada, A., et al., 2007. Evolutionary trends of phylogenetic diversity of nitrogen fixation genes in the gut community of wood-feeding termites. Mol. Ecol. 16 (18), 3768–3777. https://doi.org/10.1111/j.1365-294X.2007.03326.x.

Zientz, E., et al., 2006. Relevance of the endosymbiosis of *Blochmannia floridanus* and carpenter ants at different stages of the life cycle of the host. Appl. Environ. Microbiol. 72 (9), 6027–6033. https://doi.org/10.1128/AEM.00933-06.

CHAPTER SIX

Regulation of an insect symbiosis

Alex C.C. Wilson*
Department of Biology, University of Miami, Coral Gables, FL, United States
*Corresponding author: acwilson@miami.edu

Contents

Abstract

The best studied insect nutritional endosymbiosis is that of aphids and *Buchnera aphidicola*. Work on the aphid/*Buchnera* model has uncovered multiple regulatory systems within this symbiosis that include gene losses, genome rearrangements, complementary metabolic pathways, feedback inhibition loops involving amino acid transporters, miRNAs, small RNAs as well as a conserved growth pathway. Taken together, this work on how an endosymbiont is integrated into its host at a cellular and molecular level challenges the idea of a harmonious mutualism. Here I describe each of the known regulatory systems in the aphid/*Buchnera* model of endosymbiosis, offer some new interpretations, highlight knowledge gaps and conclude by arguing that the evolution of host/endosymbiont integration is littered with signatures of conflict.

The symbiosis between aphids and the gamma proteobacterium, *Buchnera aphidicola* (Munson et al., 1991b) is at least 160 million years old (Moran et al., 1993; Munson et al., 1991a). While two groups of aphids, members

Advances in Insect Physiology, Volume 58
ISSN 0065-2806
https://doi.org/10.1016/bs.aiip.2020.04.004

of the genus *Geophemphigus* (Chong and Moran, 2018) and members of the tribe Cerataphidini (Fukatsu et al., 1994) have lost *Buchnera*, it is currently understood that all other extant Aphididae sustain their anciently acquired association with *Buchnera*.

Vertically transmitted from one generation to the next, *Buchnera* live inside specialized aphid cells called bacteriocytes (Baumann et al., 1995). Depending on the developmental stage of the host, hundreds to thousands of *Buchnera* cells are housed in each bacteriocyte (Mira and Moran, 2002), where they are separated from the bacteriocyte cytoplasm by a host-derived symbiosomal membrane (Hinde, 1971a). Typically, each symbiosome contains a single *Buchnera* cell. Approximately 90–110 bacteriocyte cells aggregate to form the bacteriome, a bilobed organ that associates closely with both the gut and germline (Baumann et al., 1995; Douglas and Wilkinson, 1998). Bacteriomes contain two cell types; large round bacteriocytes that house *Buchnera* and small, thin, intercalating cells called sheath cells that do not house *Buchnera* (Houk and Griffiths, 1980) (Fig. 1).

Compelling questions in the study of endosymbiosis concern the evolution of mechanisms that maintain organismal homeostasis. Here, I review what is known about host/symbiont integration in the aphid/*Buchnera* model of endosymbiosis. In structuring this chapter, I will consider in turn mechanisms of genomic, molecular and cellular integration. The chapter ends with an overview of how these individual processes remarkably appear to target just a handful of amino acid inputs, suggesting that the patterns of host/endosymbiont integration discovered in aphids have been constrained by the genomic, cellular and molecular capabilities of the ancestral genome and may, therefore, be generalizable across a diversity of systems.

1. Genomic integration of host and endosymbiont

Study of endosymbiont genome evolution was pioneered in *Buchnera aphidicola* (Shigenobu et al., 2000; Tamas et al., 2002). Recent analysis of *Buchnera* genomes from 39 different aphid species paints the most complete picture to date of patterns of genome evolution in an insect endosymbiont (Chong et al., 2019) and bolsters the decade old understanding that genome evolution in insect endosymbionts is characterized by large scale gene losses, nucleotide composition biases that tend towards A + T richness, and rapid sequence evolution (Moran et al., 2008). A particularly striking feature of genome evolution in *Buchnera* is the absence of chromosomal rearrangement; all 39 genomes are syntenic in their entirety, with the exception of

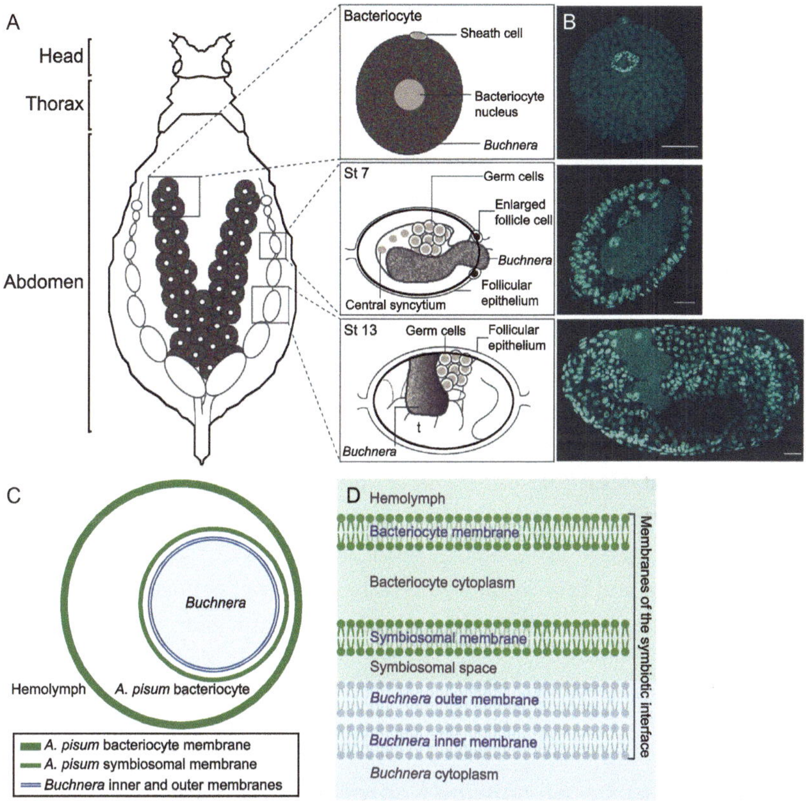

Fig. 1 The symbiotic interface. Arrangement of *Buchnera* within aphids. (A) Line drawing illustrating the arrangement of the bacteriome within an aphid abdomen in an asexual female, illustrating from top, a single bacteriocyte with sheath cell, an embryo (stage 7) at the time of *Buchnera* transmission, an embryo (stage 13) at the time of *Buchnera* cellularization; (B) DAPI stained confocal images of corresponding bacteriocyte and embryos, scale bar 20 μm; (C) Cartoon illustrating the arrangement of membranes, note that each *Buchnera* cell is surrounded by a host derived symbiosomal membrane (only one *Buchnera* cell shown); (D) Cartoon defining the membranes of the symbiotic interface. *Panel (A): modified from fig. 1k from Lu, H.-L., Price, D.R.G., Wikramanayake, A., Chang, C.-C., Wilson, A.C.C., 2016. Ontogenetic differences in localization of glutamine transporter ApGLNT1 in the pea aphid demonstrate that mechanisms of host/symbiont integration are not similar in the maternal versus embryonic bacteriome. EvoDevo 7, 1. Panel (C): fig. 1a from Price, D.R.G., Wilson, A.C.C., Luetje, C.W., 2015. Proton-dependent glutamine uptake by aphid bacteriocyte amino acid transporter ApGLNT1. Biochim. Biophys. Acta 1848, 2085–2091.*

rearrangements involving amino acid biosynthesis genes located on plasmids (Chong et al., 2019). This new analysis also revealed that rates of gene loss in *Buchnera* have varied across aphid lineages (Chong et al., 2019). In fact, some lineages of *Buchnera*, including those associated within the subfamily

Lachninae (Lamelas et al., 2011a,b; Pérez-Brocal et al., 2006) and a species of *Stegophyla* in the subfamily Phyllaphidinae (Chong et al., 2019), have lost so many genes that they now necessarily coexist with an additional obligate endosymbiont.

As studies of endosymbiont genome evolution were pioneered in *Buchnera*, it is fitting that study of host genome evolution started in a *Buchnera* host, the pea aphid, *Acyrthosiphon pisum* (International Aphid Genomics Consortium, 2010). Several features of *A. pisum* genome evolution have been argued to facilitate host/endosymbiont integration, these include loss of immunity and other microbial defence genes (Gerardo et al., 2010), loss of purine salvage (Ramsey et al., 2010) and the urea cycle (Wilson et al., 2010), the acquisition of new genes by lateral gene transfer (Nikoh et al., 2010; Nikoh and Nakabachi, 2009), the evolution of novel genes (Shigenobu and Stern, 2013) and large scale duplication of amino acid transporters (Price et al., 2011). As more host genomes are sequenced, patterns of host/endosymbiont genome evolution that are shared across multiple systems are emerging, with the most striking and best-studied pattern being that of genomic complementarity. Other patterns include the high constitutive expression of endosymbiont chaperonins, and, less well studied, the evolution of multifunctional enzymes. I will address each of these in turn.

1.1 The evolution of genomic complementarity

The evolution of genomic complementarity is facilitated by gene loss, in both endosymbionts and hosts, and sometimes also by the acquisition of new host genes by lateral gene transfer. This phenomenon was first revealed by studying the pea aphid/*Buchnera* symbiosis. In aphids, mealybugs and whiteflies, completion of pathways for the biosynthesis of five essential amino acids requires genes encoded in the host and symbiont genome (Husnik et al., 2013; International Aphid Genomics Consortium, 2010; Luan et al., 2015; Mathers et al., 2017). These amino acids include all three branched chain amino acids (leucine, isoleucine and valine), as well as methionine, and phenylalanine. Notably, in the aphid/*Buchnera* endosymbiosis, pathways for the biosynthesis of leucine, phenylalanine and methionine have been experimentally validated (Russell et al., 2013). In an elegant series of experiments that utilized recombinant enzymes, and alternate substrates, Russell et al. (2013) quantified amino acid release from isolated *Buchnera* cells and *Buchnera*-free bacteriome lysates. Those data validated the genome-based proposal that *A. pisum* and *Buchnera* collaboratively synthesize essential amino acids.

However, in the case of isoleucine and methionine the work revealed that the collaborative metabolism does not proceed in the manner predicted in the genome-based model of Wilson et al. (2010). Instead, the synthesis of 2-oxobutanoate, the aphid synthesized precursor of isoleucine, is stimulated by homoserine via cystathionine γ-lyase (E.C. 4.4.1.1) (Russell et al., 2013) and not threonine as originally proposed (Wilson et al., 2010). In addition, the synthesis of homocysteine, the aphid synthesized precursor of methionine, is stimulated by cystathionine and not cystine (Russell et al., 2013, 2014). While Russell et al. (2013) do not explicitly discuss the source of cystathionine in *A. pisum* bacteriocytes, their presentation suggests that cystine, which is upstream of cystathionine in the reverse transsulfuration pathway, is not the source. An alternative source of cystathionine in *A. pisum*, not discussed by Russell et al. (2013), could be homoserine (Fig. 2) (Matthews, 1999). Production of cystathionine from homoserine involves two steps. Here I propose that the first step from homoserine to o-phosphohomoserine is mediated by *Buchnera* genome encoded *thrB*, while the second step from o-phosphohomoserine to cystathionine is mediated in the bacteriocyte

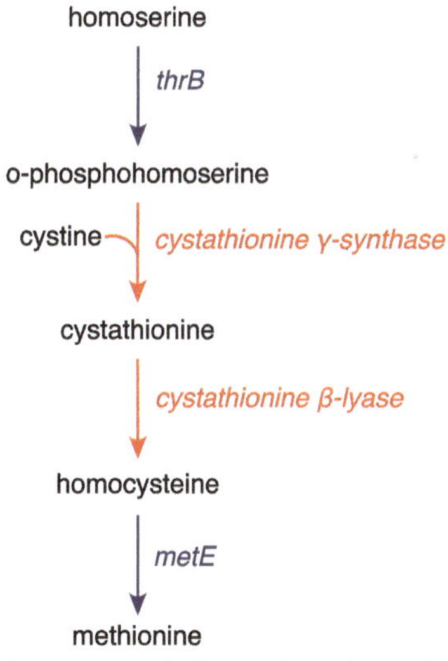

Fig. 2 Cystathionine production in aphids may be achieved through host/endosymbiont collaboration from homoserine. Enzymes in blue are encoded in the *Buchnera* genome, while enzymes in red are encoded in the *A. pisum* genome. This model requires either the shuttling of two metabolic intermediates (o-phosphohomoserine and homocysteine) and/or the shuttling of enzymes. The production of methionine from homoserine, via o-phosphohomoserine has been characterized in plants (for review see Hesse et al., 2004).

cytoplasm by *A. pisum* encoded cystathionine γ-synthase (E.C. 2.5.1.48, Fig. 2). At first glance it may appear that this new model is inconsistent with the results of Russell et al. (2013) because they reported that methionine release by isolated *Buchnera* is not stimulated by incubation with homoserine. However, since the first reaction step in the new model occurs within *Buchnera*, the result of Russell et al. (2013) is likely unsurprising because uptake of homoserine by isolated *Buchnera* would be necessary for demonstration of homoserine stimulation of methionine release. Just as collaborative biosynthesis of the branched-chain amino acids converge on a single enzyme, it is intriguing to consider that the collaborative biosynthesis of isoleucine and methionine under the new model I propose here may share both a *Buchnera* encoded metabolic precursor (homoserine) and an aphid encoded enzyme (cystathionine γ-lyase/cystathionine γ-synthase). In the context of this proposal it is important to note that the production of cystathionine from homoserine is not a well-characterized pathway. It is one that has been described in a fragmentary way in some plants (Hesse et al., 2004; Kreft et al., 2003), but one that has not made it into databases like KEGG (https://www.genome.jp/kegg/) or BRENDA (https://www.brenda-enzymes.org), thereby highlighting a need for biochemists to engage in studying host/endosymbiont metabolic collaboration.

While originally hypothesized to govern the biosynthesis of essential amino acids, collaborative pathways have also been revealed and experimentally validated for vitamin B5 (pantothenate) (Price and Wilson, 2014). *Buchnera* has lost the ability to make β-alanine, a metabolic precursor of vitamin B5, and yet some lineages of *Buchnera*, including the lineages from the pea aphid, *A. pisum*, retain *panC*, a gene encoding E.C. 6.3.2.1 (pantoate-beta-alanine ligase (AMP-forming)) that is known only to mediate the final step in B5 biosynthesis. In contrast, the aphid, *A. pisum*, is able to synthesize β-alanine by two pathways, one of which is highly expressed and up-regulated in bacteriocytes relative to whole insects (while the other has low expression in bacteriocytes and is down-regulated in bacteriocytes relative to whole body) (Price and Wilson, 2014). Once again, in the same way that the final step of branched chain amino acid biosynthesis is commonly lost from endosymbiont genomes and complemented by host insect encoded branched-chain aminotransferase (e.g., Husnik et al., 2013; Luan et al., 2015), endosymbionts across a diversity of insects that include sap-feeders and blood-feeders have lost *panD* and the ability to make β-alanine, and therefore require host provisioning of β-alanine to complete B5 biosynthesis via endosymbiont encoded *panC*, the gene that encodes the final step in B5 biosynthesis (Wilson and Duncan, 2015).

A third example of genomic complementarity in the aphid/*Buchnera* symbiosis concerns the purine salvage pathway that functions to recycle purines to purine nucleotides (Ramsey et al., 2010). In contrast to amino acid metabolism and vitamin biosynthesis, the genome-based model of collaborative purine salvage is yet to be experimentally interrogated, and further, the generality of collaborative purine salvage in insect endosymbioses beyond *A. pisum* appears not to have been explored. As suggested by Ramsey et al. (2010) the incomplete capacity of *A. pisum* for purine metabolism may explain why aposymbiotic aphids raised on diets abundant in amino acids, fail to reproduce.

1.2 The evolution of constitutive high chaperonin expression and multifunction enzymes mitigates endosymbiont genome degradation

Strictly vertically transmitted for more than 160 million years (Moran et al., 1993; Munson et al., 1991a), *Buchnera* populations are characterized by transmission bottlenecks (Buchner, 1965; Hinde, 1971b; Mira and Moran, 2002) that result in small effective population sizes (Moran, 1996), loss of DNA repair genes (Shigenobu et al., 2000; Tamas et al., 2002), and an absence of genetic recombination (Chong et al., 2019). Combined, these features result in the accumulation of mildly deleterious mutations (Moran, 1996), the evolution of codon bias as a result of increased A+T content (Woolfit and Bromham, 2003), and protein instability (McCutcheon and Moran, 2011).

Protein instability is a common feature of endosymbiont encoded proteins, a statement particularly true for ancient obligate endosymbionts like *Buchnera* (McCutcheon and Moran, 2011). Long before we were sequencing symbiont genomes and before proteomics, the proteins of *Buchnera* were being studied using polyacrylamide gel electrophoresis coupled with manipulation of *Buchnera* populations through antibiotic treatment and microinjection with radiolabelled metabolic precursors (Ishikawa, 1982, 1984). Most of this early work was led by Japanese scientist, Hajime Ishikawa. One very abundant protein of particular early interest was "symbionin" (Ishikawa, 1984). It should be noted, that care needs to be taken in interpreting some of this early work because it includes statements about "symbionin" being the *only* protein produced by *Buchnera*—which is incorrect—however, what that work did accurately draw attention to was that "symbionin" is important to symbiotic function. "Symbionin", now known to be groEL, is a molecular chaperone that is a 14 subunit homo-oligomer of 63 kDa, composed to two stacked rings of 7 subunits each (Hara and Ishikawa, 1990; Kakeda and Ishikawa, 1991). Molecular chaperones are conserved proteins that prevent misfolding of other proteins, and typically increase in abundance in cells in response to cellular stress (Dobson, 2003; Feder and Hofmann, 1999; Hartl, 1996). While the high constitutive expression of groEL was discovered in *Buchnera*, it is now understood to be a feature of host/endosymbiont coevolution and is known across many endosymbioses (for review see Kupper et al., 2014).

High constitutive groEL expression is one mechanism that mitigates the cellular impacts of endosymbiont genome degradation, another mechanism involves the evolution of multifunction enzymes. A role for multifunction enzymes in mitigating endosymbiont genome degradation was first proposed 20 years ago with publication of the first *Buchnera* genome in a landmark paper from Hajime Ishikawa's group led by Shigenobu et al. (2000). This paper was the first to report loss of *IlvE* (E.C. 2.6.1.42), the gene that encodes branched-chain-amino-acid transaminase, from an endosymbiont genome. Noting that *IlvE* was conspicuously missing from the isoleucine, leucine and valine biosynthetic pathways, Shigenobu et al. (2000) proposed that the function of *IlveE* is mediated by another aminotransferase enzyme. We now know that this "other aminotransferase" enzyme is *BCAT*, encoded by host genomes, however, an alternative explanation was originally understood; that the "other aminotransferase" was encoded by the genome of *Buchnera* as a multifunction enzyme. While the evolution of enzyme multifunctionality appears not to have facilitated gene loss of *IlveE*, it has played a role in *Buchnera* genome evolution. Many *Buchnera* lineages have lost the entire vitamin B5 biosynthesis pathway, while other *Buchnera* lineages, including that in *A. pisum*, retain an identically fragmented pathway (Price and Wilson, 2014). By heterologously expressing *Buchnera* proteins in *Escherichia coli* mutants, Price and Wilson (2014) were able to use complementation tests to examine their hypothesis that *Buchnera* maintain a functional B5 biosynthesis pathway by complementation of *panE* loss by *ilvC*. This work was the first to experimentally demonstrate that genome evolution in an insect endosymbiont has been facilitated by recruitment of a substrate-ambiguous enzyme from another metabolic pathway.

In the same way that sequencing the genomes of more and more insect endosymbionts revealed common patterns of endosymbiont genome evolution, the accumulation of paired genomes (i.e., those of a host and its endosymbiont(s) Gerardo and Wilson, 2011) has revealed that genomic complementarity in metabolic pathways and high constitutive groEL expression is central to the function of endosymbiont services in all Sternorrhynchan insects and beyond. In contrast, the role of enzyme multifunctionality in the genomic integration of hosts and endosymbionts is yet to be extensively explored.

2. Molecular integration of host and endosymbiont

A prominent question in the study of the aphid/*Buchnera* endosymbiosis in the years following development of microarray technologies concerned regulation of *Buchnera* gene expression (Moran et al., 2005; Reymond et al., 2006; Wilcox et al., 2003; Wilson et al., 2006), and in particular, regulation of essential amino acid biosynthesis. Whole genome sequencing of *Buchnera* from both *A. pisum* and the greenbug, *Schizaphis graminum*, revealed that genes for the regulation of amino acid biosynthesis have been almost entirely lost (Moran et al., 2005; Shigenobu et al., 2000; Tamas et al., 2002), raising the question of whether amino acid biosynthesis was under transcriptional control in *Buchnera* (Moran et al., 2003). More recently the role of aphid microRNAs, *Buchnera* small RNAs, and a eukaryotic signalling pathway in regulating the aphid/*Buchnera* holobiont have been studied. I will examine each of these in turn.

2.1 *Buchnera* amino acid biosynthesis gene expression is transcriptionally flat

Aphids are singular in their diet, they feed only on plant phloem sap. Phloem sap provides an abundance of carbon in the form of sugar, and nitrogen in the form of free amino acids. Notably, the total concentration and relative composition of free amino acids in phloem sap varies distinctively across plant species and amongst intraspecific plant genotypes (Sandström and Pettersson, 1994). Phloem sap amino acid concentration and composition is additionally known to vary seasonally (Weibull, 1987), and even diurnally (Kallarackal et al., 2012; Mitchell et al., 1992; Taylor et al., 2012), within a single plant genotype. Thus, aphid host plants naturally vary in their amino acid content, raising the questions of if and how essential amino acid biosynthesis is regulated in response to the nutritional needs of the symbiosis (Moran et al., 2005; Reymond et al., 2006). Using a whole genome microarray for *Buchnera* from *S. graminum* (*Buchnera* Sg), Moran et al. (2005) studied the transcriptional response to changes in amino acid provisioning using an experimental set up that involved excised young barley leaves sitting in tap water spiked with different free amino acids that included arginine, glutamine, glutamic acid, aspartic acid, leucine, and tryptophan. While the free amino acid content of the aphids fed on leaves spiked with amino acids differed from aphids fed on control diets, the whole genome microarray experiments found that only one gene, *metE*, showed a strong response to any of the treatments (Moran et al., 2005). Another gene, *argA*, showed a small but robust positive gene expression change in response to glutamine and arginine, and a small but robust negative gene expression change in response to starvation. Focusing on amino acid biosynthesis genes with respect to small but significant changes in gene expression, *trpA*, showed a minor decrease in gene expression in aphids fed on tryptophan spiked leaves, and the plasmid encoded genes (*trpEG* and *leuABCD*) slightly increased in expression in starved aphids relative to control aphids (Moran et al., 2005). In a set of similar experiments for *Buchnera* from *A. pisum* (*Buchnera* APS) only four amino acid biosynthesis genes were upregulated in aphids fed on a diet that lacked tyrosine and phenylalanine (Reymond et al., 2006). Consistent with the work in *S. graminum* (Moran et al., 2005), Reymond et al. (2006) found that the transcriptional response of *Buchnera* APS to changes in nutrition are weak and inconsistent, further suggesting that the transcriptional response of *Buchnera* does not affect rates of essential amino acid biosynthesis.

Collectively this microarray work revealed that *Buchnera* gene expression is transcriptionally flat in response to fluctuating nutrient supply and led to the suggestion that regulation of amino acid biosynthesis may be achieved by movement of precursor amino acids from host to endosymbiont (Moran et al., 2003; Reymond et al., 2006). This sort of regulation could be mediated by amino acid transporters in the membranes that separate host from endosymbiont (Fig. 1). Following completion of the *A. pisum* genome, work from my group has contributed insights into the role of amino acid transporters in regulating host/endosymbiont amino acid metabolism. I describe this work in the section of this chapter that focuses on cellular integration.

2.2 microRNAs are emerging as regulators of genes important for symbiotic function

MicroRNAs are known regulators of host/microbe interactions across a diversity of systems including host/pathogen and host/facultative symbiont. Recently, by working in a comparative framework to leverage the natural experiment of evolution, a set of 14 conserved aphid-encoded microRNAs were implicated in regulating the aphid/*Buchnera* endosymbiosis (Feng et al., 2018). Remarkably, 10 of these 14 microRNAs were implicated previously in other host/microbe interaction studies. Following identification of conserved microRNAs that are expressed in aphid bacteriocytes, Feng et al. (2018) predicted microRNA targets and found 103 target predictions to be shared between *A. pisum* and the green peach aphid, *Myzus persicae*. Cluster of Orthologous Group (COG) analysis of these 103 targets found that amino acid transport and metabolism genes, and signal transduction mechanisms were significantly over-represented in the target set relative to the genomes and bacteriome transcriptomes of both species. The biology of 4/9 amino acid transport and metabolism targets is particularly intriguing. One, ACYPI000333, is the ortholog of a low-affinity amino acid transporter, SLC38A9, that functions as a central part of the molecular machinery that activates mechanistic target of rapamycin complex 1 (mTORC1, see below). The second is *BCAT* (branched-chain-amino-acid transaminase, ACYPI008372), the aphid gene discussed earlier in this chapter in the section on genomic complementarity, that mediates the final reaction in the collaborative biosynthesis of isoleucine, leucine and valine (Hansen and Moran, 2011; Russell et al., 2013; Wilson et al., 2010). The third is orthologous to kynurenine-oxoglutarate transaminase (ACYPI001759), an enzyme that functions in tryptophan metabolism to synthesize kynurenate and xanthurenate. Notably, identification of kynurenine-oxoglutarate transaminase as a target led Feng et al. (2018) to propose a novel mechanism by which aphids possibly control glutamate transport to *Buchnera* from bacteriocyte cytoplasm (Fig. 3).

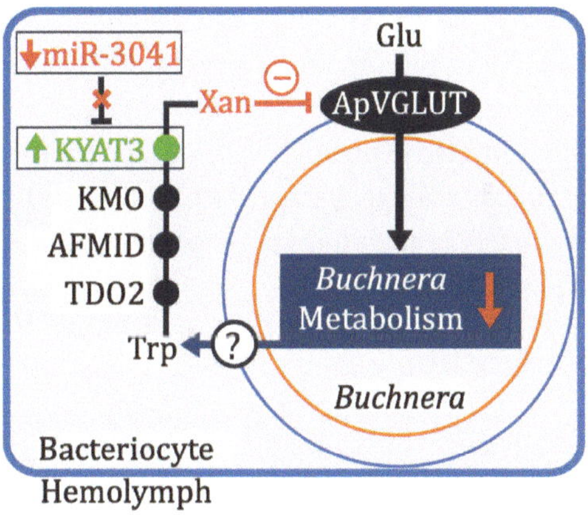

Fig. 3 Proposed mechanism for host control of glutamate transport to *Buchnera* from aphid bacteriocytes that is regulated by aphid genome encoded microRNA, miR-3041. This mechanism couples biosynthesis by *Buchnera* of the essential amino acid tryptophan with aphid supply of the nonessential amino acid glutamate that serves as an amino acid precursor and/or amino donor in the biosynthesis of essential amino acids that include histidine, isoleucine, leucine, phenylalanine, tryptophan, and valine. KYAT3, kynurenine-oxoglutarate transaminase; KMO, kynurenine 3-monooxygenase; AFMID, arylformamidase; TDO2, tryptophan 2,3-dioxygenase; Trp, tryptophan; Xan, xanthurenate; Glu, glutamate; ApVGLUT, aphid vesicular glutamate transporter. *From Feng, H., Wang, L., Wuchty, S., Wilson, A.C.C., 2018. MicroRNA regulation in an ancient obligate endosymbiosis. Mol. Ecol. 27, 1777–1793, fig. 7.*

The fourth encodes adenosylmethionine decarboxylase and is implicated in regulation of both β-alanine production by aphid hosts (see Section 1.1), and *Buchnera*-mediated biosynthesis of adenine (Feng et al., 2018).

The accessibility of whole genome and transcriptome sequencing, coupled with the availability of free annotation tools, mean that it is relatively easy to predict the regulation of genes by microRNAs in a new genome. However, the challenge comes in working to sort the interactions that are real from those that are noise. While experimental interrogation of computationally predicted microRNA::mRNA interactions would ideally be achieved in vivo, not all systems have established genetic tools. The life cycle of aphids has some features that are particularly attractive experimentally, for example, their ability to reproduce asexually facilitates manipulative experiments using genetically identical aphids replicated across multiple treatments, and yet other features, such as the relative inaccessibility of the sexual part of the aphid life cycle, have hindered the development of genetic tools in aphids. In spite of these limitations associated with in vivo validation, advances have been made in studying the biology of aphids and *Buchnera* by leveraging heterologous expression systems (Feng et al., 2019b; Price and Wilson, 2014; Price et al., 2007, 2010, 2014; Thairu et al., 2018). Using a dual luciferase assay in mouse NIH/3T3 cell culture, Feng et al. (2019b) demonstrated that *M. persicae* miR92a interacts with the microRNA target region of *M. persicae* bacteriocyte-specific secreted protein 1 (SP1, Shigenobu and Stern, 2013). This work experimentally validated the earlier computational prediction that *SP1* is regulated by miR92a in aphid bacteriocytes (Feng et al., 2018).

2.3 *Buchnera* encode a set of conserved small RNAs of unknown function

Small RNAs, known regulators of bacterial gene expression (Georg and Hess, 2011; Storz et al., 2011; Waters and Storz, 2009), likely function in regulation of the reduced genomes of intracellular bacteria such as *Buchnera* (Hansen and Degnan, 2014). An asexual female adult aphid contains two *Buchnera* populations, those contained in her bacteriome (maternal bacteriome) and those contained in the bacteriomes of her developing embryos (embryonic bacteriome) (Baumann et al., 1995). Label-free mass-spectrometry quantification of *A. pisum* maternal and embryonic *Buchnera* populations revealed differential protein regulation across the two populations, in the absence of differential mRNA expression (Hansen and Degnan, 2014). Further interrogation of small RNA expression in five *Buchnera* lineages revealed a set of anciently conserved novel intergenic and antisense small RNAs, and untranslated regions that are inferred to facilitate post-transcriptional regulation of *Buchnera* gene expression (Hansen and Degnan, 2014). In later work, Thairu et al. (2018) used an *Escherichia coli* green fluorescent protein (GFP) reporter assay to validate their earlier prediction of a candidate antisense small RNA regulating expression of carbamoyl phosphate synthase large chain (CarB), a protein involved in biosynthesis of the essential amino acid, arginine. At this time it seems clear that *Buchnera* encoded small RNAs likely play a role in regulating *Buchnera* function. Given that similar regulatory mechanisms are also known from the intracellular pathogen *M. pneumoniae*, it seems reasonable to consider that post-transcriptional mechanisms mediated by small RNAs are a feature of endosymbiont genome regulation and function. What remains to be understood is the extent of post-transcriptional regulation across taxa that house nutritional endosymbionts and the ways that these mechanisms interface with mechanisms of host control.

2.4 Expression of the amino acid sensing hub of the mechanistic target of rapamycin pathway is upregulated in aphid bacteriocytes

The mechanistic target of rapamycin (mTOR) pathway is an ancient, conserved signalling pathway that integrates extra and intracellular signals to regulate cellular metabolism, growth, proliferation, and survival (Liu and Sabatini, 2020). mTOR is regulated by two complexes: Complex 1 (mTORC1) and Complex 2 (mTORC2). Amongst other signals, mTORC1 integrates amino acid signals; while mTORC2 integrates growth factors (Saxton and Sabatini, 2017). I first learnt about mTOR when working with Dr. Hsiao-ling Lu to interpret the cytoplasmic localization of amino acid transporter ApGLNT1 during asexual aphid embryogenesis (Lu et al., 2016). When considering the *raison d'être* of the aphid/*Buchnera* endosymbiosis, the tight regulation of mTOR via amino acid availability through mTORC1 (Yuan et al., 2013) renders mTOR a compelling candidate pathway for the molecular and cellular integration of aphid hosts and their *Buchnera*.

Recently, my group annotated mTOR in *A. pisum* and *M. persicae* (James et al., 2018). Following annotation of a complete invertebrate mTOR pathway, comparison of the expression of mTOR genes in *M. persicae* whole insect, gut and bacteriome tissues from three different lineages of *M. persicae* revealed first, that mTOR was more highly expressed in bacteriocytes than it was in whole insect, and second, that genes specific to the amino acid sensing mTORC1 were significantly more highly expressed in bacteriomes than were the amino-acid insensitive mTORC2 specific genes (James et al., 2018). Two groups of amino acids are sensed by mTORC1. The first group includes arginine, glutamine and asparagine, while the second group includes leucine, methionine, isoleucine and valine (Rebsamen et al., 2015; Wolfson et al., 2016). To the reader well familiar with the aphid/*Buchnera* symbiosis, or likely any Sternorrhynchan nutritional endosymbiosis, this is a striking list of seven amino acids. The list is striking because four of the seven are collaboratively synthesized by host and endosymbiont (leucine, methionine, isoleucine and valine), and one, arginine (a non-essential amino acid in most animals), is provisioned by *Buchnera*—because aphids have lost the urea-cycle-derived capacity to synthesize arginine (International Aphid Genomics Consortium, 2010). The remaining two amino acids, glutamine and asparagine, are both found at high concentration in aphid hemolymph (Sasaki and Ishikawa, 1995) and notably, in the case of glutamine, are implicated as central to the cellular regulation of amino acid biosynthesis in aphid bacteriocytes (see Section 3.2).

3. Cellular integration of host and endosymbiont

The evolution of genomic complementarity, the complementation of endosymbiont biosynthesis pathways by genes encoded in the nuclear genome of hosts, requires concomitant evolution of cellular processes that facilitate the shuttling of metabolites and/or gene products across the intracellular and cellular membranes that separate host and endosymbiont. While the cellular arrangement of *Buchnera* within their host aphid changes through development, through most of an aphid's life (with the exception of a short time during transmission) *Buchnera* are surrounded by a host symbiosomal membrane and localize to host bacteriocyte cells, that are organized along with sheath cells to form the bacteriome (Koga et al., 2012). Thus, during the majority of an aphid's life *Buchera* are found inside bacteriocyte cells, enveloped by symbiosomal membranes. Yet, *Buchnera* also occupy other cellular spaces during transmission and early development (Hinde, 1971b; Koga et al., 2012; Miura et al., 2003). These spaces include the maternal hemolymph and the syncytium of the embryonic blastula, but not sheath cells. With respect to the nutritional function of the endosymbiosis the mature cellular arrangement of host and endosymbiont means that *Buchnera* are separated from their host by a lot of membrane barriers that have been referred to as the "symbiotic interface" (Fig. 1).

3.1 Translocation of gene products at the symbiotic interface

Within pathway metabolic collaboration requires the shuttling of metabolites or enzymes across a series of membrane barriers that separate the cytoplasm of *Buchnera* from the cytoplasm of aphid bacteriocyte cells (Fig. 1). Previously it has been argued that within pathway metabolic collaboration is not facilitated by shuttling to *Buchnera* of aphid encoded enzymes because the aphid genes lack recognizable signal sequences (Wilson et al., 2010). However, it should be noted that the question of enzyme shuttling in the context of metabolic collaboration has received scant attention beyond the *A. pisum* genome companion paper of Wilson et al. (2010) where we stated that the "enzymes of interest lack any recognizable signal sequence", without providing any annotation that this is indeed the case. Further, immunolocalization work by Nakabachi et al. (2014) of RlpA4, a laterally acquired *A. pisum* gene, demonstrated that proteins encoded in the host genome and synthesized in maternal bacteriocytes can indeed be targeted and trafficked into *Buchnera* cells, although evidence of the synthesis of RlpA4 in the maternal bacteriocytes is not included in the manuscript. In contrast, in a related nutritional endosymbiosis, that of the citrus mealybug, *Planococcus citri*, with the nested endosymbionts *Candidatus* Tremblaya princeps (*Tremblaya*) that hosts the endosymbiont *Candidatus* Moranella endobia (*Moranella*), a host genome encoded protein, MurF, was recently found to localize to both the cytoplasm of *Moranella* and to host mealybug tissue, but not to *Tremblaya* (Bublitz et al., 2019). With such precedents it seems reasonable to revisit the possibility that amino acid biosynthesis enzymes are indeed trafficked between partners at the symbiotic interface.

If within pathway collaboration is not facilitated by traffic of enzymes at the symbiotic interface then metabolic collaboration must be realized via the trafficking of metabolic intermediates. While a number of studies have sought to address the validation and development of genome based models of collaborative metabolism (MacDonald et al., 2011, 2012; Price and Wilson, 2014; Russell et al., 2013, 2014; see, e.g., Thomas et al., 2009), the question of how metabolic intermediates are trafficked between host and endosymbiont remains open.

3.2 Translocation of amino acids at the symbiotic interface

By coupling heterologous expression of aphid amino acid transporters in oocytes of the African Clawed Frog, *Xenopus laevis*, with transporter immunolocalization using custom antibodies we have now functionally characterized two amino acid transporters (Feng et al., 2019a; Price et al., 2014, 2015); both are transporters of nonessential amino acids. The first to be characterized was ApGLNT1 (*Acyrthosiphon pisum* Glutamine Transporter 1, formerly *A. pisum* gene ACYPI001018). At the symbiotic interface, ApGLNT1 transports glutamine (only glutamine) from aphid hemolymph into bacteriocyte cells (Price et al., 2014). Most notably, glutamine transport by ApGLNT1 is competitively inhibited by arginine, a *Buchnera* synthesized essential amino acid (Price et al., 2014, 2015). This inhibition of ApGLNT1 glutamine transport by arginine means that glutamine supply to bacteriocytes is regulated by synthesis of a *Buchnera* provisioned essential amino acid, arginine. The second amino acid transporter to be characterized was ApNEAAT1 (*Acyrthosiphon pisum* Nonessential Amino Acid Transporter 1, formerly *A. pisum* gene ACYPI008971). ApNEAAT1, is a transporter of small nonessential amino acids that include serine, proline, cysteine, alanine, and glycine. At the symbiotic interface ApNEAAT1 localizes to both the bacteriocyte cell membrane and the symbiosomal membrane that surrounds individual *Buchnera* cells. Rather unusually, ApNEAAT1 transport is not driven by ion gradients (it is electroneutral); transport is instead driven by amino acid concentration gradients—a feature that elegantly serves the nutritional needs of host and endosymbiont (Feng et al., 2019a). Functional characterization of just two of the 40 nutritional amino acid transporters encoded in the *A. pisum* genome (Duncan et al., 2014, 2016; Price et al., 2011) has started to provide considerable insight into regulation of amino acid biosynthesis, and the mechanisms by which host and symbiont amino acid demands are met. Just as predicted by Moran et al. (2003) the nutritional needs of the endosymbiosis do indeed appear to be regulated by host control of amino acid supply by ApGLNT1, a remarkable mechanism that may, or may not be broadly functional across related endosymbioses.

Functional characterization of the mechanisms of cellular integration of hosts and endosymbionts is yielding clear insights into the mechanisms by which nutritional endosymbioses are regulated and cellular homeostasis is maintained in these systems. The work is often slow and painstaking both because of the level of evidence required to generate for example, robust immunolocalization data, and because it requires the application of approaches that span traditional sub-disciplinary boundaries, and therefore benefits from collaboration amongst scientists with diverse skill sets. It is fortunate that the field has a strong history of such collaborations (e.g., the early collaboration of Paul and Linda Baumann with Nancy Moran).

4. Concluding remarks—A mutualism in conflict

I am captivated by the story of aphids and their nutritional endosymbiont, *Buchnera aphidicola*. When I travel, I meet new people who know a lot about biology, but nothing about the obligate nutritional endosymbionts of insects like aphids. Such biologists often assume that an ancient endosymbiosis like that of aphids and *Buchnera* is one that exists in perfect harmony, devoid of conflict. This is a tempting mistake to make. It is true that *Buchnera* have been vertically transmitted for 160 million years (Moran et al., 1993; Munson et al., 1991a), it is also true that *Buchnera* cannot live outside aphids (Shigenobu et al., 2000) and that aphids without *Buchnera* fail to reproduce (Ohtaka and Ishikawa, 1991; Sasaki et al., 1991). Yet, the fitness of aphids and *Buchnera* are incompletely aligned (Bennett and Moran, 2015) and tracks of this incomplete alignment of fitness litter the genome of the aphid/*Buchnera* holobiont. These tracks are the signatures of evolutionary conflict (Queller and Strassmann, 2018). In concluding this chapter, I highlight the tracks of conflict that jump out in the context of considering the role of mTOR in regulating holobiont function (Table 1).[a]

The mTOR signalling pathway receives amino acid inputs at two points (Fig. 4). Of the seven amino acids known to signal input into the pathway, the biosynthesis of five feature genomic tracks that mark what Bennett and Moran (2015) refer to as mutational decay of the symbiont coupled with compensation by the host. I suggest, however, that these tracks may not reflect cooperation, but rather that they result from conflict, and serve as records of the endosymbiont and host wrestling for control of mTOR, a pathway that centrally regulates the maintenance of cellular and organismal homeostasis. In the case of the three amino acids (isoleucine, methionine and valine) marked by one genomic track, all three involve complementation of metabolic pathways, historically encoded in the *Buchnera* genome, by host genes, and thereby represent a gain of host input into the biosynthesis of amino acids that are key mTOR inputs. Remarkably, in the case of isoleucine, valine and leucine (below), computational prediction of miRNA targets indicates that host control involves not just complementation of the pathway but the evolution of

[a] From the Merriam-Webster Dictionary, *Track: (a): a footprint whether recent or fossil, (b): detectable evidence (such as the wake of a ship, a line of footprints, or a wheel rut) that something has passed.* Over the course of evolution, genomes are marked with tracks that include the mutation, duplication, loss and acquisition of genes and regulatory elements. These tracks reveal the ways that past selection and genetic drift have carried an organism into the present.

Table 1 Five of the seven amino acids that provide inputs into mTOR are essential amino acids.

Amino acid	Signature of conflict	Control
Arginine	1. Loss of urea cycle from host insect genomes, complemented by retention of arginine biosynthesis pathways in endosymbiont genomes	H→S
	2. Gain of urea cycle genes by lateral gene transfer from bacteria in some host insect genomes	H→S→H
	3. Regulation of host provision of glutamine by *Buchnera* encoded amino acid, arginine via ApGLNT1	H→S
Isoleucine	4. Host/endosymbiont metabolic complementation in the biosynthesis of the branched-chain amino acids	S→H
Leucine	5. Some *Buchnera* lineages encode leucine biosynthesis genes on a plasmid	S
	6. Host/endosymbiont metabolic complementation in the biosynthesis of the branched-chain amino acids	S→H
Methionine	7. Host/endosymbiont metabolic complementation in methionine biosynthesis	S→H
Valine	8. Host/endosymbiont metabolic complementation in the biosynthesis of the branched-chain amino acids	S→H

Listed alphabetically in the table, they include arginine, the amino acid that aphids have lost the ability to synthesize, and four of the five amino acids that are commonly collaboratively synthesized by Sternorrhyncha hosts and their nutritional endosymbionts (isoleucine, leucine, methionine and valine).

finescale regulation by host encoded miRNAs (Feng et al., 2018). The remaining two essential amino acid inputs, arginine and leucine, are marked by multiple genomic tracks, that may serve as records of an on-going tug-of-war between host and endosymbiont.

Leucine biosynthesis, like that of isoleucine, methionine and valine biosynthesis involves a derived state of complementation by host genes. Yet, biosynthesis of leucine is also marked in some *Buchnera* lineages by a second track encoded entirely in the endosymbiont genome. Some lineages of *Buchnera* encode genes for leucine biosynthesis not on the bacterial chromosome but on a plasmid, pLeu (Baumann et al., 1999; Bracho et al., 1995; Chong et al., 2019). While early evidence indicated that *leuABCD* were transferred from the chromosome to pLeu only once (or very few times) (Moran et al., 2003), the most recent evidence based on whole genome sequencing demonstrates multiple rearrangements (Chong et al., 2019).

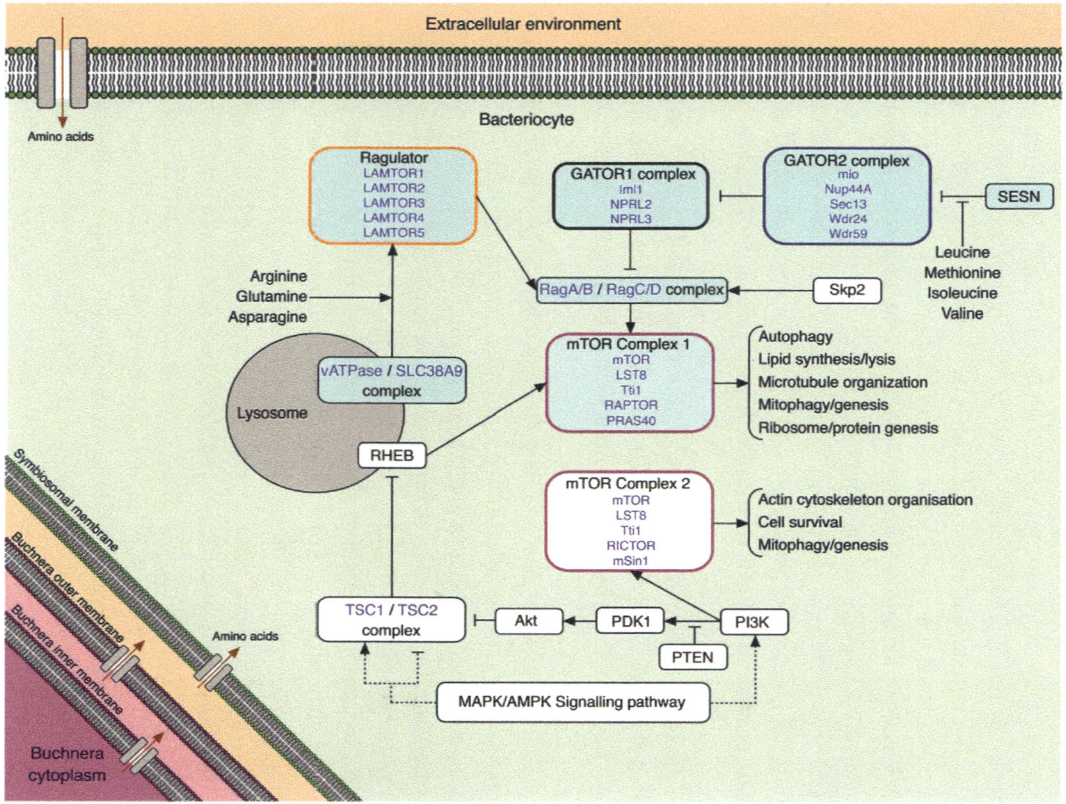

Fig. 4 The mTOR pathway within an aphid bacteriocyte. Amino acid inputs are received in two places. Arginine, glutamine, and asparagine are sensed via the vATPase/SLC38A9 complex, while leucine, methionine, isoleucine and valine are sensed by sestrin (SESN). *From James, E.B., Feng, H., Wilson, A.C.C., 2018. mTOR complex 1 implicated in aphid/Buchnera host/symbiont integration. G3 8, 3083–3091, fig. 1.*

These rearrangements include changes in gene order within the operon (from ABCD to BCDA), multiple movements of the operon from chromosome to plasmid and back to chromosome as evidenced by finding the operon in different chromosome locations across taxa (Chong et al., 2019). Historically it had been argued that the presence of amino acid biosynthesis genes on plasmids in *Buchnera* was driven by selection to meet host metabolic needs (Moran et al., 2003, 2005), and yet, experimental work did little to support that argument and so with time, it fell out of favour. Here, I suggest that the pLeu localization of leucine biosynthesis genes might be driven by conflict, and selfishness on the part of *Buchnera* to control amino acid inputs into mTOR to the benefit of self.

Arginine biosynthesis is marked by three genomic tracks that signal an on-going tug-of-war between host and endosymbiont. The first is endosymbiont complementation of arginine biosynthesis following loss of most urea cycle genes in the host genome. While this pattern was first described in the pea aphid, *A. pisum* (International Aphid Genomics Consortium, 2010; Wilson et al., 2010), it is a feature of host/endosymbiont coevolution that is common in the Sternorrhyncha—hosts lose urea cycle genes, while against a template of large-scale gene losses, endosymbionts fail to lose the ability to make arginine (Duncan, 2016). The second, known only at this time from *A. pisum*, involves the localization of amino acid transporter ApGLNT1 to the plasma membrane of bacteriocytes where it has been proposed to function to regulate amino acid biosynthesis in bacteriocytes by controlling supply of the precursor amino acid glutamine by competitive negative feedback inhibition by arginine (Price et al., 2014, 2015). This mechanism couples amino acid biosynthesis to endosymbiont provisioned essential amino acid biosynthesis, the *raison d'être* of the aphid/*Buchnera* endosymbiosis. The third, is marked by lateral gene transfer of bacterial genes into host genomes where they function to complement earlier gene losses by the host. While this track is not yet known in aphids, it is known from insects that include whiteflies and mealybugs (Husnik et al., 2013; Luan et al., 2015). Here I suggest that the repair of urea cycle function by lateral gene transfer in these systems may be driven by conflict, and selfishness on the part of host insects to regain control of the biosynthesis of a key mTOR input. Many pieces are still missing from the development of this argument about arginine, and possibly I'm putting the pieces together to tell the wrong story. However, in working to understand the role of mTOR in holobiont (host + endosymbiont) function, I find the possibility that this integration has been driven by conflict intriguing.

The idea that conflict plays a role in the evolution of endosymbiosis is not new (Bennett and Moran, 2015; Garcia and Gerardo, 2014; Husnik and Keeling, 2019; Keeling and McCutcheon, 2017; Kiers and West, 2016; McCutcheon et al., 2019; Sullivan, 2017; Wernegreen, 2004). To date, the central argument against ancient endosymbioses being harmonious mutualisms, pivots on observations of endosymbiont genome evolution. Here, in thinking about host/endosymbiont metabolic collaboration with a focus on mTOR, I shift the perspective a little by highlighting places that evolution has likely been impacted by an on-going tug-of-war between host and endosymbiont. A challenge going forward will be that definitive tests to distinguish between the role of conflict versus cooperation in the evolution of a holobiont remain elusive. We may never fully understand the dance, but it is clear to many in the field that the evolution of obligate, intracellular endosymbioses, are not solely driven by selection for cooperation, but rather that conflict, mutation and genetic drift have been, and will continue to be, powerful forces in the evolution of insect nutritional endosymbioses.

Acknowledgements

The writing of this chapter benefited from useful discussions with many people, including my current and former students, Edward James, Celeste Banfill, Honglin Feng, Rebecca Duncan, and my colleagues, J. David Van Dyken and Hsiao-ling Lu. Hsiao-ling Lu was especially generous in providing the confocal images included in Fig. 1. This chapter was additionally improved by the input of an anonymous reviewer (you know who you are—thank you), and the editors of this volume, Jake Russell and Kerry Oliver, who continue to inspire me with their abilities to pull our community together to produce work like this volume. The preparation of this review was supported by National Science Foundation Award IOS-1354154.

References

Baumann, P., Baumann, L., Lai, C.Y., Rouhbakhsh, D., Moran, N.A., Clark, M.A., 1995. Genetics, physiology, and evolutionary relationships of the genus *Buchnera*: intracellular symbionts of aphids. Annu. Rev. Microbiol. 49, 55–94.

Baumann, L., Baumann, P., Moran, N.A., Sandström, J., Thao, M.L., 1999. Genetic characterization of plasmids containing genes encoding enzymes of leucine biosynthesis in endosymbionts (*Buchnera*) of aphids. J. Mol. Evol. 48, 77–85.

Bennett, G.M., Moran, N.A., 2015. Heritable symbiosis: the advantages and perils of an evolutionary rabbit hole. Proc. Natl. Acad. Sci. U. S. A. 112, 10169–10176.

Bracho, A.M., Martínez-Torres, D., Moya, A., Latorre, A., 1995. Discovery and molecular characterization of a plasmid localized in *Buchnera* sp. bacterial endosymbiont of the aphid *Rhopalosiphum padi*. J. Mol. Evol. 41, 67–73.

Bublitz, D.C., Chadwick, G.L., Magyar, J.S., Sandoz, K.M., Brooks, D.M., Mesnage, S., Ladinsky, M.S., Garber, A.I., Bjorkman, P.J., Orphan, V.J., McCutcheon, J.P., 2019. Peptidoglycan production by an insect-bacterial mosaic. Cell 179, 703–712.e7.

Buchner, P., 1965. Endosymbiosis of Animals with Plant Microorganisms. Interscience Publishers.

Chong, R.A., Moran, N.A., 2018. Evolutionary loss and replacement of *Buchnera*, the obligate endosymbiont of aphids. ISME J. 12, 898–908.

Chong, R.A., Park, H., Moran, N.A., 2019. Genome evolution of the obligate endosymbiont *Buchnera aphidicola*. Mol. Biol. Evol. 36, 1481–1489.

Dobson, C.M., 2003. Protein folding and misfolding. Nature 426, 884–890.

Douglas, A.E., Wilkinson, T.L., 1998. Host cell allometry and regulation of the symbiosis between pea aphids, *Acyrthosiphon pisum*, and bacteria, *Buchnera*. J. Insect Physiol. 44, 629–635.

Duncan, R.P., 2016. The Role of Gene Co-option in the Evolution of Sap-Feeding Insect Endosymbioses (Hemiptera: Sternorrhyncha). University of Miami.

Duncan, R.P., Husnik, F., Van Leuven, J.T., Gilbert, D.G., Dávalos, L.M., McCutcheon, J.P., Wilson, A.C.C., 2014. Dynamic recruitment of amino acid transporters to the insect/symbiont interface. Mol. Ecol. 23, 1608–1623.

Duncan, R.P., Feng, H., Nguyen, D.M., Wilson, A.C.C., 2016. Gene family expansions in aphids maintained by endosymbiotic and nonsymbiotic traits. Genome Biol. Evol. 8, 753–764.

Feder, M.E., Hofmann, G.E., 1999. Heat-shock proteins, molecular chaperones, and the stress response: evolutionary and ecological physiology. Annu. Rev. Physiol. 61, 243–282.

Feng, H., Wang, L., Wuchty, S., Wilson, A.C.C., 2018. microRNA regulation in an ancient obligate endosymbiosis. Mol. Ecol. 27, 1777–1793.

Feng, H., Edwards, N., Anderson, C.M.H., Althaus, M., Duncan, R.P., Hsu, Y.-C., Luetje, C.W., Price, D.R.G., Wilson, A.C.C., Thwaites, D.T., 2019a. Trading amino acids at the aphid–*Buchnera* symbiotic interface. Proc. Natl. Acad. Sci. U. S. A. 116, 16003–16011.

Feng, H., Park, J.S., Zhai, R.G., Wilson, A.C.C., 2019b. microRNA-92a regulates the expression of aphid bacteriocyte-specific secreted protein 1. BMC. Res. Notes 12, 638.

Fukatsu, T., Aoki, S., Kurosu, U., Ishikawa, H., 1994. Phylogeny of Cerataphidini aphids revealed by their symbiotic microorganisms and basic structure of their galls—implications for host-symbiont coevolution and evolution of sterile soldier castes. Zoolog. Sci. 11, 613–623.

Garcia, J.R., Gerardo, N.M., 2014. The symbiont side of symbiosis: do microbes really benefit? Front. Microbiol. 5, 510.

Georg, J., Hess, W.R., 2011. Cis-antisense RNA, another level of gene regulation in bacteria. Microbiol. Mol. Biol. Rev. 75, 286–300.

Gerardo, N.M., Wilson, A.C.C., 2011. The power of paired genomes. Mol. Ecol. 20, 2038–2040.

Gerardo, N.M., Altincicek, B., Anselme, C., Atamian, H., Barribeau, S.M., de Vos, M., Duncan, E.J., Evans, J.D., Gabaldón, T., Ghanim, M., Heddi, A., Kaloshian, I., Latorre, A., Moya, A., Nakabachi, A., Parker, B.J., Pérez-Brocal, V., Pignatelli, M., Rahbé, Y., Ramsey, J.S., Spragg, C.J., Tamames, J., Tamarit, D., Tamborindeguy, C., Vincent-Monegat, C., Vilcinskas, A., 2010. Immunity and other defenses in pea aphids, *Acyrthosiphon pisum*. Genome Biol. 11, R21.

Hansen, A.K., Degnan, P.H., 2014. Widespread expression of conserved small RNAs in small symbiont genomes. ISME J. 8, 2490–2502.

Hansen, A.K., Moran, N.A., 2011. Aphid genome expression reveals host–symbiont cooperation in the production of amino acids. Proc. Natl. Acad. Sci. U. S. A. 108, 2849–2854.

Hara, E., Ishikawa, H., 1990. Purification and partial characterization of symbionin, an aphid endosymbiont-specific protein. Insect Biochem. 20, 421–427.

Hartl, F.U., 1996. Molecular chaperones in cellular protein folding. Nature 381, 571–579.

Hesse, H., Kreft, O., Maimann, S., Zeh, M., Hoefgen, R., 2004. Current understanding of the regulation of methionine biosynthesis in plants. J. Exp. Bot. 55, 1799–1808.
Hinde, R., 1971a. The fine structure of the mycetome symbiotes of the aphids *Brevicoryne brassicae*, *Myzus persicae*, and *Macrosiphum rosae*. J. Insect Physiol. 17, 2035–2050.
Hinde, R., 1971b. The control of the mycetome symbiotes of the aphids *Brevicoryne brassicae*, *Myzus persicae*, and *Macrosiphum rosae*. J. Insect Physiol. 17, 1791–1800.
Houk, E.J., Griffiths, G.W., 1980. Intracellular symbiotes of the Homoptera. Annu. Rev. Entomol. 25, 161–187.
Husnik, F., Keeling, P.J., 2019. The fate of obligate endosymbionts: reduction, integration, or extinction. Curr. Opin. Genet. Dev. 58-59, 1–8.
Husnik, F., Nikoh, N., Koga, R., Ross, L., Duncan, R.P., Fujie, M., Tanaka, M., Satoh, N., Bachtrog, D., Wilson, A.C.C., von Dohlen, C.D., Fukatsu, T., McCutcheon, J.P., 2013. Horizontal gene transfer from diverse bacteria to an insect genome enables a tripartite nested mealybug symbiosis. Cell 153, 1567–1578.
International Aphid Genomics Consortium, 2010. Genome sequence of the pea aphid *Acyrthosiphon pisum*. PLoS Biol. 8, e1000313.
Ishikawa, H., 1982. Host-symbiont interactions in the protein synthesis in the pea aphid, *Acyrthosiphon pisum*. Insect Biochem. 12, 613–622.
Ishikawa, H., 1984. Characterization of the protein species synthesized *in vivo* and *in vitro* by an aphid endosymbiont. Insect Biochem. 14, 417–425.
James, E.B., Feng, H., Wilson, A.C.C., 2018. mTOR complex 1 implicated in aphid/*Buchnera* host/symbiont integration. G3 8, 3083–3091.
Kakeda, K., Ishikawa, H., 1991. Molecular chaperon produced by an intracellular symbiont. J. Biochem. 110, 583–587.
Kallarackal, J., Bauer, S.N., Nowak, H., Hajirezaei, M.-R., Komor, E., 2012. Diurnal changes in assimilate concentrations and fluxes in the phloem of castor bean (*Ricinus communis* L.) and tansy (*Tanacetum vulgare* L.). Planta 236, 209–223.
Keeling, P.J., McCutcheon, J.P., 2017. Endosymbiosis: the feeling is not mutual. J. Theor. Biol. 434, 75–79.
Kiers, E.T., West, S.A., 2016. Evolution: welcome to symbiont prison. Curr. Biol. 26, R66–R68.
Koga, R., Meng, X.-Y., Tsuchida, T., Fukatsu, T., 2012. Cellular mechanism for selective vertical transmission of an obligate insect symbiont at the bacteriocyte-embryo interface. Proc. Natl. Acad. Sci. U. S. A. 109, E1230–E1237.
Kreft, O., Hoefgen, R., Hesse, H., 2003. Functional analysis of cystathionine gamma-synthase in genetically engineered potato plants. Plant Physiol. 131, 1843–1854.
Kupper, M., Gupta, S.K., Feldhaar, H., Gross, R., 2014. Versatile roles of the chaperonin GroEL in microorganism-insect interactions. FEMS Microbiol. Lett. 353, 1–10.
Lamelas, A., Gosalbes, M.J., Manzano-Marín, A., Peretó, J., Moya, A., Latorre, A., 2011a. *Serratia symbiotica* from the aphid *Cinara cedri*: a missing link from facultative to obligate insect endosymbiont. PLoS Genet. 7, e1002357.
Lamelas, A., Gosalbes, M.J., Moya, A., Latorre, A., 2011b. New clues about the evolutionary history of metabolic losses in bacterial endosymbionts, provided by the genome of *Buchnera aphidicola* from the aphid *Cinara tujafilina*. Appl. Environ. Microbiol. 77, 4446–4454.
Liu, G.Y., Sabatini, D.M., 2020. mTOR at the nexus of nutrition, growth, ageing and disease. Nat. Rev. Mol. Cell Biol. 21, 183–203.
Lu, H.-L., Price, D.R.G., Wikramanayake, A., Chang, C.-C., Wilson, A.C.C., 2016. Ontogenetic differences in localization of glutamine transporter ApGLNT1 in the pea aphid demonstrate that mechanisms of host/symbiont integration are not similar in the maternal versus embryonic bacteriome. EvoDevo 7, 1.

Luan, J.-B., Chen, W., Hasegawa, D.K., Simmons, A.M., Wintermantel, W.M., Ling, K.-S., Fei, Z., Liu, S.-S., Douglas, A.E., 2015. Metabolic coevolution in the bacterial Symbiosis of whiteflies and related plant sap-feeding insects. Genome Biol. Evol. 7, 2635–2647.

MacDonald, S.J., Thomas, G.H., Douglas, A.E., 2011. Genetic and metabolic determinants of nutritional phenotype in an insect-bacterial symbiosis. Mol. Ecol. 20, 2073–2084.

Macdonald, S.J., Lin, G.G., Russell, C.W., Thomas, G.H., Douglas, A.E., 2012. The central role of the host cell in symbiotic nitrogen metabolism. Proc. Biol. Sci. 279, 2965–2973.

Mathers, T.C., Chen, Y., Kaithakottil, G., Legeai, F., Mugford, S.T., Baa-Puyoulet, P., Bretaudeau, A., Clavijo, B., Colella, S., Collin, O., Dalmay, T., Derrien, T., Feng, H., Gabaldón, T., Jordan, A., Julca, I., Kettles, G.J., Kowitwanich, K., Lavenier, D., Lenzi, P., Lopez-Gomollon, S., Loska, D., Mapleson, D., Maumus, F., Moxon, S., Price, D.R.G., Sugio, A., van Munster, M., Uzest, M., Waite, D., Jander, G., Tagu, D., Wilson, A.C.C., van Oosterhout, C., Swarbreck, D., Hogenhout, S.A., 2017. Rapid transcriptional plasticity of duplicated gene clusters enables a clonally reproducing aphid to colonise diverse plant species. Genome Biol. 18, 27.

Matthews, B.F., 1999. Lysine, threonine and methionine biosynthesis. In: Singh, B.K. (Ed.), Plant Amino Acids: Biochemistry and Biotechnology. Marcel Dekker, New York, pp. 205–225.

McCutcheon, J.P., Moran, N.A., 2011. Extreme genome reduction in symbiotic bacteria. Nat. Rev. Microbiol. 10, 13–26.

McCutcheon, J.P., Boyd, B.M., Dale, C., 2019. The life of an insect endosymbiont from the cradle to the grave. Curr. Biol. 29, R485–R495.

Mira, A., Moran, N.A., 2002. Estimating population size and transmission bottlenecks in maternally transmitted endosymbiotic bacteria. Microb. Ecol. 44, 137–143.

Mitchell, D.E., Gadus, M.V., Madore, M.A., 1992. Patterns of assimilate production and translocation in muskmelon (Cucumis melo L.). Plant Physiol. 99, 959–965.

Miura, T., Braendle, C., Shingleton, A., Sisk, G., Kambhampati, S., Stern, D.L., 2003. A comparison of parthenogenetic and sexual embryogenesis of the pea aphid *Acyrthosiphon pisum* (Hemiptera: Aphidoidea). J. Exp. Zool. B Mol. Dev. Evol. 295, 59–81.

Moran, N.A., 1996. Accelerated evolution and Muller's rachet in endosymbiotic bacteria. Proc. Natl. Acad. Sci. U. S. A. 93, 2873–2878.

Moran, N.A., Munson, M.A., Baumann, P., Ishikawa, H., 1993. A molecular clock in endosymbiotic bacteria is calibrated using the insect hosts. Proc. R. Soc. Lond. B Biol. Sci. 253, 167–171.

Moran, N.A., Plague, G.R., Sandstrom, J.P., Wilcox, J.L., 2003. A genomic perspective on nutrient provisioning by bacterial symbionts of insects. Proc. Natl. Acad. Sci. U. S. A. 100, 14543–14548.

Moran, N.A., Dunbar, H.E., Wilcox, J.L., 2005. Regulation of transcription in a reduced bacterial genome: nutrient-provisioning genes of the obligate symbiont *Buchnera aphidicola*. J. Bacteriol. 187, 4229–4237.

Moran, N.A., McCutcheon, J.P., Nakabachi, A., 2008. Genomics and evolution of heritable bacterial symbionts. Annu. Rev. Genet. 42, 165–190.

Munson, M.A., Baumann, P., Clark, M.A., Baumann, L., Moran, N.A., Voegtlin, D.J., Campbell, B.C., 1991a. Evidence for the establishment of aphid-eubacterium endosymbiosis in an ancestor of four aphid families. J. Bacteriol. 173, 6321–6324.

Munson, M.A., Baumann, P., Kinsey, M.G., 1991b. *Buchnera* gen. nov. and *Buchnera aphidicola* sp. nov., a taxon consisting of the mycetocyte-associated, primary endosymbionts of aphids. Int. J. Syst. Bacteriol. 41, 566–568.

Nakabachi, A., Ishida, K., Hongoh, Y., Ohkuma, M., Miyagishima, S.-Y., 2014. Aphid gene of bacterial origin encodes a protein transported to an obligate endosymbiont. Curr. Biol. 24, R640–R641.

Nikoh, N., Nakabachi, A., 2009. Aphids acquired symbiotic genes via lateral gene transfer. BMC Biol. 7, 12.

Nikoh, N., McCutcheon, J.P., Kudo, T., Miyagishima, S.-Y., Moran, N.A., Nakabachi, A., 2010. Bacterial genes in the aphid genome: absence of functional gene transfer from *Buchnera* to its host. PLoS Genet. 6, e1000827.

Ohtaka, C., Ishikawa, H., 1991. Effects of heat treatment on the symbiotic system of an aphid mycetocyte. Symbiosis 11, 19–30.

Pérez-Brocal, V., Gil, R., Ramos, S., Lamelas, A., Postigo, M., Michelena, J.M., Silva, F.J., Moya, A., Latorre, A., 2006. A small microbial genome: the end of a long symbiotic relationship? Science 314, 312–313.

Price, D.R.G., Wilson, A.C.C., 2014. A substrate ambiguous enzyme facilitates genome reduction in an intracellular symbiont. BMC Biol. 12, 110.

Price, D.R.G., Karley, A.J., Ashford, D.A., Isaacs, H.V., Pownall, M.E., Wilkinson, H.S., Gatehouse, J.A., Douglas, A.E., 2007. Molecular characterisation of a candidate gut sucrase in the pea aphid, *Acyrthosiphon pisum*. Insect Biochem. Mol. Biol. 37, 307–317.

Price, D.R.G., Tibbles, K., Shigenobu, S., Smertenko, A., Russell, C.W., Douglas, A.E., Fitches, E., Gatehouse, A.M.R., Gatehouse, J.A., 2010. Sugar transporters of the major facilitator superfamily in aphids; from gene prediction to functional characterization. Insect Mol. Biol. 19 (Suppl. 2), 97–112.

Price, D.R.G., Duncan, R.P., Shigenobu, S., Wilson, A.C.C., 2011. Genome expansion and differential expression of amino acid transporters at the aphid/*Buchnera* symbiotic interface. Mol. Biol. Evol. 28, 3113–3126.

Price, D.R.G., Feng, H., Baker, J.D., Bavan, S., Luetje, C.W., Wilson, A.C.C., 2014. Aphid amino acid transporter regulates glutamine supply to intracellular bacterial symbionts. Proc. Natl. Acad. Sci. U. S. A. 111, 320–325.

Price, D.R.G., Wilson, A.C.C., Luetje, C.W., 2015. Proton-dependent glutamine uptake by aphid bacteriocyte amino acid transporter ApGLNT1. Biochim. Biophys. Acta 1848, 2085–2091.

Queller, D.C., Strassmann, J.E., 2018. Evolutionary conflict. Annu. Rev. Ecol. Evol. Syst. 49, 73–93.

Ramsey, J.S., MacDonald, S.J., Jander, G., Nakabachi, A., Thomas, G.H., Douglas, A.E., 2010. Genomic evidence for complementary purine metabolism in the pea aphid, *Acyrthosiphon pisum*, and its symbiotic bacterium *Buchnera aphidicola*. Insect Mol. Biol. 19 (Suppl. 2), 241–248.

Rebsamen, M., Pochini, L., Stasyk, T., de Araújo, M.E.G., Galluccio, M., Kandasamy, R.K., Snijder, B., Fauster, A., Rudashevskaya, E.L., Bruckner, M., Scorzoni, S., Filipek, P.A., Huber, K.V.M., Bigenzahn, J.W., Heinz, L.X., Kraft, C., Bennett, K.L., Indiveri, C., Huber, L.A., Superti-Furga, G., 2015. SLC38A9 is a component of the lysosomal amino acid sensing machinery that controls mTORC1. Nature 519, 477–481.

Reymond, N., Calevro, F., Viñuelas, J., Morin, N., Rahbé, Y., Febvay, G., Laugier, C., Douglas, A., Fayard, J.-M., Charles, H., 2006. Different levels of transcriptional regulation due to trophic constraints in the reduced genome of *Buchnera aphidicola* APS. Appl. Environ. Microbiol. 72, 7760–7766.

Russell, C.W., Bouvaine, S., Newell, P.D., Douglas, A.E., 2013. Shared metabolic pathways in a coevolved insect-bacterial symbiosis. Appl. Environ. Microbiol. 79, 6117–6123.

Russell, C.W., Poliakov, A., Haribal, M., Jander, G., van Wijk, K.J., Douglas, A.E., 2014. Matching the supply of bacterial nutrients to the nutritional demand of the animal host. Proc. Biol. Sci. 281, 20141163.

Sandström, J., Pettersson, J., 1994. Amino acid composition of phloem sap and the relation to intraspecific variation in pea aphid (*Acyrthosiphon pisum*) performance. J. Insect Physiol. 40, 947–955.

Sasaki, T., Ishikawa, H., 1995. Production of essential amino acids from glutamate by mycetocyte symbionts of the pea aphid, *Acyrthosiphon pisum*. J. Insect Physiol. 41, 41–46.
Sasaki, T., Hayashi, H., Ishikawa, H., 1991. Growth and reproduction of the symbiotic and aposymbiotic pea aphids, *Acyrthosiphon pisum* maintained on artificial diets. J. Insect Physiol. 37, 749–756.
Saxton, R.A., Sabatini, D.M., 2017. mTOR signaling in growth, metabolism, and disease. Cell 169, 361–371.
Shigenobu, S., Stern, D.L., 2013. Aphids evolved novel secreted proteins for symbiosis with bacterial endosymbiont. Proc. Biol. Sci. 280, 20121952.
Shigenobu, S., Watanabe, H., Hattori, M., Sakaki, Y., Ishikawa, H., 2000. Genome sequence of the endocellular bacterial symbiont of aphids *Buchnera* sp. APS. Nature 407, 81–86.
Storz, G., Vogel, J., Wassarman, K.M., 2011. Regulation by small RNAs in bacteria: expanding frontiers. Mol. Cell 43, 880–891.
Sullivan, W., 2017. Wolbachia, bottled water, and the dark side of symbiosis. Mol. Biol. Cell 28, 2343–2346.
Tamas, I., Klasson, L., Canbäck, B., Näslund, A.K., Eriksson, A.-S., Wernegreen, J.J., Sandström, J.P., Moran, N.A., Andersson, S.G.E., 2002. 50 million years of genomic stasis in endosymbiotic bacteria. Science 296, 2376–2379.
Taylor, S.H., Parker, W.E., Douglas, A.E., 2012. Patterns in aphid honeydew production parallel diurnal shifts in phloem sap composition. Entomol. Exp. Appl. 142, 121–129.
Thairu, M.W., Cheng, S., Hansen, A.K., 2018. A sRNA in a reduced mutualistic symbiont genome regulates its own gene expression. Mol. Ecol. 27, 1766–1776.
Thomas, G.H., Zucker, J., Macdonald, S.J., Sorokin, A., Goryanin, I., Douglas, A.E., 2009. A fragile metabolic network adapted for cooperation in the symbiotic bacterium *Buchnera aphidicola*. BMC Syst. Biol. 3, 24.
Waters, L.S., Storz, G., 2009. Regulatory RNAs in bacteria. Cell 136, 615–628.
Weibull, J., 1987. Seasonal changes in the free amino acids of oat and barley phloem sap in relation to plant growth stage and growth of *Rhopalosiphum padi*. Ann. Appl. Biol. 111, 729–737.
Wernegreen, J.J., 2004. Endosymbiosis: lessons in conflict resolution. PLoS Biol. 2, E68.
Wilcox, J.L., Dunbar, H.E., Wolfinger, R.D., Moran, N.A., 2003. Consequences of reductive evolution for gene expression in an obligate endosymbiont. Mol. Microbiol. 48, 1491–1500.
Wilson, A.C.C., Duncan, R.P., 2015. Signatures of host/symbiont genome coevolution in insect nutritional endosymbioses. Proc. Natl. Acad. Sci. U. S. A. 112, 10255–10261.
Wilson, A.C.C., Dunbar, H.E., Davis, G.K., Hunter, W.B., Stern, D.L., Moran, N.A., 2006. A dual-genome microarray for the pea aphid, *Acyrthosiphon pisum*, and its obligate bacterial symbiont, *Buchnera aphidicola*. BMC Genomics 7, 50.
Wilson, A.C.C., Ashton, P.D., Calevro, F., Charles, H., Colella, S., Febvay, G., Jander, G., Kushlan, P.F., Macdonald, S.J., Schwartz, J.F., Thomas, G.H., Douglas, A.E., 2010. Genomic insight into the amino acid relations of the pea aphid, *Acyrthosiphon pisum*, with its symbiotic bacterium *Buchnera aphidicola*. Insect Mol. Biol. 19 (Suppl. 2), 249–258.
Wolfson, R.L., Chantranupong, L., Saxton, R.A., Shen, K., Scaria, S.M., Cantor, J.R., Sabatini, D.M., 2016. Sestrin2 is a leucine sensor for the mTORC1 pathway. Science 351, 43–48.
Woolfit, M., Bromham, L., 2003. Increased rates of sequence evolution in endosymbiotic bacteria and fungi with small effective population sizes. Mol. Biol. Evol. 20, 1545–1555.
Yuan, H.-X., Xiong, Y., Guan, K.-L., 2013. Nutrient sensing, metabolism, and cell growth control. Mol. Cell 49, 379–387.

CHAPTER SEVEN

Functions and mechanisms of symbionts of insect disease vectors

Kevin J. Vogel[a], Kerri L. Coon[b]
[a]Department of Entomology, The University of Georgia, Athens, GA, United States
[b]Department of Entomology, The University of Wisconsin-Madison, Madison, WI, United States

Contents

Abstract

Insects are the most significant vectors of human disease. Disease-transmitting insects often rely on vertebrate blood (haematophagy) either as a sole food source or as a significant component of their diet. While blood is a nutrient-rich resource, it also lacks key

Advances in Insect Physiology, Volume 58
ISSN 0065-2806
https://doi.org/10.1016/bs.aiip.2020.03.004

vitamins necessary for host development. Many insects which feed obligately on blood rely on bacterial symbionts to overcome this dietary limitation. Microbes associated with vectors have also been demonstrated to play essential roles in host development, immune system modulation, and vector competency. These microbes range from irreversibly specialized symbionts that are vertically transmitted from mother to offspring, to facultatively symbiotic, free-living microbes acquired from the local environment. Some may be found at variable frequencies and yet still shape numerous aspects of host physiology. Here we focus our attention on three groups of insect disease vectors and their varied associations with symbiotic microbes. This review will examine tsetse flies (*Glossina*), vectors of *Trypanosoma* species responsible for human and animal African trypanosomiasis, kissing bugs (Triatominae), the vectors of *T. cruzi* that causes Chagas' disease, and mosquitoes (Culicidae), vectors of a wide array of pathogens from *Plasmodium* to arboviruses. We explore the rapidly growing understanding of how microbes contribute to the development, physiology, and pathogen transmission of these insects.

1. Introduction

Five orders of insects contain members that feed on vertebrate blood including Diptera (flies), Hemiptera (true bugs), Psocodea (lice), Siphonaptera (fleas), and Lepidoptera (moths and butterflies) (Lehane, 2005). Haematophagy evolved between 6 and 18 times in insects, and several more times in other metazoans (Husník et al., 2011; Lehane, 2005; Wiegmann et al., 2011). The distribution of this habit is highly uneven among these orders, e.g. with 11 families of Diptera including haematophagous species, and only a single genus of Lepidoptera (*Calyptra*) being known to feed on vertebrate blood (Zaspel and Hoy, 2008). Blood feeding insects can be broadly divided into two groups based on their dietary fidelity. The obligately haematophagous insects consume blood exclusively, while facultatively haematophagous insects consume blood along with other nutrients or only during certain periods of development. Blood feeding in facultative lineages is often but not always limited to adult females and associated with reproduction. In these groups, blood proteins and lipid are converted into yolk proteins, which are packaged into developing oocytes. In contrast, both sexes of the obligately haematophagous groups rely on blood as a sole or primary nutrient source throughout development, though it also plays a major role in the production of eggs.

Feeding on vertebrate blood has both benefits and drawbacks for arthropods. The primary advantage of blood is its high protein content, though it also contains significant amounts of lipids and cholesterol, which are common limiting nutrients in insect food sources. Given the relatively modest number of species (~14,000) that have evolved this feeding habit (Lehane, 2005), it would appear that blood feeding presents a number of evolutionary hurdles. First is feeding on a host, which requires specialized mouthparts to pierce host skin and consume blood while avoiding retaliation by the host. Salivary proteins from the insect modulate a number of physiological processes to facilitate blood feeding, including suppression of host clotting and immune responses (Abdeladhim et al., 2014; Santiago et al., 2020; Valenzuela et al., 2003). Second, most blood feeding insects take large meals that may dramatically increase the weight of the insect. Many species have evolved highly efficient mechanisms to eliminate and expel excess water from the blood meal via diuresis (Piermarini, 2016). This diuresis also concentrates salts and urea in the blood meal, which requires additional detoxification mechanisms (Pennington et al., 2003). Digestion of blood itself releases large amounts of haeme-containing molecules. Haeme, if left unmodified, creates reactive oxygen species (ROS) which are toxic to the insect, leading different species to evolve various means to sequester or detoxify haeme (Graça-Souza et al., 2006). In addition to these costs of blood feeding, the acquisition of pathogens from the vertebrate host is a potentially significant cost to the insect. While much attention is rightly paid to the consequences of insect-vectored diseases in vertebrate hosts, these pathogens often impart costs to their insect vectors as well. Blood-acquired parasites may reduce host fitness in a variety of ways including disrupted feeding, immune activation, tissue damage, and reduced fecundity (Ahmed et al., 2002; Peterson et al., 2015). These hurdles have been repeatedly overcome not only by evolution of novel host solutions to the consequences of haematophagy, but also by associations with microorganisms.

Although haematophagous insects are widely known for their ability to transmit pathogenic microorganisms to humans and other vertebrates, these insects maintain a diverse array of relationships with microbes beyond these vectored pathogens. The roles of these microorganisms vary from beneficial: promoting host development, nutrient provisioning, and pathogen blocking; to detrimental: manipulating reproduction and inducing disease. In this review we focus on the beneficial interactions between three groups of haematophagous insects and their microbiota: tsetse flies, triatomine kissing bugs, and mosquitoes (Fig. 1). Tsetse flies and triatomine kissing bugs are obligate blood feeders which have evolved distinct relationships with their

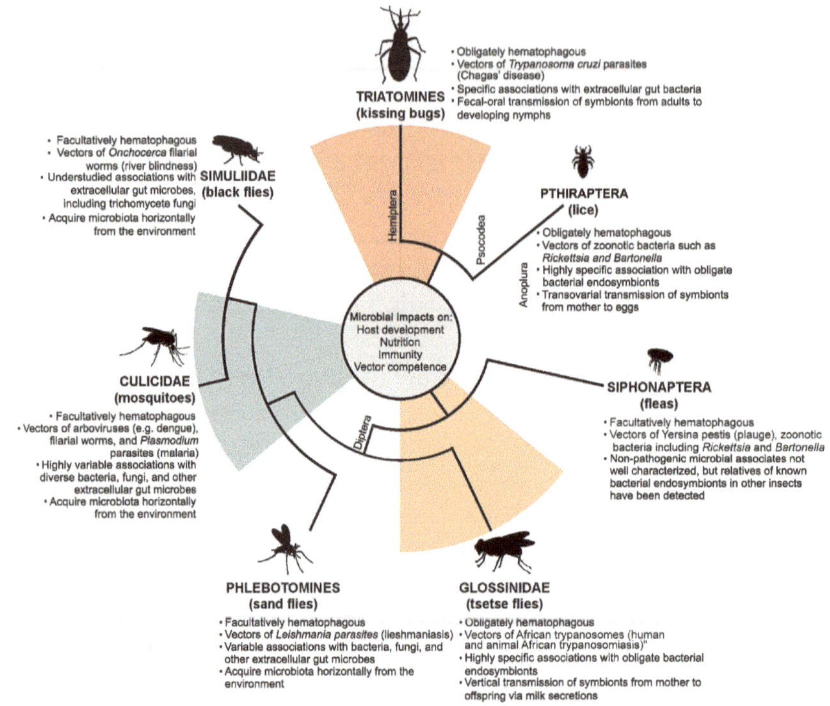

Fig. 1 Circular cladogram showing the relationships between major insect disease vectors, including the groups of focus in this review (highlighted in red, yellow, or blue). Information regarding each group's life history, primary vectored pathogens, and associations with non-pathogenic microbes is also provided.

symbiotic microbes, while mosquitoes are facultative haematophages engaging in well-studied relationships with opportunistic symbionts. We will consider the function of these symbionts in host development and homeostasis as well as their role in shaping the vectorial capacity of these insects.

2. Obligately haematophagous insects and their symbionts

The use of blood as a primary or exclusive food source has evolved in three orders of insects—Diptera (four families), Hemiptera (three families), and Psocodea (one suborder) (Lehane, 2005; Wiegmann et al., 2011). These insects generally require a blood meal to complete each juvenile moult, and may require an additional blood meal for each cycle of reproduction. This diet provides insufficient B vitamins to support insect development and reproduction (Brecher and Wigglesworth, 1944). Since insects are incapable of *de novo* B vitamin synthesis, haematophagous species rely on microbial partners for these essential nutrients.

For most obligately haematophagous insects, their bacterial partners are the archetypical endosymbionts—ancient, metabolically-integrated bacteria that reside in specialized host cells (Buchner, 1965). The genomes of these bacteria have been degraded through genetic drift and elevated mutation rates (Akman et al., 2002; Hosokawa et al., 2010). Many genes once thought to be necessary for bacteria are absent in such symbionts; consequently, these bacteria are unable to grow outside their specialized host environments and rely on maternal transmission to offspring. This mode of transmission results in severe population bottlenecking, reducing the effective population size of the symbiont and consequently elevating the fixation of deleterious mutations (Moran, 1996). Despite encoding a fraction of the genes of free-living bacteria, these symbionts have independently maintained genes necessary for B vitamin biosynthesis while having lost a number of genes common to most free-living bacteria (Akman et al., 2002; Boyd et al., 2017; Hosokawa et al., 2010; Nováková et al., 2015). The commonalities seen in genomes of the obligate symbionts of insects such as lice, bedbugs, and tsetse flies represent a stunning case of convergent evolution.

3. Glossinidae

3.1 Hippoboscoidea, a group of unusual dipterans

The superfamily Hippoboscoidea are a monophyletic group encompassing four families of obligate, blood feeding, ectoparasitic flies: tsetse flies (Glossinidae), louse flies (Hippoboscidae), and bat flies (Streblidae and Nycteribiidae). Molecular phylogenetic studies suggest that the Glossinidae are an outgroup to the remaining families, which constitute the monophyletic Pupipara. Glossinidae are large flies found throughout sub-Saharan Africa. They rely on sight and olfaction to identify mammalian hosts from which they feed (Buxton, 1955; Petersen et al., 2007).

Though they exhibit varied dispersal and host seeking, the Hippoboscoidea are united by their obligate haematophagy and the evolution of adenotrophic viviparity, in which a single larva develops entirely within the mother's uterus and is supported through the production of protein—and phospholipid-rich "milk" produced by the insect's milk glands (Benoit et al., 2015). This milk supports the development of a single larva within the female's uterus. Fully developed larvae are then deposited in surrounding soil and immediately pupate. In some groups including the Glossinidae and most Hippoboscidae, these pupae eclose into adult stages that conform to a recognizable dipteran body plan. Other groups, most notably the bat flies but also some hippoboscids, lack wings and halteres, and have highly developed legs which they use to attach to their hosts, where they remain until death or forced removal.

The monogeneric Glossinidae contains approximately 23 species exclusively found in sub-Saharan Africa. They feed solely on blood from a wide variety of mammalian hosts and occupy a variety of habitats as adults. Of these, members of the *Glossina morsitans* species group are the most economically significant. Their transmission of animal African trypanosomiasis, caused by *Trypanosoma vivax*, *T. congolense* and *T. brucei*, has impeded farming of cattle, pigs, and other livestock in a large area of sub-Saharan Africa and is considered a major factor limiting economic development in this region (Lehane, 2005). The causative agent of human African trypanosomiasis, or sleeping sickness, *Trypanosoma brucei*, is transmitted by several *Glossina* species, but members of the *G. palpalis* species group are the most medically significant vectors. Along with transmission by tsetse flies, *T. brucei* may also be transmitted through the bites of other blood feeding flies (Tabanidae and *Stomoxys* sp.; Desquesnes and Dia, 2004; Mihok et al., 1995). Both the human- and animal-infecting *Trypanosoma* sp. are ingested by the fly during a blood meal. In the gut, they differentiate before traversing the midgut epithelium as infective metacyclic trypanosomes that invade the fly's salivary glands (*T. brucei*) or the proboscis (*T. congolense*). Parasites are then transmitted to new hosts through the fly's bite (Gibson et al., 2017). Though significant gains have recently been made in limiting human African trypanosomiasis, sylvatic transmission of the parasite among wild animal hosts may subvert efforts to completely eliminate the disease (Mehlitz and Molyneux, 2019).

3.2 The obligate symbiont of tsetse flies, *Wigglesworthia glossinidia*

Two symbionts of tsetse flies have received a large amount of attention within this group: the obligate symbiont *Wigglesworthia glossinidia*, found in all tsetse flies, and the facultative symbiont *Sodalis glossinidius,* with variable distribution across species and populations. Both are found in a bacteriome organ composed of enlarged cells lining the anterior midgut (Rio et al., 2012). They are also found extracellularly in host milk secretions (Ma and Denlinger, 1974; Pais et al., 2008), which facilitate their transmission to developing offspring (Rio et al., 2012). The facultative symbiont *S. glossinidius* is further found in the host midgut and occasionally in the haemolymph (Cheng and Aksoy, 1999).

Wigglesworthia is often compared to the obligate symbiont of plant phloem-feeding aphids, *Buchnera aphidicola*. Both are obligate nutritional symbionts that reside in specialized host tissue and have genomes of similar size (~400–700 kb, Chong et al., 2019; Rio et al., 2012). Genome sequencing studies have demonstrated that *Wigglesworthia* is subject to strong population bottlenecks, strong genetic drift, and accelerated sequence evolution—the same mixture of demographic processes, dominant evolutionary forces, and rates of change that characterize *Buchnera* (Akman et al., 2002) and symbionts of other sap-feeding insects (McCutcheon and Moran, 2012; Moran, 1996). When coupled with a deletional mutation bias in bacterial genomes (Mira et al., 2001), the relatively large influence of drift relative to purifying selection has dramatically reduced genome size for haematophages' symbionts, with those from *Wigglesworthia* (i.e. 703–719 kb) comprising a small fraction of sizes seen among related free-living Gammaproteobacteria (Akman et al., 2002; Rio et al., 2012). While these symbionts have clearly lost many of the genes present in free-living Gammaproteobacteria, their genomes are still larger than those of many other obligate symbionts (McCutcheon and Moran, 2012), displaying fewer of the bizarre features characterizing the most highly eroded genomes (Lukasik et al., 2018; McCutcheon et al., 2009).

Though some of the differences between *Wigglesworthia* and *Buchnera* are likely due to differential selection pressures owing to the particular nature of the symbioses, some of these may be consequences of the younger age of the *Wigglesworthia*-*Glossina* relationship (50–80 my; Rio et al., 2012) relative to the *Buchnera*-Aphididae relationship (160–280 my; Moran et al., 1993). Genome sequences of *Wigglesworthia* from different *Glossina* hosts have additional features distinguishing them from genomes of other, more ancient endosymbionts. Comparisons of the two sequenced genomes from hosts *Glossina morsitans* and *G. brevipalpis*, revealed a major genomic rearrangement, something not observed in *Buchnera* or other obligate intracellular symbionts (Chong et al., 2019; Rio et al., 2012). *Wigglesworthia* also appears to be more transcriptionally dynamic than *Buchnera* (Bing et al., 2017; Munoz et al., 2017; Rio et al., 2012; Wilcox et al., 2003), exhibiting, for instance, stage-specific expression of flagellar proteins of hypothesized importance in facilitating maternal transmission. In contrast to other well-studied bacteriocyte-associated symbionts, *Wigglesworthia* encodes genes necessary for synthesis of cell wall components including lipopolysaccharide and peptidoglycan. The presence of these genes likely allows *Wigglesworthia* to persist in multiple locations within its host. This symbiont is not enveloped in a host membrane within the cytosol of *Glossina* cells, as seen in many other obligate symbionts (Rio et al., 2012). Furthermore, unlike *Buchnera*, which is only found outside host cells during early embryogenesis and egg formation (Braendle et al., 2003), *Wigglesworthia* is found extracellularly in the milk secretions of the mother and is not transovarially transmitted (Cheng and Aksoy, 1999; Ma and Denlinger, 1974; Pais et al., 2008). Presumably, this extracellular stage requires a functional cell wall and has strengthened selection to retain these genes.

Inferences from the genomes of *Wigglesworthia* have supported this symbiont's long-hypothesized role: permitting obligate haematophagy through the synthesis and provisioning of B vitamins that are scarce in blood diets. In particular, the genomes from both strains encode machinery for synthesis of vitamins B_1, B_2, B_5, B_6, and B_7 (Akman et al., 2002; Rio et al., 2012). These findings support experimental evidence of *Wigglesworthia*'s role wherein removing the bacteria from the insect with antibiotics sterilizes the insect . Some of the effects of symbiont removal can be ameliorated by provisioning the fly with either B vitamin mixtures or *Wigglesworthia* extracted from untreated flies (Michalkova et al., 2014; Pais et al., 2008). Comparative genomics between *Wigglesworthia* of *G. morsitans* and *G. brevipalpis* demonstrate that B vitamin synthesis varies between strains, with the symbiont of *G. morsitans* retaining the ability to synthesize folate (B_9) from chorismate, while *Wigglesworthia* of *G. brevipalpis* lacks the genes required for folate synthesis (Snyder and Rio, 2015). A number of hypotheses have been put forward as to how *G. brevipalpis* acquires folate, though none have been explicitly tested.

Recent work has begun to illuminate how *Wigglesworthia* influences *Glossina* biology beyond direct nutrient provisioning. Weiss et al. (2011, 2012) examined the role of *Wigglesworthia* in the development of the host immune system and in shaping responses to pathogen challenges. Selective clearance of *Wigglesworthia*, but not *Sodalis*, resulted in an impaired immune response to *Escherichia coli* when aposymbiotic flies cured of *Wigglesworthia* were injected with the *E. coli*. Reduced expression of antimicrobial peptides (AMPs) and prophenoloxidase (PPO) was seen in the flies without *Wigglesworthia* when subjected to *E. coli* challenge relative to controls. Furthermore, the loss of *Wigglesworthia* also impaired phagocytosis of *E. coli*, suggesting that the impact of aposymbiosis on the host affects a wide range of host immune processes. Transplantation of haemocytes or *Wigglesworthia* extracts from wild-type *G. morsitans* successfully restored immune function in aposymbiotic flies (Weiss et al., 2013, 2019). The host immune system also appears to influence *Wigglesworthia*, as a fly immune protein, PGRP-LB is highly abundant in tsetse milk. This amidase-containing protein is hypothesized to degrade *Wigglesworthia* peptidoglycan, preventing the bacteria from stimulating a host immune response in the milk gland. Knockdown of PGRP-LB, by RNAi results in increased synthesis of AMPs, decreased titres of *Wigglesworthia*, and reduced fecundity (Wang and Aksoy, 2012; Wang et al., 2009). The obligate symbiont was also shown to be essential for proper maturation of the melanization-related immune response of *G. morsitans*. Clearance of *Wigglesworthia* from the fly resulted in downregulation of an odorant binding protein, OBP6. Silencing of this gene during larval development resulted in increased mortality of adult flies challenged with *E. coli*. This phenotype was determined to result from loss of crystal cells during development, which are necessary for proper melanization. Without these cells flies could not successfully melanize wounds and died at high rates from simple needle punctures (Benoit et al., 2017). While annotated as an odorant binding protein, olfactory-related genes have been identified as important for haematopoiesis in other insects (Shim et al., 2013; Thomas et al., 2016).

Both the immune and nutritional function of *Wigglesworthia* may play important roles in the ability of tsetse flies to transmit *T. brucei*. One of the few major differences between *Wigglesworthia* from *G. morsitans* and *G. brevipalpis* is that the *G. morsitans* symbiont encodes the ability to synthesize folate (B_9) from chorismate (Rio et al., 2012). Snyder and Rio (2015) tested whether inactivation of chorismate biosynthesis in *Wigglesworthia* had significant effects on infection susceptibility of *G. morsitans* with *T. brucei*. Their results demonstrate that folate production by *Wigglesworthia* has a large role in the success of the auxotrophic *T. brucei* to establish in tsetse flies. Further work has shown that *G. brevipalpis*, which harbours a *Wigglesworthia* strain unable to produce folate, is a less efficient vector of *T. brucei* than *G. morsitans*. Experimental supplementation of *G. brevipalpis*' diet with folate increases the proportion of flies with *T. brucei* infection relative to non-supplemented control flies, further establishing the important role of symbiont-produced folate in the vector competency of tsetse flies (Rio et al., 2019).

The immune response of tsetse flies has been shown to vary when challenged with *T. brucei* depending on age and presence of *Wigglesworthia*. *T. brucei* is most successful at colonizing recently-eclosed flies relative to older flies, likely due to increased immune effector expression in older flies (Weiss et al., 2013). In *G. morsitans* experimentally deprived of their *Wigglesworthia*, this adult refractoriness to *T. brucei* infection is lost, possibly due to degradation of the peritrophic matrix lining the adult gut and reduced expression of AMPs in older adults (Pais et al., 2008; Weiss et al., 2013). Together with nutritional provisioning experiments, a clear picture has emerged that *Wigglesworthia* is essential for host development and reproduction while also a significant contributor to its vectorial capacity.

3.3 *Sodalis*, a secondary symbiont influencing vector competence

In addition to *Wigglesworthia*, many *Glossina* species harbour a secondary facultative symbiont, *Sodalis glossinidius*, the prevalence of which varies among populations of tsetse flies. Members of this genus have repeatedly and independently established as symbionts of insects (Boyd et al., 2017; Koga and Moran, 2014; Santos-Garcia et al., 2015), including as a symbiont of the bat fly *Melophagus ovinus* (Chrudimský et al., 2012). In contrast to many symbiotic bacterial lineages, a closely related, free-living member of this genus, *S. praecaptivus,* has also been identified and cultured (Clayton et al., 2012). Several species of *Sodalis* have been subject to genome sequencing, revealing a fascinating continuum of symbiont integration with the host. In some insects, *Sodalis* has repeatedly been captured as an obligate symbiont, including in the spittlebug *Philaenus spumarius*, the lygaeoid *Henestaris halophilus*, and the seal louse *Proechinophthirus fluctus*. These *Sodalis* are bacteriocyte-associated, have reduced genome size, limited pseudogenes, and no evidence of active mobile genetic elements (Boyd et al., 2017; Koga and Moran, 2014; Santos-Garcia et al., 2015). In contrast, the *Sodalis* associated with both *Glossina* species and *Sitophilus oryzae* weevils have typical bacterial genome sizes (4.0–4.5 Mb) but also abundant pseudogenes and active mobile genetic elements (Belda et al., 2010; Oakeson et al., 2014). *S. glossinidius* exhibits a very low coding density, harbouring over 2200 pseudogenes comprising upwards of 50% of the genome (Belda et al., 2010; Goodhead et al., 2020). Several pseudogenes are dynamically expressed *in vitro* (Goodhead et al., 2020), suggestive of a recently captured symbiont in the early stages of genome degradation. Despite these abundant pseudogenes, computational analyses suggest that *S. glossinidius* has not lost major metabolic pathways necessary for life outside the host (Belda et al., 2012). Such findings suggest that these symbionts are subject to the same evolutionary pressures that lead to symbiont genome erosion (Moran, 1996), although *Sodalis* from *Glossina* and *S. oryzae* are likely at an earlier stage of symbiotic integration than other members of the genus.

S. glossinidius varies in abundance among *Glossina* species and geographic locations, being fixed in some locations (Dennis et al., 2014; Farikou et al., 2010; Maudlin et al., 1990; Tagueu et al., 2018; Wamwiri et al., 2013). Epidemiological evidence supports a role for *S. glossinidius* in the vector competency of *Glossina* sp. Multiple surveys have reported a positive association between *T. brucei* infection and presence of *Sodalis* in the fly (Doudoumis et al., 2017; Geiger et al., 2007; Griffith et al., 2018), although other studies have found no association (Channumsin et al., 2018; Odeniran et al., 2019). Mechanistically, *S. glossinidius* is hypothesized to support *T. brucei* infection through modulation of the fly's immune response to the parasite. Digestion of fly chitin by *S. glossinidius* produces *N*-acetyl-D-glucosamine, a sugar that interferes with a host lectin with trypanocidal activity (Dale and Maudlin, 1999; Dale and Welburn, 2001; Hall et al., 2019; Welburn et al., 1993).

Recent studies have established lab populations of *G. morsitans* which have been cured of *S. glossinidius* (Trappeniers et al., 2019). Studies on these flies revealed that they elicited a strong immune reaction to challenges with both *E. coli* and the free-living relative of *S. glossinidius*, *S. praecaptivus*. In contrast, when challenged with re-infection of *S. glossinidius*, a minimal immune reaction was observed. The lack of an immune response to the symbiont suggests that either the host immune system does not recognize *S. glossinidius* or that the immune response is suppressed. This may be due to the observed resistance of *S. glossinidius* to host PGRP-LB, a highly abundant immune protein in adults and in tsetse milk (Trappeniers et al., 2019).

The ability to culture *S. glossinidius in vitro* has prompted interest in exploiting the symbiont for paratransgenic control of trypanosomiasis (Dale and Maudlin, 1999). The bacterium has been shown to be amenable to transformation with both green fluorescent protein (Cheng and Aksoy, 1999) as well as nanobody fragments (i.e. single-domain antibodies) targeting *T. brucei* (De Vooght et al., 2012, 2015). The ability of recombinant *S. glossinidius* to be transmitted through both males and females likely increases the potential of a paratransgenic strain to spread through a population, an important factor for any such strategy (De Vooght et al., 2015, 2018). Currently applied technologies to control *Glossina* populations include the sterile insect technique (SIT), in which radiation is used to produce flies incapable of producing viable offspring. An intriguing study demonstrated that *S. glossinidius* can survive radiation treatment of fly pupae and recover to wild-type levels, suggesting that SIT and paratransgenesis could be paired to improve control of *Glossina* transmission of *T. brucei* (Demirbas-Uzel et al., 2018).

3.4 Future directions

Symbionts of tsetse flies play important nutritional, immunological, and developmental roles in the biology of the host. *Wigglesworthia* is key to the ability of tsetse flies to transmit trypanosomes, while *Sodalis* appears to play a more complicated role in trypanosome transmission. Although transmission of human African trypanosomaisis and animal African trypanosomiasis has been dramatically reduced in recent years (Franco et al., 2014), sylvatic transmission of the pathogens may hamper eradication of these diseases (Cordon-Obras et al., 2015). Paratransgenic approaches leveraging *Sodalis* to control human and animal African trypanosomaisis are appealing, especially in limiting sylvatic transmission of the parasites. The success of such approaches will be contingent upon a deep understanding of the interactions between *Sodalis*, *Wigglesworthia*, the insect host, and parasite.

4. Triatomines: Vectors of *Trypanosoma cruzi*

Triatomines, colloquially known as kissing bugs, are a group of large, obligately blood feeding hemipterans that evolved from predatory Reduviidae an estimated 35–40 million years ago (Monteiro et al., 2018). Evidence suggests that these insects began feeding on blood after establishing a close association with vertebrate nests, likely due to the stable environment and abundance of prey arthropods (Schofield, 2000). The piercing mouthparts of their insectivorous ancestors may have been easily coopted for blood feeding. The promiscuous probing behaviour of closely related predatory reduviids suggests that an ancestral, nest-associated reduviid may have discovered an abundant, nutrient-rich food source waiting to be tapped (Monteiro et al., 2018). The group has subsequently evolved into some 130 species with most diversity in the neotropics.

Triatomine kissing bugs are notable vectors of *Trypanosoma cruzi*. The parasite and insect were identified by Carlos Chagas as the organism responsible for Chagas' disease in 1909. This neglected tropical disease continues to be a significant pathogen of humans, infecting an estimated 8 million people and resulting in 10,000 deaths annually (WHO, 2015). *T. cruzi* is acquired by kissing bugs through ingestion of an infected blood meal. The parasite then differentiates in the midgut, eventually migrating to the hindgut, developing into infectious parasites that are excreted along with the faeces. The parasites then enter the vertebrate host through skin lesions or mucus membranes, where they infect host cells and tissues. The parasites complete their life cycle in the vertebrate host, where they are present in the host's bloodstream, ready to be consumed by feeding triatomines.

Vector competence varies among triatomines likely due to differences in host use and behaviour. Several species defecate on the host during or immediately after feeding, increasing the opportunities for the parasites to enter the host through the bite wound or nearby mucous membranes such as the mouth (Reisenman et al., 2011; Wood, 1951). Effective vectors also maintain a peridomestic habitat, feeding in and around human domiciles, whereas several other species that are not important vectors display a strongly sylvatic habitat (Montgomery et al., 2016). Multiple other factors, including the blood meal source, the abiotic environment (humidity, temperature), the host immune system, gut microbiota, and genetic variation within the parasite appear to have large effects on vector competence (de Fuentes-Vicente et al., 2018).

4.1 Early studies: *Rhodococcus rhodnii* and other symbionts

Unlike other obligately haematophagous insects, members of the Triatominae lack the intracellular, strictly vertically transmitted symbionts seen in other groups. Rather, these insects maintain associations with bacteria in their midgut. Early studies on the model kissing bug *Rhodnius prolixus* identified these as a *Bacillus* (Duncan, 1926), which were subsequently re-classified as *Nocardia rhodnii*, and finally as *Rhodococcus rhodnii*. Several studies (Brecher and Wigglesworth, 1944; Wigglesworth, 1936) established that this symbiont was not transferred within the eggs of the insect but rather via coprophagy of faeces by nymphs, as removal of eggs from adult habitats followed by surface sterilization and rearing in pristine containers produced aposymbiotic nymphs lacking *Rhodococcus rhodnii* (Brecher and Wigglesworth, 1944). Wigglesworth noted, though no evidence was provided, that these bacteria were also found in the guts of *Triatoma rubrofasciata*, *Triatoma infestans*, and *Eutriatoma flavida* (Wigglesworth, 1936). While aposymbiotic nymphs initially displayed minimal changes in development during the early instars compared to conventional nymphs, they exhibited dramatically lengthened durations of later instars, ultimately failing to moult to adulthood (Baines, 1956; Brecher and Wigglesworth, 1944; Harington,

1960). Thus, it was established that *Rhodnius prolixus* and later *Triatoma infestans* (Goodchild, 1955) were reliant on these gut bacteria for successful development to adulthood.

The functional role of *Rhodococcus rhodnii* for *Rhodnius prolixus* development was initially hypothesized to be provisioning of B vitamins as was suggested for other symbionts of obligate haematophagous insects (Wigglesworth, 1926, 1936). Demonstrating this proved difficult and several approaches were used, sometimes producing conflicting results. Experiments injecting mice with B vitamin solutions then feeding aposymbiotic *Rhodnius prolixus* on these mice suggested that loss of any B vitamin was detrimental to *Rhodnius prolixus* development (Baines, 1956). Removal of pantothenic acid (B_5) had the most significant impact on developmental times while removal of folate (B_9) or cobalamin (B_{12}) from the vitamin cocktail injected into the mice had the smallest effect on development (Baines, 1956). Later studies produced *Rhodococcus rhodnii* strains that were auxotrophic for various B vitamins. When reared with an auxotrophic lineage of *Rhodococcus rhodnii*, *Rhodnius prolixus* hosts developed significantly faster than when reared aposymbiotically. From this, the authors' concluded that the bacteria are likely providing some additional service to the host beyond vitamin provisioning (Hill et al., 1976).

Early studies suggested a monotypic association of *Rhodnius prolixus* and *Rhodococcus rhodnii* (Bewig and Schwartz, 1954, 1955; Brecher and Wigglesworth, 1944; Duncan, 1926; Lake and Friend, 1967; Wigglesworth, 1936), but others found distinct microbes in the guts of other species of kissing bugs and often times multiple species of bacteria were found within an individual or population (Cavanagh and Marsden, 1969; Goodchild, 1955; Gumpert and Schwartz, 1962; Marchette and Hatie, 1965; Nyirady, 1973; Weurman, 1946). Cultivated gut bacteria were typically members of the Actinobacteria, with multiple isolations of *Corynebacterium* spp. by independent researchers (Goodchild, 1955; Gumpert and Schwartz, 1962). These studies pre-dated the use of culture-independent methods to identify and enumerate bacteria, which potentially resulted in underestimating the diversity of bacteria or overstating the association of specific bacteria and kissing bugs.

Both *Wolbachia* and *Arsenophonus* have been identified in triatomine bugs. These bacteria are distributed widely across diverse arthropods and can behave as both reproductive manipulators and also as nutritional symbionts (Engelstadter and Hurst, 2009; Hosokawa et al., 2010). *Arsenophonus triatomae* was one of the earliest microbes identified via molecular methods (Hypša and Dale, 1997), though an early microscopy study identified a transovarially transmitted, intracellular, Gram-negative bacterium in *Triatoma infestans* that may have been *A. triatomae* (Hypša, 1993). Though no function has yet been ascribed, this symbiont has repeatedly been found in both wild and lab colonies of different kissing bug species (Diaz et al., 2016; Hernández-Rodríguez, 2006; Lima et al., 2018; Šorfová et al., 2008). Interestingly, *A. triatomae* was initially described as having the highest abundance in glial cells of the central nervous system along with the salivary glands (Hypša and Dale, 1997). The presence of *A. triatomae* in the salivary glands of *Triatoma infestans* was recently confirmed through 16S rRNA gene sequencing (Lima et al., 2018), but the significance of this bacteria to the biology of the insect remains unclear. *Wolbachia* has also been detected in several surveys of the microbial communities within various kissing bugs (Espino et al., 2009; Kieran et al., 2019; Waltmann et al., 2019), but no evidence of its effect on triatomines has been reported to date.

4.2 Next generation sequencing and the microbiome of Triatominae

The application of culture-independent methods revealed a previously hidden diversity of gut microbes in kissing bugs. The earliest study applied restriction-fragment length polymorphism (RFLP) or denaturing gel gradient electrophoresis (DGGE) to study kissing bug microbiota in a culture-independent fashion and revealed a more complex picture of the microbiota (da Mota et al., 2012). Notably, this study failed to recover *Rhodococcus rhodnii* from *Rhodnius prolixus*, finding that its microbiome was dominated by *Serratia marcescens*. Due to the limitations of DGGE, it is difficult to definitively say whether the absence of *Rhodococcus rhodnii* from the samples is an artefact of the method or reflects a true absence of *Rhodococcus rhodnii*. The authors found evidence of *Rohrkolberia cinguli*, an obligate intracellular symbiont of the bull rush bug *Chialis typhae*, in the kissing bugs *Dipetalogaster maximus* and *Rhodnius prolixus*. Similar to earlier studies, the authors recovered *A. triatomae* from samples of *Triatoma infestans* along with *S. marcescens*.

Large-scale sampling of kissing bug microbiota via 16S rRNA sequencing with next-generation technologies has continued to reveal a diverse, variable community within kissing bugs. These studies have often, but not always, identified members from the phylum Actinobacteria as abundant members of the gut community, especially in species within the genus *Rhodnius* but also other triatomines (Carels et al., 2017; Diaz et al., 2016; Dumonteil et al., 2018; Gumiel et al., 2015; Montoya-Porras et al., 2018;

Oliveira et al., 2018; Waltmann et al., 2019). Within this phylum, bacteria within the Nocardiaceae were prevalent, including those from the genus *Rhodococcus*, suggesting this group to be common symbionts in kissing bugs. Firmicutes have also been recovered as abundant or dominant members of the microbiome in multiple studies, especially members of the genus *Enterococcus* (Oliveira et al., 2018; Rodríguez-Ruano et al., 2018). Other groups frequently observed include *Serratia marcescens* (Azambuja et al., 2017; da Mota et al., 2012) and various other Proteobacteria (Montoya-Porras et al., 2018). *Arsenophonus* is commonly found in *Triatoma* species, and *Wolbachia* is found at variable frequencies among populations (Espino et al., 2009).

The function of these microbes in kissing bug biology remains understudied. Nutritional provisioning of B vitamins is generally accepted as the major function of kissing bug microbiota and is supported by several aforementioned studies (Baines, 1956; Hill et al., 1976; Lake and Friend, 1968; Nyirady, 1973). However, these studies were limited in their ability to isolate effects of B vitamin deficiency due to symbionts as well as their ability to supplement B vitamins in the diet. Using strains of *Rhodococcus rhodnii* that were auxotrophic for individual B vitamins, Hill et al. (1976) found that *Rhodnius prolixus* inoculated with the auxotrophic strains developed faster than aposymbiotic *Rhodnius prolixus* and from this concluded that the role of symbionts in *Rhodnius prolixus* development is likely more than B vitamin provisioning. Furthermore, several studies indicated that the absence of bacteria had a minimal effect on development of *Rhodnius prolixus* when reared on germ-free mice rather than rabbits (Auden, 1974; Baines, 1956; Nyirady, 1973). This could potentially be explained by differences in serum B vitamin levels between mice and rabbits, although this was not addressed. While vitamin synthesis may not be the sole function of *Rhodococcus rhodnii* in kissing bugs, the genome sequence of the bacteria reveals that it does encode the ability to synthesize all B vitamins de novo (Pachebat et al., 2013). Blood meal digestion appears to be somewhat affected by absence of microbes (Gumpert and Schwartz, 1962), though when assayed, *Rhodococcus rhodnii* appears to have minimal haemolytic properties (Vogel, unpublished).

Beyond nutritional provisioning, symbionts of many insects interact with the immune system of the host. In most insects, including mosquitoes and tsetse flies, a chitinous peritrophic matrix is present in the midgut and functions as a barrier to microbial entry into the haemocoel through the epithelial membrane (Aksoy, 2019; Valzania et al., 2018a). Hemipterans, including triatomines, lack this peritrophic matrix, instead having a lipid-protein perimicrovillar matrix separating the midgut lumen from the membrane surface of the midgut (Billingsley, 1990; Gutiérrez-Cabrera et al., 2016). The interaction of symbionts with the perimicrovillar matrix and how this shapes the host immune response to gut bacteria remains unknown, though changes in the thickness of this membrane during feeding and starvation (Gutiérrez-Cabrera et al., 2016) may allow more direct contact of symbionts with the host membranes.

While it has long been suggested that hemipterans lack many components of the IMD and Toll immune pathways (Zumaya-Estrada et al., 2018), recent evidence suggests that these pathways are more complete than previously assumed (Nishide et al., 2019; Salcedo-Porras and Lowenberger, 2019). Strong evidence has emerged that the triatomine immune system is shaped by and shapes the microbial community. Silencing of the IMD immune transcription factor *relish* by RNA-interference resulted in an increase in the titre of *Rhodococcus rhodnii* symbionts within *Rhodnius prolixus* (Mesquita et al., 2015) suggesting the host actively modulates its endogenous microbiome. Immune effector expression can be induced by feeding bacteria to triatomines, though the location (gut or haemolymph) and impacts of antimicrobial effects appear to be specific to different bacterial groups (Vieira et al., 2014). Further evidence that the host immune system interacts with the microbiota was provided by inhibiting the NF-kB pathway. Suppression of this immune pathway with the drug IMD-0354 resulted in decreased AMP expression and increased bacterial abundance in the anterior midgut (Vieira et al., 2018).

4.3 *Trypanosomes*, the microbiome, and paratransgenesis

A major focus of investigations into the microbiome of kissing bugs has been the interaction of the gut bacteria with parasites. Previously, it was widely believed that *T. cruzi* had little effect on the insect host. However, recent shown that depending on the strain of *T. cruzi* examined, the parasite can be pathogenic, reducing insect survival and slowing development (Peterson et al., 2015). In addition to *T. cruzi*, triatomines are hosts to other trypanosomatids, including *T. rangeli* and *Blastocrithidia triatomae*. While *T. rangeli* and *B. triatomae* parasites are not pathogenic to mammals, they do cause disease in *Rhodnius prolixus* and *Triatoma infestans* kissing bugs, respectively (Eichler and Schaub, 2002). Unlike *T. cruzi*, which only occupies the digestive tract of triatomines, *T. rangeli* and *B. triatomae* penetrate the midgut and infect the haemocoel, eventually establishing an infection in the salivary glands. From there they are transmitted through the bite of a triatomine to a vertebrate host, rather than through the faeces. Studies of *T. rangeli* and *B. triatomae* suggest that there is an antagonistic interaction between these parasites and gut symbionts, resulting in reduced symbiont titre (Eichler and Schaub, 2002). *T. cruzi* however, had no impact on the gut microbiota. It has been proposed that since these trypanosomatids are unable to produce B vitamins, they may rely on vitamin synthesis by symbionts for successful infection (Eichler and Schaub, 2002).

The obligate requirement of bacterial symbionts for kissing bug development along with the ability to culture and genetically manipulate *Rhodococcus rhodnii* has made this system a model for paratransgenesis. Initial efforts were able to transform *Rhodococcus rhodnii* with a shuttle plasmid that was maintained for up to six months without antibiotic selection (Beard et al., 1992). This was followed by the construction of an expression plasmid encoding the antimicrobial peptide Cecropin A, which was shown to be effective at killing *T. cruzi* in the insect gut. Transformed *Rhodococcus rhodnii* expressing the peptide were provided to insects via an artificial faeces formulation (Cruziguard), leading to the establishment of the transgenic *Rhodococcus rhodnii* in *Rhodnius prolixus*. Insects harbouring the transgenic symbiont were frequently, though not universally, cleared of *T. cruzi* (Durvasula et al., 1997). Further development of this paratransgenesis system included creation of a chromosomally-integrating vector for *Rhodococcus rhodnii* to ensure stability of the expression construct in the genome independently of antibiotic selection (Dotson et al., 2003) along with symbionts expressing alternative AMPs (Fieck et al., 2010). An alternative symbiont vector, a *Corynebacterium* capable of infecting other triatomines, was also transformed to express antibody fragments which could be designed to target *T. cruzi* (Durvasula et al., 2008).

4.4 Future directions

Detailed comparisons among microbiome studies of kissing bugs remains difficult due to variable study species, geographic locations, wild versus lab populations, and tissue sampling techniques (whole body versus gut). Despite this, several commonalities emerge. First is that previous culture-dependent methods used to study the microbiome failed to capture the true diversity of the gut microbiota. This is unsurprising given estimates that a large proportion of microbes have never been successfully cultured (Steen et al., 2019). Furthermore, culture-based efforts to sample the kissing bug microbiome often relied on one or a few types of media, further reducing the likelihood of capturing the total diversity present.

Second, the microbial communities are variable within and among species sampled at different geographic locations. This feature had been established prior to culture-independent studies (Goodchild, 1955; Nyirady, 1973) but is now well supported based on 16S rRNA gene surveys of different populations (Kieran et al., 2019; Rodríguez-Ruano et al., 2018). While it has long been hypothesized that coprophagy is the main route of transmission of triatomine microbiota, these 16S rRNA gene surveys suggest that environmental microbes are frequently acquired and lost by kissing bugs. The functional implications of this diversity remain unclear but suggest that multiple bacteria are capable of functioning as symbionts.

Third, the microbiome of triatomines changes across insect ontogeny, with diversity declining as the insects develop (Kieran et al., 2019; Rodríguez-Ruano et al., 2018). Several factors could explain the loss of diversity seen in later instars. The immune system appears to actively regulate microbial populations (Mesquita et al., 2015; Vieira et al., 2018), and this may select for specific, tolerant microbes that can then stably colonize the gut. Additional studies are necessary to illuminate the mechanisms that favour certain microbes over others during triatomine development.

Fourth, despite the variance in kissing bug microbiomes, several groups of bacteria are repeatedly recovered. These groups include Nocardiaceae (including *Rhodococcus rhodnii*), *Serratia marcescens*, *Enterococcus*, and various Proteobacteria (Azambuja et al., 2017; da Mota et al., 2012; Montoya-Porras et al., 2018; Rodríguez-Ruano et al., 2018; Waltmann et al., 2019). These results suggest that there is some degree of specificity to the community, but whether this is shaped by differential acquisition by the host or ability of the bacteria to colonize the gut remains unclear.

Finally, diversity is highest in insects sampled from wild populations as opposed to lab colonies (Waltmann et al., 2019). In wild populations, these insects may have increased opportunities to acquire microbes from the environment that are not encountered in a lab setting. Given the reduction in diversity during insect development, the limited diversity of lab populations may be a function of bottlenecking of the microbial population limiting which groups are likely to be transmitted to subsequent offspring. This may be particularly true in lab colonies maintained over many generations. In conjunction with an inability to characterize un-culturable bacteria, this could explain the early findings of monotypic associations of kissing bugs such as *Rhodnius prolixus* with *Rhodococcus rhodnii* (Brecher and Wigglesworth, 1944; Wigglesworth, 1936).

The emerging picture of the triatomine microbiome suggests much remains unknown. Are bacteria providing more than B vitamins to the insect? What bacterial factors are required for successful colonization? What host genes are involved in regulating symbionts? If strategies such as paratransgenesis are to be successfully deployed to reduce Chagas' disease transmission by triatomines, then a full understanding of the host, microbial, and environmental factors that govern community establishment and stability is essential.

5. Facultatively haematophagous insects

Facultative blood feeding is found in a number of groups, though most species are concentrated in the Diptera. These organisms often rely on blood for reproductive purposes: females consume blood and convert the nutrients into yolk for the production of eggs. In most of these insects, only females are haematophagous and only consume blood during the adult life stage of development. Interestingly, the facultative blood feeders are predominantly holometabolous, with larvae that are often aquatic or soil-dwelling and not host-associated. Molecular phylogenetic studies suggest that facultative haematophagy evolved at least 12 times independently within the Diptera (Wiegmann et al., 2011), with separate origins within Lepidoptera and Siphonaptera.

Unlike most obligately haematophagous insects, the symbiotic microbes of facultatively blood feeding insects appear to be environmentally acquired rather than vertically transmitted (Coon et al., 2014, 2016a; Díaz-Nieto et al., 2016). The microbiota is generally restricted to the gut, though *Wolbachia* and other reproductive manipulators are capable of infecting various tissues of the insects. The various impacts of these microbes in host biology are still being elucidated; but they have so far been implicated in development (Coon et al., 2014, 2017), nutrition (Chouaia et al., 2012; Díaz-Nieto et al., 2016; Mitraka et al., 2013; Wotton et al., 1997), and host-parasite interactions (Boissière et al., 2012; Cappelli et al., 2019; Dennison et al., 2014; Eappen et al., 2013). Although there are several groups of facultative blood feeders that are significant vectors of diseases affecting humans (e.g. Simuliidae and Phlebotiminae), the vast majority of work on interactions between symbionts and facultative microbes has been performed in the Culicidae and is the focus of the remainder of our review.

6. Functions of gut microbiota in the Culicidae

Perhaps the best-known insect vectors of disease are mosquitoes, which transmit the causative agents of many important human diseases, including dengue, malaria, chikungunya, yellow fever, and Zika. All mosquitoes belong to the family Culicidae (Order: Diptera), which to date contains about 3500 species and subspecies that are distributed throughout most temperate, tropical and/or subtropical regions of the world (Clements, 1992). It has long been known that both immature and adult stage mosquitoes host communities of microbes in their digestive tract (Chao and Wistreich, 1959; Ferguson and Micks, 1961; Hinman, 1930; Jones and DeLong, 1961; Rozeboom, 1935). The last few decades have seen the publication of numerous studies on mosquito gut microbiota, ranging from culture-dependent and culture-independent surveys of the composition and diversity of mosquito gut microbial communities to molecular studies on their function. Here, we provide a brief synthesis of our current knowledge on this topic, with an emphasis on more recent studies on the impact of gut microbiota on mosquito development, immunity, and vector competence.

6.1 Acquisition and transmission of mosquito gut microbiota

Mosquito larvae hatch from eggs deposited on or near the surface of standing water. Larvae moult through four aquatic instars before pupating and emerging as adults (Clements, 1992). Larvae feed on detritus, microorganisms, and small invertebrates in their aquatic habitat (Merritt et al., 1992), while adults of both sexes are terrestrial and feed on sugar sources (Foster, 1995). Adult female mosquitoes also must usually feed on blood from a vertebrate host to produce eggs, which is how mosquitoes acquire and transmit disease-causing organisms (Briegel, 2003). Culture-dependent and culture-independent approaches have been used to isolate and identify members of the gut microbiota from several mosquito species collected in the laboratory or field (Strand, 2017, 2018). Most of these studies have focused on the characterization of bacteria via 16S rRNA gene sequencing (Minard et al., 2013). These taxonomic surveys indicate that adult mosquitoes contain low diversity (<20 species) bacterial communities, which vary tremendously across species, among individuals, and as a function of collection site (Bascuñan et al., 2018; Boissière et al., 2012; Buck et al., 2016; Coon et al., 2014; Dickson et al., 2018; Duguma et al., 2015; Gimonneau et al., 2014; Muturi et al., 2016b, 2017, 2018; Osei-Poku et al., 2012; Tchioffo et al., 2015; Thongsripong et al., 2018; Wang et al., 2011, 2018; Zouache et al., 2011). In contrast, the bacteria present in larvae consistently overlap with the bacteria present in their aquatic environment, and bacterial diversity in larvae is consistently much higher (~200 species) than in adults (Bascuñan et al., 2018; Coon et al., 2014, 2016b; Gimonneau et al., 2014; Wang et al., 2018). Experimental data also suggest that a subset of the bacteria present in larvae is able to persist to the adult stage through transstadial transmission (Coon et al., 2014), although bacterial abundance is extremely low in newly emerged adults (Coon et al., 2016a; Moll et al., 2001). Newly emerged adults also immediately imbibe water from their larval habitat, which reseeds the gut with environmental microbes (Lindh et al., 2008a).

6.2 Roles in growth and development

Recent studies indicate that larvae of *Aedes aegypti* and several other mosquito species require living gut bacteria for successful development under normal conditions (Coon et al., 2014, 2016b, 2020; Valzania et al., 2018). The presence of gut bacteria reduces gut oxygen levels, which is associated with stabilization of specific transcription factors, called hypoxia-inducible factors (HIFs), in the midgut (Coon et al., 2017, 2020; Valzania et al., 2018a, b). In turn, HIF stabilization correlates with activation of several pathways with functions essential for growth and moulting, including the target of rapamycin, adenosine monophosphate-activated protein kinase, and insulin/insulin growth factor signalling pathways (Valzania et al., 2018a; Vogel et al., 2017). Recent studies also indicate that many different types of microorganisms individually induce gut hypoxia, HIF stabilization, larval growth, and moulting (Valzania et al., 2018b), and that several members of the gut microbiota in *A. aegypti* mosquitoes and *E. coli* can individually support the development of larvae into adults that are comparable in size and fecundity to conventional larvae colonized by a mixed bacterial community (Coon et al., 2014, 2016a).

Bacteria and other microorganisms can also serve as an important source of nutrition for developing larvae (Merritt et al., 1992; Romoli et al., 2019). Microbial abundance is lower in aquatic habitats where larvae are present (Kaufman et al., 1999; Walker and Merritt, 1991), and there are dramatic shifts in the composition and structure of bacterial and fungal communities in the water where larvae feed (Kaufman et al., 1999). Results from several studies indicate that growth and survival rates can vary substantially between larvae reared in the presence of different bacteria and other microorganisms (Coon et al., 2016a; Dickson et al., 2017; Valzania et al., 2018b). Microbiota composition present in the larval environment also strongly determines adult fitness traits such as individual body size, teneral reserves (nutritional reserves available upon emergence), fecundity, longevity, and vector competency (Coon et al., 2016a; Dickson et al., 2017). Recent studies with *A. aegypti* also report that yeast extract and heat-killed bacteria can promote larval growth, but only when provided at very high doses with other nutrient-rich dietary components in darkness (Correa et al., 2018). Thus, while the presence of living microbes in the gut is sufficient to induce gut hypoxia and moulting, certain microbial species and assemblages also appear to provide photolabile nutrients or other factors that improve mosquito fitness.

6.3 Roles in reproduction

While obligately haematophagous insects harbour symbionts that play critical roles in host development and reproduction by providing essential B vitamins that are deficient in the vertebrate blood they consume (see above), no mosquitoes are known to harbour obligate symbionts. However, some evidence suggests that the mosquito gut microbiota is involved in certain nutritional processes important for reproduction. First, feeding antibiotics to *A. aegypti* adult females reduces the abundance of bacteria in the midgut after a blood meal, which correlates with reduced blood meal digestion and egg production (Gaio Ade et al., 2011; Gusmão et al., 2010). Some evidence also supports roles for gut microbiota in nutrient acquisition by autogenous mosquitoes, which produce eggs without blood feeding. Coon et al. (2016a) assessed the effects of specific members of the gut microbiota in *A. aegypti* and *Aedes atropalpus* larvae on female fitness traits including egg production. A unique feature of the life history of *A. atropalpus* is that adult females produce a first clutch of eggs without blood feeding (Clements, 1992; O'Meara, 1985), which is facilitated by enhanced nutrient acquisition by larvae and the availability of larger teneral reserves that provide resources for egg production (Chambers and Klowden, 1994; Merritt et al., 1992; Su and Mulla, 1997; Telang et al., 2006; Telang and Wells, 2004). Interestingly, several bacterial isolates supported normal development and reproduction of *A. aegypti*, but only partially rescued development and reproduction by *A. atropalpus* (Coon et al., 2016a). Specific bacterial isolates also contributed differently to glycogen, protein, and lipid stores in newly emerged *A. atropalpus* females (Coon et al., 2016a). Thus, an important feature of autogeny in *A. atropalpus* could be greater dependence on the composition of the gut microbiota for development and reproduction. In contrast, impacts of gut bacteria on nutrient reserves in *A. aegypti* are likely compensated for by the added nutrients females obtain through blood feeding.

6.4 Impacts on vector competence

Gut microbiota can reduce susceptibility of mosquitoes to infection by human pathogens, either through activation of mosquito immune responses, resource competition, or the production of anti-pathogen molecules (Bahia et al., 2014; Boissière et al., 2012; Cirimotich et al., 2011; Dong et al., 2009; Eappen et al., 2013; Gendrin et al., 2015; Gonzalez-Ceron et al., 2003; Joyce et al., 2011; Pumpuni et al., 1993, 1996; Ramirez et al., 2012, 2014; Xi et al., 2008). These studies have been extensively reviewed elsewhere (Briegel, 2003; Dennison et al., 2014; Hegde et al., 2015; Minard et al., 2013; Rodgers et al., 2017a; Van Tol and Dimopoulos, 2016). More recently, the gut microbiota has also been shown to induce proper formation of the peritrophic matrix, which serves as the first barrier to pathogen entry (Rodgers et al., 2017b; Song et al., 2018). However, the complex variation in gut microbial communities has made it difficult to identify specific taxa that consistently limit pathogen infection and transmission. Different strains of taxa within the Enterobacteriaceae (e.g. *Enterobacter, Pantoea, Proteus,* and *Serratia*), for example, vary substantially in their ability to inhibit pathogen development *in vivo*, with some strains even enhancing pathogen infectivity (Apte-Deshpande et al., 2012, 2014; Bahia et al., 2014; Bando et al., 2013; Carissimo et al., 2015; Cirimotich et al., 2011; Dennison et al., 2016; Dickson et al., 2017; Ramirez et al., 2012; Tchioffo et al., 2013, 2016; Wu et al., 2019).

Gut microbiota may also influence the efficacy of existing mosquito control agents such as insecticides and entomopathogenic bacteria and fungi (Dada et al., 2018; Ramirez et al., 2014; Wei et al., 2017). Naturally occurring bacteria present in environments where mosquitoes persist also have the capacity to impact their host seeking and oviposition behaviours (Lindh et al., 2008b; Trexler et al., 2003; Verhulst et al., 2009, 2010, 2011). In this way, microbes can also indirectly affect pathogen transmission by altering the rate at which infected mosquitoes come into contact with susceptible hosts, or the likelihood that they will locate oviposition sites that support development of their offspring.

6.5 Interactions with *Wolbachia*

Several mosquito species are also often infected by the intracellular bacterium *Wolbachia*, which is maternally transmitted by many arthropods, including other haematophagous insects (see above) (Hegde et al., 2015). This bacterium is not considered a member of the gut microbiota but is present at high titres in the reproductive tracts of adult mosquitoes and to a lesser extent the gut and other tissues (Bourtzis et al., 2014). *Wolbachia* infections have the potential to rapidly invade mosquito populations by inducing a reproductive phenotype known as cytoplasmic incompatibility, which leads to early embryonic death when *Wolbachia*-infected males mate with uninfected females (Hoffmann and Turelli, 1988; Turelli and Hoffmann, 1991). Infected females, in contrast, produce viable embryos when mated with either infected or uninfected males and therefore have a reproductive advantage over uninfected females. *Wolbachia* can also provide infected individuals with other fitness benefits, including nutrient provisioning (Brownlie et al., 2009; Hosokawa et al., 2010) and protection against pathogens (Hedges et al., 2008; Teixeira et al., 2008). Certain *Wolbachia* strains have been demonstrated to suppress propagation and transmission of dengue virus and malaria parasites in *Wolbachia*-infected mosquitoes (reviewed in Bourtzis et al., 2014; Caragata et al., 2016; Flores and O'Neill, 2018). Studies have also reported the stable introduction of *Wolbachia* into uninfected mosquitoes and the trans-generational spread of *Wolbachia* infection with little to no fitness costs (Bian et al., 2013; Dutra et al., 2016; Hoffmann et al., 2011; Joshi et al., 2014; Joubert et al., 2016; McMeniman et al., 2009; Schmidt et al., 2017; Walker et al., 2011; Xi et al., 2005). However, successful invasion of *Wolbachia* into naturally occurring mosquito populations also depends on the initial frequency of infection (Turelli, 2010), and recent studies suggest that interactions between *Wolbachia* and members of the native gut microbiota may also be important for its establishment and persistence in mosquito hosts (Asgari, 2017). Members of the gut microbiota have been found to negatively affect *Wolbachia* transmission in *Anopheles stephensi* mosquitoes (Hughes et al., 2014; Rossi et al., 2015), although more recent studies suggest that these patterns may vary between different *Wolbachia* strains and mosquito species combinations (Audsley et al., 2017, 2018; Chen et al., 2016).

The presence of *Wolbachia* could also impact gut microbiota composition and therefore host development, fecundity, and vector competence. Audsley et al. (2017) reported that *Wolbachia* infection significantly altered the relative abundance—but not species diversity—of resident bacteria in field-collected adult *A. aegypti* mosquitoes, although the affected taxa represented only a small proportion of the gut bacterial community present in the individuals they sampled. They also examined field-collected larvae, in which *Wolbachia* infection had little effect on resident bacteria (Audsley et al., 2017). Studies in laboratory reared mosquitoes support these findings (Chen et al., 2016), which suggests a low potential for *Wolbachia* to impact host fitness through reciprocal interactions with the native gut microbiota. However, more work is needed to determine whether these observations extend to other mosquito species or mosquitoes collected from different geographic locations.

6.6 Future directions

Due to their impacts on mosquito growth and development, reproduction, and vector competence, members of the gut microbiota of mosquitoes are emerging as novel targets for the control of a number of mosquito-borne pathogens, and paratransgenic approaches using genetically modified bacteria have already shown promise in *Anopheles* mosquitoes for the control of malaria (Wang and Jacobs-Lorena, 2013). Efforts to identify and disseminate unmodified bacteria that naturally inhibit pathogen infection or transmission are also ongoing. However, despite recent progress in our understanding of mosquito-microbiota interactions, a number of important basic questions remain underexplored:

1. *How do various microorganisms, including non-bacterial members of the gut microbiota, affect the vectorial capacity of mosquitoes?* Numerous types of microorganisms, including bacteria, algae, protists, fungi, rotifers, and viruses, have been identified from the mosquito gut and have the potential to produce anti-pathogen molecules and/or alter host immunity or fitness (Belda et al., 2017; Bolling et al., 2015; Bozic et al., 2017; Chandler et al., 2015; Demaio et al., 1996; Hinman, 1930; Merritt et al., 1990; Muturi et al., 2016a; Roundy et al., 2017; Steyn et al., 2016; Tawidian et al., 2019; Thongsripong et al., 2018; Walker et al., 1988). However, studies to date have almost exclusively focused on bacteria and few have stepped beyond taxonomic characterization to provide insights into the mechanisms underlying how gut microbiota impact mosquito survival, reproduction, and/or growth and development. Future experimentation using axenic (microbe-free) and gnotobiotic mosquitoes colonized by a defined gut microbiota will be

necessary to identify specific taxa with functions of interest (e.g. the ability to antagonize human pathogens).

2. *How does the gut environment shape the acquisition and maintenance of key microbial species?* Successful implementation of microbe-based control strategies will rely on our ability to effectively disseminate candidate microbes into mosquito populations in the field, via attractive sugar baits to target adult mosquitoes (Bilgo et al., 2018; Mancini et al., 2016) or introduction into larval habitats. Thus, a better understanding of how microbes persist in environments where mosquitoes feed and the mechanisms underlying how they colonize and persist in mosquito hosts is essential.
3. *How do microbial interactions impact the stability and function of mosquito gut microbial communities?* Finally, heterogeneity in gut microbiota composition between individuals and across mosquito populations could have unpredictable effects on introduced microbes due to competition or other interactions (Bahia et al., 2014; Coon et al., 2014; Hegde et al., 2018; Terenius et al., 2012). As previously noted, the gut microbiota may also influence the efficacy of other mosquito and pathogen control agents, including entomopathogens and *Wolbachia*. The ability of *Wolbachia* to spread rapidly through mosquito populations makes characterizing the interactions between stable *Wolbachia* infection and gut microbiota a particularly high priority for future research.

7. Concluding Remarks

Like all other metazoans, the multiple origins of haematophagy in insects occurred in a world filled with microbes. These microbes shaped the evolution of these insects, unsurprisingly resulting in the reliance of many on symbiotic bacteria, whether specialized or opportunistic, for their development. The relationships are highly varied among disease vectors: obligately haematophagous insects generally rely on vertically transmitted bacteria to provision them with nutrients absent in their diet, while facultatively haematophagous insects generally acquire bacteria from the environment that perform important developmental roles for the host. In both groups, microbes appear to impact vector competence. Our understanding of how these microbes interface with the host and the pathogens they transmit is still in its early stages. Despite this, significant progress has been made in exploiting these symbiotic relationships to control pathogen transmission, though there are vast holes in our understanding of the nature of these relationships.

Acknowledgements

This work was supported by funding from the National Institutes of Health (award K22AI127849 to KJV), United States Department of Agriculture (award 2018-67012-29991 to KLC), The University of Georgia College of Agricultural and Environmental Sciences Experiment Station, and The University of Wisconsin.

References

Abdeladhim, M., Kamhawi, S., Valenzuela, J.G., 2014. What's behind a sand fly bite? The profound effect of sand fly saliva on host hemostasis, inflammation and immunity. Infect. Genet. Evol. 28, 691–703.

Ahmed, A.M., Baggott, S.L., Maingon, R., Hurd, H., 2002. The costs of mounting an immune response are reflected in the reproductive fitness of the mosquito *Anopheles gambiae*. Oikos 97, 371–377.

Akman, L., Yamashita, A., Watanabe, H., Oshima, K., Shiba, T., Hattori, M., Aksoy, S., 2002. Genome sequence of the endocellular obligate symbiont of tsetse flies, *Wigglesworthia glossinidia*. Nat. Genet. 32, 402–407.

Aksoy, S., 2019. Tsetse peritrophic matrix influences for trypanosome transmission. J. Insect Physiol. 118, 103919.

Apte-Deshpande, A., Paingankar, M., Gokhale, M.D., Deobagkar, D.N., 2012. *Serratia odorifera* a midgut inhabitant of *Aedes aegypti* mosquito enhances its susceptibility to dengue-2 virus. PLoS One 7, e40401.

Apte-Deshpande, A.D., Paingankar, M.S., Gokhale, M.D., Deobagkar, D.N., 2014. *Serratia odorifera* mediated enhancement in susceptibility of *Aedes aegypti* for chikungunya virus. Indian J. Med. Res. 139, 762–768.

Asgari, S., 2017. Chapter 10—Host-microbe interactions: a case for *Wolbachia* dialogue. In: Wikel, S.K., Aksoy, S., Dimopoulos, G. (Eds.), Arthropod Vector: Controller of Disease Transmission. Academic Press, London.

Auden, D.T., 1974. Studies on the development of *Rhodnius prolixus* and the effects of its symbiote *Nocardia rhodnii*. J. Med. Entomol. 11, 68–71.

Audsley, M.D., Ye, Y.H., McGraw, E.A., 2017. The microbiome composition of *Aedes aegypti* is not critical for *Wolbachia*-mediated inhibition of dengue virus. PLoS Negl. Trop. Dis. 11, e0005426.

Audsley, M.D., Seleznev, A., Joubert, D.A., Woolfit, M., O'Neill, S.L., McGraw, E.A., 2018. *Wolbachia* infection alters the relative abundance of resident bacteria in adult *Aedes aegypti* mosquitoes, but not larvae. Mol. Ecol. 27, 297–309.

Azambuja, P., Garcia, E.S., Waniek, P.J., Vieira, C.S., Figueiredo, M.B., Gonzalez, M.S., Mello, C.B., Castro, D.P., Ratcliffe, N.A., 2017. *Rhodnius prolixus*: from physiology by Wigglesworth to recent studies of immune system modulation by *Trypanosoma cruzi* and *Trypanosoma rangeli*. J. Insect Physiol. 97, 45–65.

Bahia, A.C., Dong, Y., Blumberg, B.J., Mlambo, G., Tripathi, A., Benmarzouk-Hidalgo, O.J., Chandra, R., Dimopoulos, G., 2014. Exploring *Anopheles* gut bacteria for *Plasmodium* blocking activity. Environ. Microbiol. 16, 2980–2994.

Baines, S., 1956. The role of the symbiotic bacteria in the nutrition of *Rhodnius prolixus* (Hemiptera). J. Exp. Biol. 33, 533–541.

Bando, H., Okado, K., Guelbeogo, W.M., Badolo, A., Aonuma, H., Nelson, B., Fukumoto, S., Xuan, X., Sagnon, N., Kanuka, H., 2013. Intra-specific diversity of *Serratia marcescens* in *Anopheles* mosquito midgut defines *Plasmodium* transmission capacity. Sci. Rep. 3, 1641.

Bascuñan, P., Niño-Garcia, J.P., Galeano-Castañeda, Y., Serre, D., Correa, M.M., 2018. Factors shaping the gut bacterial community assembly in two main Colombian malaria vectors. Microbiome 6, 148.

Beard, C.B., Mason, P.W., Aksoy, S., Tesh, R.B., Richards, F.F., 1992. Transformation of an insect symbiont and expression of a foreign gene in the Chagas' disease vector *Rhodnius prolixus*. Am. J. Trop. Med. Hyg. 46, 195–200.

Belda, E., Moya, A., Bentley, S., Silva, F.J., 2010. Mobile genetic element proliferation and gene inactivation impact over the genome structure and metabolic capabilities of *Sodalis glossinidius*, the secondary endosymbiont of tsetse flies. BMC Genomics 11, 449.

Belda, E., Silva, F.J., Peretó, J., Moya, A., 2012. Metabolic networks of *Sodalis glossinidius*: a systems biology approach to reductive evolution. PLoS One 7, e30652.

Belda, E., Coulibaly, B., Fofana, A., Beavogui, A.H., Traore, S.F., Gohl, D.M., Vernick, K.D., Riehle, M.M., 2017. Preferential suppression of *Anopheles gambiae* host sequences allows detection of the mosquito eukaryotic microbiome. Sci. Rep. 7, 3241.

Benoit, J.B., Attardo, G.M., Baumann, A.A., Michalkova, V., Aksoy, S., 2015. Adenotrophic viviparity in tsetse flies: potential for population control and as an insect model for lactation. Annu. Rev. Entomol. 60, 351–371.

Benoit, J.B., Vigneron, A., Broderick, N.A., Wu, Y., Sun, J.S., Carlson, J.R., Aksoy, S., Weiss, B.L., 2017. Symbiont-induced odorant binding proteins mediate insect host hematopoiesis. eLife 6, e19535.

Bewig, F., Schwartz, W., 1954. Untersuchungen über die symbiose der triatomiden *Rhodnius prolixus* Stål und *Triatoma infestans* Klug. Naturwissenschaften 41, 435.

Bewig, F., Schwartz, W., 1955. Über die symbiose von triatomiden mit *Nocardia rhodnii* (Erikson) Waksman und Henrici. Naturwissenschaften 42, 423–424.

Bian, G., Joshi, D., Dong, Y., Lu, P., Zhou, G., Pan, X., Xu, Y., Dimopoulos, G., Xi, Z., 2013. *Wolbachia* invades *Anopheles stephensi* populations and induces refractoriness to *Plasmodium* infection. Science 340, 748–751.

Bilgo, E., Vantaux, A., Sanon, A., Ilboudo, S., Dabiré, R.K., Jacobs-Lorena, M., Diabate, A., 2018. Field assessment of potential sugar feeding stations for disseminating bacteria in a paratransgenic approach to control malaria. Malar. J. 17, 367.

Billingsley, P.F., 1990. The midgut ultrastructure of hematophagous insects. Annu. Rev. Entomol. 35, 219–248.

Bing, X.L., Attardo, G.M., Vigneron, A., Aksoy, E., Scolari, F., Malacrida, A., Weiss, B.L., Aksoy, S., 2017. Unravelling the relationship between the tsetse fly and its obligate symbiont *Wigglesworthia*: transcriptomic and metabolomic landscapes reveal highly integrated physiological networks. Proc. R. Soc. B Biol. Sci. 284, 20170360.

Boissière, A., Tchioffo, M.T., Bachar, D., Abate, L., Marie, A., Nsango, S.E., Shahbazkia, H.R., Awono-Ambene, P.H., Levashina, E.A., Christen, R., Morlais, I., 2012. Midgut microbiota of the malaria mosquito vector *Anopheles gambiae* and interactions with *Plasmodium falciparum* infection. PLoS Pathog. 8, e1002742.

Bolling, B.G., Weaver, S.C., Tesh, R.B., Vasilakis, N., 2015. Insect-specific virus discovery: significance for the arbovirus community. Viruses 7, 4911–4928.

Bourtzis, K., Dobson, S.L., Xi, Z., Rasgon, J.L., Calvitti, M., Moreira, L.A., Bossin, H.C., Moretti, R., Baton, L.A., Hughes, G.L., Mavingui, P., Gilles, J.R., 2014. Harnessing mosquito-*Wolbachia* symbiosis for vector and disease control. Acta Trop. 132 (Suppl), S150–S163.

Boyd, B.M., Allen, J.M., Nguyen, N.P., Vachaspati, P., Quicksall, Z.S., Warnow, T., Mugisha, L., Johnson, K.P., Reed, D.L., 2017. Primates, lice and bacteria: speciation and genome evolution in the symbionts of hominid lice. Mol. Biol. Evol. 34, 1743–1757.

Bozic, J., Capone, A., Pediconi, D., Mensah, P., Cappelli, A., Valzano, M., Mancini, M.V., Scuppa, P., Martin, E., Epis, S., Rossi, P., Favia, G., Ricci, I., 2017. Mosquitoes can harbour yeasts of clinical significance and contribute to their environmental dissemination. Environ. Microbiol. Rep. 9, 642–648.

Braendle, C., Miura, T., Bickel, R., Shingleton, A.W., Kambhampati, S., Stern, D.L., 2003. Developmental origin and evolution of bacteriocytes in the aphid-*Buchnera* symbiosis. PLoS Biol. 1, E21.

Brecher, G., Wigglesworth, V.B., 1944. The transmission of *Actinomyces rhodnii* Erikson in *Rhodnius prolixus* Stål (Hemiptera) and its influence on the growth of the host. Parasitology 35, 220–224.

Briegel, H., 2003. Physiological bases of mosquito ecology. J. Vector Ecol. 28, 1–11.

Brownlie, J.C., Cass, B.N., Riegler, M., Witsenburg, J.J., Iturbe-Ormaetxe, I., McGraw, E.A., O'Neill, S.L., 2009. Evidence for metabolic provisioning by a common invertebrate endosymbiont, *Wolbachia pipientis*, during periods of nutritional stress. PLoS Pathog. 5, e1000368.

Buchner, P., 1965. Endosymbiosis of Animals With Plant Microorganisms. John Wiley & Sons, New York.

Buck, M., Nilsson, L.K., Brunius, C., Dabire, R.K., Hopkins, R., Terenius, O., 2016. Bacterial associations reveal spatial population dynamics in *Anopheles gambiae* mosquitoes. Sci. Rep. 6, 22806.

Buxton, P.A., 1955. The Natural History of Tsetse Flies. H.K. Lewis & Co. Ltd, London.

Cappelli, A., Damiani, C., Mancini, M.V., Valzano, M., Rossi, P., Serrao, A., Ricci, I., Favia, G., 2019. *Asaia* activates immune genes in mosquito eliciting an anti-*Plasmodium* response: implications in malaria control. Front. Genet. 10, 836.

Caragata, E.P., Dutra, H.L.C., Moreira, L.A., 2016. Exploiting intimate relationships: controlling mosquito-transmitted disease with *Wolbachia*. Trends Parasitol. 32, 207–218.

Carels, N., Gumiel, M., da Mota, F.F., Moreira, C.J.D., Azambuja, P., 2017. A metagenomic analysis of bacterial microbiota in the digestive tract of triatomines. Bioinform. Biol. Insights 11, 1–19.

Carissimo, G., Pondeville, E., McFarlane, M., Dietrich, I., Mitri, C., Bischoff, E., Antoniewski, C., Bourgouin, C., Failloux, A.B., Kohl, A., Vernick, K.D., 2015. Antiviral immunity of *Anopheles gambiae* is highly compartmentalized, with distinct roles for RNA interference and gut microbiota. Proc. Natl. Acad. Sci. U. S. A. 112, E176–E185.

Cavanagh, P., Marsden, P.D., 1969. Bacteria isolated from gut of some reduviid bugs. Trans. R. Soc. Trop. Med. Hyg. 63, 415–416.

Chambers, G.M., Klowden, M.J., 1994. Nutritional reserves of autogenous and anautogenous selected strains of *Aedes albopictus* (Diptera: Culicidae). J. Med. Entomol. 31, 554–560.

Chandler, J.A., Liu, R.M., Bennett, S.N., 2015. RNA shotgun metagenomic sequencing of northern California (USA) mosquitoes uncovers viruses, bacteria, and fungi. Front. Microbiol. 6, 185.

Channumsin, M., Ciosi, M., Masiga, D., Turner, C.M.R., Mable, B.K., 2018. *Sodalis glossinidius* presence in wild tsetse is only associated with presence of trypanosomes in complex interactions with other tsetse-specific factors. BMC Microbiol. 18, 163.

Chao, J., Wistreich, G.A., 1959. Microbial isolations from the midgut of *Culex tarsalis* Coquillet. J. Insect Pathol. 1, 311–318.

Chen, S., Zhao, J., Joshi, D., Xi, Z., Norman, B., Walker, E.D., 2016. Persistent infection by *Wolbachia* wAlbB has no effect on composition of the gut microbiota in adult female *Anopheles stephensi*. Front. Microbiol. 7, 1485.

Cheng, Q., Aksoy, S., 1999. Tissue tropism, transmission and expression of foreign genes *in vivo* in midgut symbionts of tsetse flies. Insect Mol. Biol. 8, 125–132.

Chong, R.A., Park, H., Moran, N.A., 2019. Genome evolution of the obligate endosymbiont *Buchnera aphidicola*. Mol. Biol. Evol. 36, 1481–1489.

Chouaia, B., Rossi, P., Epis, S., Mosca, M., Ricci, I., Damiani, C., Ulissi, U., Crotti, E., Daffonchio, D., Bandi, C., Favia, G., 2012. Delayed larval development in *Anopheles* mosquitoes deprived of *Asaia* bacterial symbionts. BMC Microbiol. 12 (Suppl. 1), S2.

Chrudimský, T., Husník, F., Nováková, E., Hypša, V., 2012. Candidatus *Sodalis melophagi* sp nov.: phylogenetically independent comparative model to the tsetse fly symbiont *Sodalis glossinidius*. PLoS One 7, e40354.

Cirimotich, C.M., Dong, Y., Clayton, A.M., Sandiford, S.L., Souza-Neto, J.A., Mulenga, M., Dimopoulos, G., 2011. Natural microbe-mediated refractoriness to *Plasmodium* infection in *Anopheles gambiae*. Science 332, 855–858.

Clayton, A.L., Oakeson, K.F., Gutin, M., Pontes, A., Dunn, D.M., Von Niederhausern, A.C., Weiss, R.B., Fisher, M., Dale, C., 2012. A novel human-infection-derived bacterium provides insights into the evolutionary origins of mutualistic insect-bacterial symbioses. PLoS Genet. 8, e1002990.

Clements, A.N., 1992. The Biology of Mosquitoes, Volume 1: Development, Nutrition, and Reproduction. Chapman & Hall, New York.

Coon, K.L., Vogel, K.J., Brown, M.R., Strand, M.R., 2014. Mosquitoes rely on their gut microbiota for development. Mol. Ecol. 23, 2727–2739.

Coon, K.L., Brown, M.R., Strand, M.R., 2016a. Gut bacteria differentially affect egg production in the anautogenous mosquito *Aedes aegypti* and facultatively autogenous mosquito *Aedes atropalpus* (Diptera: Culicidae). Parasit. Vectors 9, 375.

Coon, K.L., Brown, M.R., Strand, M.R., 2016b. Mosquitoes host communities of bacteria that are essential for development but vary greatly between local habitats. Mol. Ecol. 25, 5806–5826.

Coon, K.L., Valzania, L., McKinney, D.A., Vogel, K.J., Brown, M.R., Strand, M.R., 2017. Bacteria-mediated hypoxia functions as a signal for mosquito development. Proc. Natl. Acad. Sci. U. S. A. 114, E5362–E5369.

Coon, K.L., Valzania, L., Brown, M.R., Strand, M.R., 2020. Predaceous *Toxorhynchites* mosquitoes require a living gut microbiota to develop. Proc. R. Soc. B Biol. Sci. 287, 20192705.

Cordon-Obras, C., Rodriguez, Y.F., Fernandez-Martinez, A., Cano, J., Ndong-Mabale, N., Ncogo-Ada, P., Ndongo-Asumu, P., Aparicio, P., Navarro, M., Benito, A., Bart, J.M., 2015. Molecular evidence of a *Trypanosoma brucei gambiense* sylvatic cycle in the human African trypanosomiasis foci of Equatorial Guinea. Front. Microbiol. 6, 765.

Correa, M.A., Matusovsky, B., Brackney, D.E., Steven, B., 2018. Generation of axenic *Aedes aegypti* demonstrate live bacteria are not required for mosquito development. Nat. Commun. 9, 4464.

da Mota, F.F., Marinho, L.P., Moreira, C.J.D., Lima, M.M., Mello, C.B., Garcia, E.S., Carels, N., Azambuja, P., 2012. Cultivation-independent methods reveal differences among bacterial gut microbiota in triatomine vectors of chagas disease. PLoS Negl. Trop. Dis. 6, e1631.

Dada, N., Sheth, M., Liebman, K., Pinto, J., Lenhart, A., 2018. Whole metagenome sequencing reveals links between mosquito microbiota and insecticide resistance in malaria vectors. Sci. Rep. 8, 2084.

Dale, C., Maudlin, I., 1999. *Sodalis* gen. nov. and *Sodalis glossinidius* sp. nov., a microaerophilic secondary endosymbiont of the tsetse fly *Glossina morsitans morsitans*. Int. J. Syst. Bacteriol. 49 (Pt. 1), 267–275.

Dale, C., Welburn, S.C., 2001. The endosymbionts of tsetse flies: manipulating host-parasite interactions. Int. J. Parasitol. 31, 628–631.

de Fuentes-Vicente, J.A., Gutiérrez-Cabrera, A.E., Flores-Villegas, A.L., Lowenberger, C., Benelli, G., Salazar-Schettino, P.M., Córdoba-Aguilar, A., 2018. What makes an effective Chagas disease vector? Factors underlying *Trypanosoma cruzi*-triatomine interactions. Acta Trop. 183, 23–31.

De Vooght, L., Caljon, G., Stijlemans, B., De Baetselier, P., Coosemans, M., Van Den Abbeele, J., 2012. Expression and extracellular release of a functional anti-trypanosome nanobody in *Sodalis glossinidius*, a bacterial symbiont of the tsetse fly. Microb. Cell Factories 11, 23.

De Vooght, L., Caljon, G., Van Hees, J., Van Den Abbeele, J., 2015. Paternal transmission of a secondary symbiont during mating in the viviparous tsetse fly. Mol. Biol. Evol. 32, 1977–1980.

De Vooght, L., Van Keer, S., Van Den Abbeele, J., 2018. Towards improving tsetse fly paratransgenesis: stable colonization of *Glossina morsitans morsitans* with genetically modified *Sodalis*. BMC Microbiol. 18, 165.

Demaio, J., Pumpuni, C.B., Kent, M., Beier, J.C., 1996. The midgut bacterial flora of wild *Aedes triseriatus*, *Culex pipiens*, and *Psorophora columbiae* mosquitoes. Am. J. Trop. Med. Hyg. 54, 219–223.

Demirbas-Uzel, G., De Vooght, L., Parker, A.G., Vreysen, M.J.B., Mach, R.L., Van Den Abbeele, J., Abd-Alla, A.M.M., 2018. Combining paratransgenesis with SIT: impact of ionizing radiation on the DNA copy number of *Sodalis glossinidius* in tsetse flies. BMC Microbiol. 18, 160.

Dennis, J.W., Durkin, S.M., Downie, J.E.H., Hamill, L.C., Anderson, N.E., Macleod, E.T., 2014. *Sodalis glossinidius* prevalence and trypanosome presence in tsetse from Luambe National Park, Zambia. Parasit. Vectors 7, 378.

Dennison, N.J., Jupatanakul, N., Dimopoulos, G., 2014. The mosquito microbiota influences vector competence for human pathogens. Curr. Opin. Insect Sci. 3, 6–13.

Dennison, N.J., Saraiva, R.G., Cirimotich, C.M., Mlambo, G., Mongodin, E.F., Dimopoulos, G., 2016. Functional genomic analyses of *Enterobacter*, *Anopheles* and *Plasmodium* reciprocal interactions that impact vector competence. Malar. J. 15, 425.

Desquesnes, M., Dia, M.L., 2004. Mechanical transmission of *Trypanosoma vivax* in cattle by the African tabanid *Atylotus fuscipes*. Vet. Parasitol. 119, 9–19.

Diaz, S., Villavicencio, B., Correia, N., Costa, J., Haag, K.L., 2016. Triatomine bugs, their microbiota and *Trypanosoma cruzi*: asymmetric responses of bacteria to an infected blood meal. Parasit. Vectors 9, 636.

Díaz-Nieto, L.M., D'Alessio, C., Perotti, M.A., Berón, C.M., 2016. *Culex pipiens* development is greatly influenced by native bacteria and exogenous yeast. PLoS One 11, e0153133.

Dickson, L.B., Jiolle, D., Minard, G., Moltini-Conclois, I., Volant, S., Ghozlane, A., Bouchier, C., Ayala, D., Paupy, C., Moro, C.V., Lambrechts, L., 2017. Carryover effects of larval exposure to different environmental bacteria drive adult trait variation in a mosquito vector. Sci. Adv. 3, e1700585.

Dickson, L.B., Ghozlane, A., Volant, S., Bouchier, C., Ma, L., Vega-Rua, A., Dusfour, I., Jiolle, D., Paupy, C., Mayanja, M.N., Kohl, A., Lutwama, J.J., Duong, V., Lambrechts, L., 2018. Diverse laboratory colonies of *Aedes aegypti* harbor the same adult midgut bacterial microbiome. Parasit. Vectors 11, 207.

Dong, Y., Manfredini, F., Dimopoulos, G., 2009. Implication of the mosquito midgut microbiota in the defense against malaria parasites. PLoS Pathog. 5, e1000423.

Dotson, E.M., Plikaytis, B., Shinnick, T.M., Durvasula, R.V., Beard, C.B., 2003. Transformation of *Rhodococcus rhodnii*, a symbiont of the Chagas disease vector *Rhodnius prolixus*, with integrative elements of the L1 mycobacteriophage. Infect. Genet. Evol. 3, 103–109.

Doudoumis, V., Blow, F., Saridaki, A., Augustinos, A., Dyer, N.A., Goodhead, I., Solano, P., Rayaisse, J.B., Takac, P., Mekonnen, S., Parker, A.G., Abd-Alla, A.M.M., Darby, A., Bourtzis, K., Tsiamis, G., 2017. Challenging the *Wigglesworthia, Sodalis, Wolbachia* symbiosis dogma in tsetse flies: *Spiroplasma* is present in both laboratory and natural populations. Sci. Rep. 7, 4699.

Duguma, D., Hall, M.W., Rugman-Jones, P., Stouthamer, R., Terenius, O., Neufeld, J.D., Walton, W.E., 2015. Developmental succession of the microbiome of *Culex* mosquitoes. BMC Microbiol. 15, 140.

Dumonteil, E., Ramirez-Sierra, M.J., Perez-Carrillo, S., Teh-Poot, C., Herrera, C., Gourbière, S., Waleckx, E., 2018. Detailed ecological associations of triatomines revealed by metabarcoding and next-generation sequencing: implications for triatomine behavior and *Trypanosoma cruzi* transmission cycles. Sci. Rep. 8, 4140.

Duncan, J.T., 1926. On a bactericidal principle present in the alimentary canal of insects and arachnids. Parasitology 18, 238–252.

Durvasula, R.V., Gumbs, A., Panackal, A., Kruglov, O., Aksoy, S., Merrifield, R.B., Richards, F.F., Beard, C.B., 1997. Prevention of insect-borne disease: an approach using transgenic symbiotic bacteria. Proc. Natl. Acad. Sci. U. S. A. 94, 3274–3278.

Durvasula, R.V., Sundaram, R.K., Kirsch, P., Hurwitz, I., Crawford, C.V., Dotson, E., Beard, C.B., 2008. Genetic transformation of a Corynebacterial symbiont from the Chagas disease vector *Triatoma infestans*. Exp. Parasitol. 119, 94–98.

Dutra, H.L., Rocha, M.N., Dias, F.B., Mansur, S.B., Caragata, E.P., Moreira, L.A., 2016. *Wolbachia* blocks currently circulating Zika virus isolates in Brazilian *Aedes aegypti* mosquitoes. Cell Host Microbe 19, 771–774.

Eappen, A.G., Smith, R.C., Jacobs-Lorena, M., 2013. *Enterobacter*-activated mosquito immune responses to *Plasmodium* involve activation of SRPN6 in *Anopheles stephensi*. PLoS One 8, e62937.

Eichler, S., Schaub, G.A., 2002. Development of symbionts in triatomine bugs and the effects of infections with trypanosomatids. Exp. Parasitol. 100, 17–27.

Engelstadter, J., Hurst, G.D.D., 2009. The ecology and evolution of microbes that manipulate host reproduction. Annu. Rev. Ecol. Evol. Syst. 40, 127–149.

Espino, C.I., Gómez, T., González, G., Do Santos, M.F.B., Solano, J., Sousa, O., Moreno, N., Windsor, D., Ying, A., Vilchez, S., Osuna, A., 2009. Detection of *Wolbachia* bacteria in multiple organs and feces of the triatomine insect *Rhodnius pallescens* (Hemiptera, Reduviidae). Appl. Environ. Microbiol. 75, 547–550.

Farikou, O., Njiokou, F., Mbida, J.A.M., Njitchouang, G.R., Djeunga, H.N., Asonganyi, T., Simarro, P.P., Cuny, G., Geiger, A., 2010. Tripartite interactions between tsetse flies, *Sodalis glossinidius* and trypanosomes—an epidemiological approach in two historical human African trypanosomiasis foci in Cameroon. Infect. Genet. Evol. 10, 115–121.

Ferguson, M.J., Micks, D.W., 1961. Microorganisms associated with mosquitoes: I. Bacteria isolated from the midgut of adult *Culex fatigans* Wiedemann. J. Insect Pathol. 3, 112–119.

Fieck, A., Hurwitz, I., Kang, A.S., Durvasula, R., 2010. *Trypanosoma cruzi*: synergistic cytotoxicity of multiple amphipathic anti-microbial peptides to *T. cruzi* and potential bacterial hosts. Exp. Parasitol. 125, 342–347.

Flores, H.A., O'Neill, S.L., 2018. Controlling vector-borne diseases by releasing modified mosquitoes. Nat. Rev. Microbiol. 16, 508–518.

Foster, W.A., 1995. Mosquito sugar feeding and reproductive energetics. Annu. Rev. Entomol. 40, 443–474.

Franco, J.R., Simarro, P.P., Diarra, A., Jannin, J.G., 2014. Epidemiology of human African trypanosomiasis. Clin. Epidemiol. 6, 257–275.

Gaio Ade, O., Gusmão, D.S., Santos, A.V., Berbert-Molina, M.A., Pimenta, P.F., Lemos, F.J., 2011. Contribution of midgut bacteria to blood digestion and egg production in *Aedes aegypti* (Diptera: Culicidae) (L.). Parasit. Vectors 4, 105.

Geiger, A., Ravel, S., Mateille, T., Janelle, J., Patrel, D., Cuny, G., Frutos, R., 2007. Vector competence of *Glossina palpalis gambiensis* for *Trypanosoma brucei* s.l. and genetic diversity of the symbiont *Sodalis glossinidius*. Mol. Biol. Evol. 24, 102–109.

Gendrin, M., Rodgers, F.H., Yerbanga, R.S., Ouédraogo, J.B., Basáñez, M.G., Cohuet, A., Christophides, G.K., 2015. Antibiotics in ingested human blood affect the mosquito microbiota and capacity to transmit malaria. Nat. Commun. 6, 5921.

Gibson, W., Kay, C., Peacock, L., 2017. *Trypanosoma congolense*: molecular toolkit and resources for studying a major livestock pathogen and model trypanosome. Adv. Parasitol. 98, 283–309.

Gimonneau, G., Tchioffo, M.T., Abate, L., Boissière, A., Awono-Ambene, P.H., Nsango, S.E., Christen, R., Morlais, I., 2014. Composition of *Anopheles coluzzii* and *Anopheles gambiae* microbiota from larval to adult stages. Infect. Genet. Evol. 28, 715–724.

Gonzalez-Ceron, L., Santillan, F., Rodriguez, M.H., Mendez, D., Hernandez-Avila, J.E., 2003. Bacteria in midguts of field-collected *Anopheles albimanus* block *Plasmodium vivax* sporogonic development. J. Med. Entomol. 40, 371–374.

Goodchild, A.J.P., 1955. The bacteria associated with *Triatoma infestans* and some other species of Reduviidae. Parasitology 45, 441–448.

Goodhead, I., Blow, F., Brownridge, P., Hughes, M., Kenny, J., Krishna, R., Mclean, L., Pongchauikul, P., Beynon, R., Darby, A.C., 2020. Large-scale and significant expression from pseudogenes in *Sodalis glossinidius*—a facultative bacterial endosymbiont. Microb. Genom. 6, 3–15.

Graça-Souza, A.V., Maya-Monteìro, C., Paiva-Silva, G.O., Braz, G.R.C., Paes, M.C., Sorgine, M.H.F., Oliveira, M.F., Oliveira, P.L., 2006. Adaptations against heme toxicity in blood-feeding arthropods. Insect Biochem. Mol. Biol. 36, 322–335.

Griffith, B.C., Weiss, B.L., Aksoy, E., Mireji, P.O., Auma, J.E., Wamwiri, F.N., Echodu, R., Murilla, G., Aksoy, S., 2018. Analysis of the gut-specific microbiome from field-captured tsetse flies, and its potential relevance to host trypanosome vector competence. BMC Microbiol. 18, 146.

Gumiel, M., da Mota, F.F., Rizzo, V.D., Sarquis, O., De Castro, D.P., Lima, M.M., Garcia, E.D., Carels, N., Azambuja, P., 2015. Characterization of the microbiota in the guts of *Triatoma brasiliensis* and *Triatoma pseudomaculata* infected by *Trypanosoma cruzi* in natural conditions using culture independent methods. Parasit. Vectors 8, 245.

Gumpert, J., Schwartz, W., 1962. Untersuchungen über die symbiose von tieren mit pilzen und bakterien. Zeitschrift fur Allgemeine Mikrobiologie 2, 209–225.

Gusmão, D.S., Santos, A.V., Marini, D.C., Bacci Jr., M., Berbert-Molina, M.A., Lemos, F.J., 2010. Culture-dependent and culture-independent characterization of microorganisms associated with *Aedes aegypti* (Diptera: Culicidae) (L.) and dynamics of bacterial colonization in the midgut. Acta Trop. 115, 275–281.

Gutiérrez-Cabrera, A.E., Córdoba-Aguilar, A., Zenteno, E., Lowenberger, C., Espinoza, B., 2016. Origin, evolution and function of the hemipteran perimicrovillar membrane with emphasis on Reduviidae that transmit Chagas disease. Bull. Entomol. Res. 106, 279–291.

Hall, R.J., Flanagan, L.A., Bottery, M.J., Springthorpe, V., Thorpe, S., Darby, A.C., Wood, A.J., Thomas, G.H., 2019. A tale of three species: adaptation of *Sodalis glossinidius* to tsetse biology, *Wigglesworthia* metabolism, and host diet. mBio 10, e02106–e02118.

Harington, J.S., 1960. Studies on *Rhodnius prolixus*—growth and development of normal and sterile bugs, and the symbiotic relationship. Parasitology 50, 279–286.

Hedges, L.M., Brownlie, J.C., O'Neill, S.L., Johnson, K.N., 2008. *Wolbachia* and virus protection in insects. Science 322, 702.

Hegde, S., Rasgon, J.L., Hughes, G.L., 2015. The microbiome modulates arbovirus transmission in mosquitoes. Curr. Opin. Virol. 15, 97–102.

Hegde, S., Khanipov, K., Albayrak, L., Golovko, G., Pimenova, M., Saldñaa, M.A., Rojas, M.M., Hornett, E.A., Motl, G.C., Fredregill, C.L., Dennett, J.A., Debboun, M., Fofanov, Y., Hughes, G.L., 2018. Microbiome interaction networks and community structure from laboratory-reared and field-collected *Aedes aegypti*, *Aedes albopictus*, and *Culex quinquefasciatus* mosquito vectors. Front. Microbiol. 9, 2160.

Hernández-Rodríguez, C., 2006. Las comunidades microbianas: Diferentes modelos y una cierta unidad metodológica. Rev. Latinoam. Microbiol. 48, 211–225.

Hill, P., Campbell, J.A., Petrie, I.A., 1976. *Rhodnius prolixus* and its symbiotic Actinomycete—microbiological, physiological and behavioral study. Proc. R. Soc. B Biol. Sci. 194, 501–525.

Hinman, E.H., 1930. A study of the food of mosquito larvae. Am. J. Epidemiol. 12, 238–270.

Hoffmann, A.A., Turelli, M., 1988. Unidirectional incompatibility in *Drosophila simulans*: inheritance, geographic variation and fitness effects. Genetics 119, 435–444.

Hoffmann, A.A., Montgomery, B.L., Popovici, J., Iturbe-Ormaetxe, I., Johnson, P.H., Muzzi, F., Greenfield, M., Durkan, M., Leong, Y.S., Dong, Y., Cook, H., Axford, J., Callahan, A.G., Kenny, N., Omodei, C., McGraw, E.A., Ryan, P.A., Ritchie, S.A., Turelli, M., O'Neill, S.L., 2011. Successful establishment of *Wolbachia* in *Aedes* populations to suppress dengue transmission. Nature 476, 454–457.

Hosokawa, T., Koga, R., Kikuchi, Y., Meng, X.Y., Fukatsu, T., 2010. *Wolbachia* as a bacteriocyte-associated nutritional mutualist. Proc. Natl. Acad. Sci. U. S. A. 107, 769–774.

Hughes, G.L., Dodson, B.L., Johnson, R.M., Murdock, C.C., Tsujimoto, H., Suzuki, Y., Patt, A.A., Cui, L., Nossa, C.W., Barry, R.M., Sakamoto, J.M., Hornett, E.A., Rasgon, J.L., 2014. Native microbiome impedes vertical transmission of *Wolbachia* in *Anopheles* mosquitoes. Proc. Natl. Acad. Sci. U. S. A. 111, 12498–12503.

Husník, F., Chrudimský, T., Hypša, V., 2011. Multiple origins of endosymbiosis within the Enterobacteriaceae (γ-Proteobacteria): convergence of complex phylogenetic approaches. BMC Biol. 9, 87.

Hypša, V., 1993. Endocytobionts of *Triatoma infestans*—distribution and transmission. J. Invertebr. Pathol. 61, 32–38.

Hypša, V., Dale, C., 1997. *In vitro* culture and phylogenetic analysis of "Candidatus *Arsenophonus triatominarum*," an intracellular bacterium from the triatomine bug, *Triatoma infestans*. Int. J. Syst. Bacteriol. 47, 1140–1144.

Jones, W.L., DeLong, D.M., 1961. A simplified technique for sterilizing the surface of *Aedes aegypti* eggs. J. Econ. Entomol. 54, 813–814.

Joshi, D., McFadden, M.J., Bevins, D., Zhang, F., Xi, Z., 2014. *Wolbachia* strain wAlbB confers both fitness costs and benefit on *Anopheles stephensi*. Parasit. Vectors 7, 336.

Joubert, D.A., Walker, T., Carrington, L.B., De Bruyne, J.T., Kien, D.H., Hoang Nle, T., Chau, N.V., Iturbe-Ormaetxe, I., Simmons, C.P., O'Neill, S.L., 2016. Establishment of a *Wolbachia* superinfection in *Aedes aegypti* mosquitoes as a potential approach for future resistance management. PLoS Pathog. 12, e1005434.

Joyce, J.D., Nogueira, J.R., Bales, A.A., Pittman, K.E., Anderson, J.R., 2011. Interactions between La Crosse virus and bacteria isolated from the digestive tract of *Aedes albopictus* (Diptera: Culicidae). J. Med. Entomol. 48, 389–394.

Kaufman, M.G., Walker, E.D., Smith, T.W., Merritt, R.W., Klug, M.J., 1999. Effects of larval mosquitoes (*Aedes triseriatus*) and stemflow on microbial community dynamics in container habitats. Appl. Environ. Entomol. 65, 2661–2673.

Kieran, T.J., Arnold, K.M.H., Thomas, J.C., Varian, C.P., Saldaña, A., Calzada, J.E., Glenn, T.C., Gottdenker, N.L., 2019. Regional biogeography of microbiota composition in the Chagas disease vector *Rhodnius pallescens*. Parasit. Vectors 12, 504.

Koga, R., Moran, N.A., 2014. Swapping symbionts in spittlebugs: evolutionary replacement of a reduced genome symbiont. ISME J. 8, 1237–1246.

Lake, P., Friend, W.G., 1967. A monoxenic relationship *Nocardia rhodnii* Erikson in gut of *Rhodnius Prolixus* Stahl (Hemiptera—Reduviidae). Proc. Entomol. Soc. Ont. 98, 53–57.

Lake, P., Friend, W.G., 1968. Use of artificial diets to determine some of effects of *Nocardia rhodnii* on development of *Rhodnius prolixus*. J. Insect Physiol. 14, 543–562.

Lehane, M.J., 2005. The Biology of Blood Sucking in Insects. Cambridge University Press, New York.

Lima, M.S., Laport, M.S., Lorosa, E.S., Jurberg, J., Dos Santos, K.R.N., Da Silva Neto, M.A.C., Rachid, C., Atella, G.C., 2018. Bacterial community composition in the salivary glands of triatomines (Hemiptera: Reduviidae). PLoS Negl. Trop. Dis. 12, e0006739.

Lindh, J.M., Borg-Karlson, A.K., Faye, I., 2008a. Transstadial and horizontal transfer of bacteria within a colony of *Anopheles gambiae* (Diptera: Culicidae) and oviposition response to bacteria-containing water. Acta Trop. 107, 242–250.

Lindh, J.M., Kannaste, A., Knols, B.G., Faye, I., Borg-Karlson, A.K., 2008b. Oviposition responses of *Anopheles gambiae* s.s. (Diptera: Culicidae) and identification of volatiles from bacteria-containing solutions. J. Med. Entomol. 45, 1039–1049.

Lukasik, P., Nazario, K., Van Leuven, J.T., Campbell, M.A., Meyer, M., Michalik, A., Pessacq, P., Simon, C., Veloso, C., McCutcheon, J.P., 2018. Multiple origins of interdependent endosymbiotic complexes in a genus of cicadas. Proc. Natl. Acad. Sci. U. S. A. 115, E226–E235.

Ma, W.C., Denlinger, D.L., 1974. Secretory discharge and microflora of milk gland in tsetse flies. Nature 247, 301–303.

Mancini, M.V., Spaccapelo, R., Damiani, C., Accoti, A., Tallarita, M., Petraglia, E., Rossi, P., Cappelli, A., Capone, A., Peruzzi, G., Valzano, M., Picciolini, M., Diabaté, A., Facchinelli, L., Ricci, I., Favia, G., 2016. Paratransgenesis to control malaria vectors: a semi-field pilot study. Parasit. Vectors 9, 140.

Marchette, N.J., Hatie, C., 1965. Microbial isolates from digestive tract of *Triatoma protracta* Uhler (Reduviidae). J. Invertebr. Pathol. 7, 45–48.

Maudlin, I., Welburn, S.C., Mehlitz, D., 1990. The relationship between *Rickettsia*-like-organisms and trypanosome infections in natural-populations of tsetse in Liberia. Trop. Med. Parasitol. 41, 265–267.

McCutcheon, J.P., Moran, N.A., 2012. Extreme genome reduction in symbiotic bacteria. Nat. Rev. Microbiol. 10, 13–26.

McCutcheon, J.P., McDonald, B.R., Moran, N.A., 2009. Origin of an alternative genetic code in the extremely small and GC-rich genome of a bacterial symbiont. PLoS Genet. 5, e1000565.

McMeniman, C.J., Lane, R.V., Cass, B.N., Fong, A.W., Sidhu, M., Wang, Y.F., O'Neill, S.L., 2009. Stable introduction of a life-shortening *Wolbachia* infection into the mosquito *Aedes aegypti*. Science 323, 141–144.

Mehlitz, D., Molyneux, D.H., 2019. The elimination of *Trypanosoma brucei gambiense*? Challenges of reservoir hosts and transmission cycles: expect the unexpected. Parasite Epidemiol. Control 6, e00113.

Merritt, R.W., Olds, E.J., Walker, E.D., 1990. Natural food and feeding behavior of *Coquillettidia perturbans* larvae. J. Am. Mosquito Contr Assoc. 6, 35–42.

Merritt, R.W., Dadd, R.H., Walker, E.D., 1992. Feeding behavior, natural food, and nutritional relationships of larval mosquitoes. Annu. Rev. Entomol. 37, 349–376.

Mesquita, R.D., Vionette-Amaral, R.J., Lowenberger, C., Rivera-Pomar, R., Monteiro, F.A., Minx, P., Spieth, J., Carvalho, A.B., Panzera, F., Lawson, D., Torres, A.Q., Ribeiro, J.M.C., Sorgine, M.H.F., Waterhouse, R.M., Montague, M.J., Abad-Franch, F., Alves-Bezerra, M., Amaral, L.R., Araujo, H.M., Araujo, R.N., Aravind, L., Atella, G.C., Azambuja, P., Berni, M., Bittencourt-Cunha, P.R., Braz, G.R.C., Calderón-Fernández, G., Carareto, C.M.A., Christensen, M.B., Costa, I.R., Costa, S.G., Dansa, M., Daumas, C.R.O., De-Paula, I.F., Dias, F.A., Dimopoulos, G., Emrich, S.J., Esponda-Behrens, N., Fampa, P., Fernandez-Medina, R.D., Da Fonseca, R.N., Fontenele, M., Fronick, C., Fulton, L.A., Gandara, A.C., Garcia, E.S., Genta, F.A., Giraldo-Calderón, G.I.,

Gomes, B., Gondim, K.C., Granzotto, A., Guarneri, A.A., Guigó, R., Harry, M., Hughes, D.S.T., Jablonka, W., Jacquin-Joly, E., Juarez, M.P., Koerich, L.B., Latorre-Estivalis, J.M., Lavore, A., Lawrence, G.G., Lazoski, C., Lazzari, C.R., Lopes, R.R., Lorenzo, M.G., Lugon, M.D., Majerowicz, D., Marcet, P.L., Mariotti, M., Masuda, H., Megy, K., Melo, A.C.A., Missirlis, F., Mota, T., Noriega, F.G., Nouzova, M., Nunes, R.D., Oliveira, R.L.L., Oliveira-Silveira, G., Ons, S., Pagola, L., Paiva-Silva, G.O., Pascual, A., Pavan, M.G., Pedrini, N., Peixoto, A.A., Pereira, M.H., Pike, A., Polycarpo, C., Prosdocimi, F., Ribeiro-Rodrigues, R., Robertson, H.M., Salerno, A.P., Salmon, D., Santesmasses, D., Schama, R., Seabra, E.S., Silva-Cardoso, L., Silva-Neto, M.A.C., et al., 2015. Genome of *Rhodnius prolixus*, an insect vector of Chagas disease, reveals unique adaptations to hematophagy and parasite infection. Proc. Natl. Acad. Sci. U. S. A. 112, 14936–14941.

Michalkova, V., Benoit, J.B., Weiss, B.L., Attardo, G.M., Aksoy, S., 2014. Vitamin B-6 generated by obligate symbionts is critical for maintaining proline homeostasis and fecundity in tsetse flies. Appl. Environ. Microbiol. 80, 5844–5853.

Mihok, S., Maramba, O., Munyoki, E., Kagoiya, J., 1995. Mechanical transmission of *Trypanosoma* spp by African Stomoxyinae (Diptera, Muscidae). Trop. Med. Parasitol. 46, 103–105.

Minard, G., Mavingui, P., Moro, C.V., 2013. Diversity and function of bacterial microbiota in the mosquito holobiont. Parasit. Vectors 6, 146.

Mira, A., Ochman, H., Moran, N.A., 2001. Deletional bias and the evolution of bacterial genomes. Trends Genet. 17, 589–596.

Mitraka, E., Stathopoulos, S., Siden-Kiamos, I., Christophides, G.K., Louis, C., 2013. *Asaia* accelerates larval development of *Anopheles gambiae*. Pathog. Glob. Health 107, 305–311.

Moll, R.M., Romoser, W.S., Modrzakowski, M.C., Moncayo, A.C., Lerdthusnee, K., 2001. Meconial peritrophic membranes and the fate of midgut bacteria during mosquito (Diptera: Culicidae) metamorphosis. J. Med. Entomol. 38, 29–32.

Monteiro, F.A., Weirauch, C., Felix, M., Lazoski, C., Abad-Franch, F., 2018. Evolution, systematics, and biogeography of the triatominae, vectors of Chagas disease. Adv. Parasitol. 99, 265–344.

Montgomery, S.P., Parise, M.E., Dotson, E.M., Bialek, S.R., 2016. What do we know about Chagas disease in the United States? Am. J. Trop. Med. Hyg. 95, 1225–1227.

Montoya-Porras, L.M., Omar, T.C., Alzate, J.F., Moreno-Herrera, C.X., Cadavid-Restrepo, G.E., 2018. 16S rRNA gene amplicon sequencing reveals dominance of Actinobacteria in *Rhodnius pallescens* compared to *Triatoma maculata* midgut microbiota in natural populations of vector insects from Colombia. Acta Trop. 178, 327–332.

Moran, N.A., 1996. Accelerated evolution and Muller's rachet in endosymbiotic bacteria. Proc. Natl. Acad. Sci. U. S. A. 93, 2873–2878.

Moran, N.A., Munson, M.A., Baumann, P., Ishikawa, H., 1993. A molecular clock in endosymbiotic bacteria is calibrated using the insect hosts. Proc. R. Soc. B Biol. Sci. 253, 167–171.

Munoz, M.M., Pollio, A.R., White, H.L., Rio, R.V.M., 2017. Into the wild: parallel transcriptomics of the tsetse-*Wigglesworthia* mutualism within Kenyan populations. Genome Biol. Evol. 9, 2276–2291.

Muturi, E.J., Bara, J.J., Rooney, A.P., Hansen, A.K., 2016a. Midgut fungal and bacterial microbiota of *Aedes triseriatus* and *Aedes japonicus* shift in response to La Crosse virus infection. Mol. Ecol. 25, 4075–4090.

Muturi, E.J., Kim, C.H., Bara, J., Bach, E.M., Siddappaji, M.H., 2016b. *Culex pipiens* and *Culex restuans* mosquitoes harbor distinct microbiota dominated by few bacterial taxa. Parasit. Vectors 9, 18.

Muturi, E.J., Ramirez, J.L., Rooney, A.P., Kim, C.H., 2017. Comparative analysis of gut microbiota of mosquito communities in central Illinois. PLoS Negl. Trop. Dis. 11, e0005377.

Muturi, E.J., Lagos-Kutz, D., Dunlap, C., Ramirez, J.L., Rooney, A.P., Hartman, G.L., Fields, C.J., Rendon, G., Kim, C.H., 2018. Mosquito microbiota cluster by host sampling location. Parasit. Vectors 11, 468.

Nishide, Y., Kageyama, D., Yokoi, K., Jouraku, A., Tanaka, H., Futahashi, R., Fukatsu, T., 2019. Functional crosstalk across IMD and Toll pathways: insight into the evolution of incomplete immune cascades. Proc. R. Soc. B Biol. Sci. 286, 20182207.

Nováková, E., Husník, F., Šochová, E., Hypša, V., 2015. *Arsenophonus* and *Sodalis* symbionts in louse flies: an analogy to the *Wigglesworthia* and *Sodalis* system in tsetse flies. Appl. Environ. Microbiol. 81, 6189–6199.

Nyirady, S.A., 1973. Germfree culture of 3 species of Triatominae—*Triatoma protracta* (Uhler), *Triatoma rubida* (Uhler) and *Rhodnius prolixus* Stäl. J. Med. Entomol. 10, 417–448.

O'Meara, G.F., 1985. Gonotrophic interactions in mosquitoes—kicking the blood-feeding habit. Fla. Entomol. 68, 122–133.

Oakeson, K.F., Gil, R., Clayton, A.L., Dunn, D.M., Von Niederhausern, A.C., Hamil, C., Aoyagi, A., Duval, B., Baca, A., Silva, F.J., Vallier, A., Jackson, D.G., Latorre, A., Weiss, R.B., Heddi, A., Moya, A., Dale, C., 2014. Genome degeneration and adaptation in a nascent stage of symbiosis. Genome Biol. Evol. 6, 76–93.

Odeniran, P.O., Macleod, E.T., Ademola, I.O., Welburn, S.C., 2019. Endosymbionts interaction with trypanosomes in Palpalis group of *Glossina* captured in southwest Nigeria. Parasitol. Int. 70, 64–69.

Oliveira, J.L., Cury, J.C., Gurgel-Gonçalves, R., Bahia, A.C., Monteiro, F.A., 2018. Field-collected *Triatoma sordida* from central Brazil display high microbiota diversity that varies with regard to developmental stage and intestinal segmentation. PLoS Negl. Trop. Dis. 12, e0006709.

Osei-Poku, J., Mbogo, C.M., Palmer, W.J., Jiggins, F.M., 2012. Deep sequencing reveals extensive variation in the gut microbiota of wild mosquitoes from Kenya. Mol. Ecol. 21, 5138–5150.

Pachebat, J.A., Van Keulen, G., Whitten, M.M., Girdwood, S., Del Sol, R., Dyson, P.J., Facey, P.D., 2013. Draft genome sequence of *Rhodococcus rhodnii* strain LMG5362, a symbiont of *Rhodnius prolixus* (Hemiptera, Reduviidae, Triatominae), the principle vector of *Trypanosoma cruzi*. Genome Announc 1, e00329-13.

Pais, R., Lohs, C., Wu, Y.N., Wang, J.W., Aksoy, S., 2008. The obligate mutualist *Wigglesworthia glossinidia* influences reproduction, digestion, and immunity processes of its host, the tsetse fly. Appl. Environ. Microbiol. 74, 5965–5974.

Pennington, J.E., Goldstrohm, D.A., Wells, M.A., 2003. The role of hemolymph proline as a nitrogen sink during blood meal digestion by the mosquito *Aedes aegypti*. J. Insect Physiol. 49, 115–121.

Petersen, F.T., Meier, R., Kutty, S.N., Wiegmann, B.M., 2007. The phylogeny and evolution of host choice in the Hippoboscoidea (Diptera) as reconstructed using four molecular markers. Mol. Phylogenet. Evol. 45, 111–122.

Peterson, J.K., Graham, A.L., Dobson, A.P., Chávez, O.T., 2015. *Rhodnius prolixus* life history outcomes differ when infected with different *Trypanosoma cruzi* I strains. Am. J. Trop. Med. Hyg. 93, 564–572.

Piermarini, P.M., 2016. Renal excretory processes in mosquitoes. Adv. Insect Physiol. 51, 393–433.

Pumpuni, C.B., Beier, M.S., Nataro, J.P., Guers, L.D., Davis, J.R., 1993. *Plasmodium falciparum*: inhibition of sporogonic development in *Anopheles stephensi* by Gram-negative bacteria. Exp. Parasitol. 77, 195–199.

Pumpuni, C.B., Demaio, J., Kent, M., Davis, J.R., Beier, J.C., 1996. Bacterial population dynamics in three anopheline species: the impact on *Plasmodium* sporogonic development. Am. J. Trop. Med. Hyg. 54, 214–218.

Ramirez, J.L., Souza-Neto, J., Torres Cosme, R., Rovira, J., Ortiz, A., Pascale, J.M., Dimopoulos, G., 2012. Reciprocal tripartite interactions between the *Aedes aegypti* midgut microbiota, innate immune system and dengue virus influences vector competence. PLoS Neglect. Trop. Dis. 6, e1561.

Ramirez, J.L., Short, S.M., Bahia, A.C., Saraiva, R.G., Dong, Y., Kang, S., Tripathi, A., Mlambo, G., Dimopoulos, G., 2014. *Chromobacterium* Csp_P reduces malaria and dengue infection in vector mosquitoes and has entomopathogenic and *in vitro* anti-pathogen activities. PLoS Pathog. 10, e1004398.

Reisenman, C.E., Gregory, T., Guerenstein, P.G., Hildebrand, J.G., 2011. Feeding and defecation behavior of *Triatoma rubida* (Uhler, 1894) (Hemiptera: Reduviidae) under laboratory conditions, and its potential role as a vector of Chagas disease in Arizona, USA. Am. J. Trop. Med. Hyg. 85, 648–656.

Rio, R.V., Symula, R.E., Wang, J., Lohs, C., Wu, Y.N., Snyder, A.K., Bjornson, R.D., Oshima, K., Biehl, B.S., Perna, N.T., Hattori, M., Aksoy, S., 2012. Insight into the transmission biology and species-specific functional capabilities of tsetse (Diptera: glossinidae) obligate symbiont *Wigglesworthia*. mBio 3, e00240–e00311.

Rio, R.V.M., Jozwick, A.K.S., Savage, A.F., Sabet, A., Vigneron, A., Wu, Y.N., Aksoy, S., Weiss, B.L., 2019. Mutualist-provisioned resources impact vector competency. mBio 10, e00018-19.

Rodgers, F.H., Gendrin, M., Christophides, G.K., 2017a. Chapter 6—The mosquito immune system and its interactions with the microbiota: implications for disease transmission. In: Wikel, S.K., Aksoy, S., Dimopoulos, G. (Eds.), Arthropod Vector: Controller of Disease Transmission. vol. 1. Academic Press, London.

Rodgers, F.H., Gendrin, M., Wyer, C.A.S., Christophides, G.K., 2017b. Microbiota-induced peritrophic matrix regulates midgut homeostasis and prevents systemic infection of malaria vector mosquitoes. PLoS Pathog. 13, e1006391.

Rodríguez-Ruano, S.M., Škochová, V., Rego, R.O.M., Schmidt, J.O., Roachell, W., Hypša, V., Nováková, E., 2018. Microbiomes of North American Triatominae: the grounds for Chagas disease epidemiology. Front. Microbiol. 9, 1167.

Romoli, O., Schönbeck, J.C., Hapfelmeier, S., Gendrin, M., 2019. Bacteria contribute to mosquito development via folate biosynthesis and energy storage. bioRxiv. https://doi.org/10.1101/2019.12.31.891671.

Rossi, P., Ricci, I., Cappelli, A., Damiani, C., Ulissi, U., Mancini, M.V., Valzano, M., Capone, A., Epis, S., Crotti, E., Chouaia, B., Scuppa, P., Joshi, D., Xi, Z., Mandrioli, M., Sacchi, L., O'Neill, S.L., Favia, G., 2015. Mutual exclusion of *Asaia* and *Wolbachia* in the reproductive organs of mosquito vectors. Parasit. Vectors 8, 278.

Roundy, C.M., Azar, S.R., Rossi, S.L., Weaver, S.C., Vasilakis, N., 2017. Insect-specific viruses: a historical overview and recent developments. Adv. Virus Res. 98, 119–146.

Rozeboom, L.E., 1935. The relation of bacteria and bacterial filtrates to the development of mosquito larvae. Am. J. Epidemiol. 21, 167–179.

Salcedo-Porras, N., Lowenberger, C., 2019. The innate immune system of kissing bugs, vectors of chagas disease. Dev. Comp. Immunol. 98, 119–128.

Santiago, P.B., De Araújo, C.N., Charneau, S., Praça, Y.R., Bastos, I.M.D., Ribeiro, J.M.C., Santana, J.M., 2020. The pharmacopea within triatomine salivary glands. Trends Parasitol. 36, 250–265.

Santos-Garcia, D., Vargas-Chavez, C., Moya, A., Latorre, A., Silva, F.J., 2015. Genome evolution in the primary endosymbiont of whiteflies sheds light on their divergence. Genome Biol. Evol. 7, 873–888.

Schmidt, T.L., Barton, N.H., Rašić, G., Turley, A.P., Montgomery, B.L., Iturbe-Ormaetxe, I., Cook, P.E., Ryan, P.A., Ritchie, S.A., Hoffmann, A.A., O'Neill, S.L., Turelli, M., 2017. Local introduction and heterogeneous spatial spread of dengue-suppressing *Wolbachia* through an urban population of *Aedes aegypti*. PLoS Biol. 15, e2001894.

Schofield, C.J., 2000. *Trypanosoma cruzi*—The vector-parasite paradox. Memorias Do Instituto Oswaldo Cruz 95, 535–544.

Shim, J., Mukherjee, T., Mondal, B.C., Liu, T., Young, G.C., Wijewarnasuriya, D.P., Banerjee, U., 2013. Olfactory control of blood progenitor maintenance. Cell 155, 1141–1153.

Snyder, A.K., Rio, R.V.M., 2015. "*Wigglesworthia morsitans*" folate (vitamin B-9) biosynthesis contributes to tsetse host fitness. Appl. Environ. Microbiol. 81, 5375–5386.

Song, X., Wang, M., Dong, L., Zhu, H., Wang, J., 2018. PGRP-LD mediates *A. stephensi* vector competency by regulating homeostasis of microbiota-induced peritrophic matrix synthesis. PLoS Pathog. 14, e1006899.

Šorfová, P., Škeříková, A., Hypša, V., 2008. An effect of 16S rRNA intercistronic variability on coevolutionary analysis in symbiotic bacteria: molecular phylogeny of *Arsenophonus triatominarum*. Syst. Appl. Microbiol. 31, 88–100.

Steen, A.D., Crits-Christoph, A., Carini, P., Deangelis, K.M., Fierer, N., Lloyd, K.G., Cameron Thrash, J., 2019. High proportions of bacteria and Archaea across most biomes remain uncultured. ISME J. 13, 3126–3130.

Steyn, A., Roets, F., Botha, A., 2016. Yeasts associated with *Culex pipiens* and *Culex theileri* mosquito larvae and the effect of selected yeast strains on the ontogeny of *Culex pipiens*. Microb. Ecol. 71, 747–760.

Strand, M.R., 2017. Chapter 11—The gut microbiota of mosquitoes: diversity and function. In: Wikel, S.K., Aksoy, S., Dimopoulos, G. (Eds.), Arthropod Vector: Controller of Disease Transmission. vol. 1. Academic Press, London.

Strand, M.R., 2018. Composition and functional roles of the gut microbiota in mosquitoes. Curr. Opin. Insect Sci. 28, 59–65.

Su, T., Mulla, M.S., 1997. Nutritional reserves, body weight, and starvation tolerance of autogenous and anautogenous strains of *Culex tarsalis* (Diptera:Culicidae). J. Med. Entomol. 34, 68–73.

Tagueu, S.K., Farikou, O., Njiokou, F., Simo, G., 2018. Prevalence of *Sodalis glossinidius* and different trypanosome species in *Glossina palpalis palpalis* caught in the Fontem sleeping sickness focus of the southern Cameroon. Parasite 25, 44.

Tawidian, P., Rhodes, V.L., Michel, K., 2019. Mosquito-fungus interactions and antifungal immunity. Insect Biochem. Mol. Biol. 111, 103182.

Tchioffo, M.T., Boissière, A., Churcher, T.S., Abate, L., Gimonneau, G., Nsango, S.E., Awono-Ambene, P.H., Christen, R., Berry, A., Morlais, I., 2013. Modulation of malaria infection in *Anopheles gambiae* mosquitoes exposed to natural midgut bacteria. PLoS One 8, e81663.

Tchioffo, M.T., Boissière, A., Abate, L., Nsango, S.E., Bayibeki, A.N., Awono-Ambene, P.H., Christen, R., Gimonneau, G., Morlais, I., 2015. Dynamics of bacterial community composition in the malaria mosquito's epithelia. Front. Microbiol. 6, 1500.

Tchioffo, M.T., Abate, L., Boissière, A., Nsango, S.E., Gimonneau, G., Berry, A., Oswald, E., Dubois, D., Morlais, I., 2016. An epidemiologically successful *Escherichia coli* sequence type modulates *Plasmodium falciparum* infection in the mosquito midgut. Infect. Genet. Evol. 43, 22–30.

Teixeira, L., Ferreira, A., Ashburner, M., 2008. The bacterial symbiont *Wolbachia* induces resistance to RNA viral infections in *Drosophila melanogaster*. PLoS Biol. 6, e2.

Telang, A., Wells, M.A., 2004. The effect of larval and adult nutrition on successful autogenous egg production by a mosquito. J. Insect Physiol. 50, 677–685.

Telang, A., Li, Y., Noriega, F.G., Brown, M.R., 2006. Effects of larval nutrition on the endocrinology of mosquito egg development. J. Exp. Biol. 209, 645–655.

Terenius, O., Lindh, J.M., Eriksson-Gonzales, K., Bussiére, L., Laugen, A.T., Bergquist, H., Titanji, K., Faye, I., 2012. Midgut bacterial dynamics in *Aedes aegypti*. FEMS Microbiol. Ecol. 80, 556–565.

Thomas, T., De, T.D., Sharma, P., Lata, S., Saraswat, P., Pandey, K.C., Dixit, R., 2016. Hemocytome: deep sequencing analysis of mosquito blood cells in Indian malarial vector *Anopheles stephensi*. Gene 585, 177–190.

Thongsripong, P., Chandler, J.A., Green, A.B., Kittayapong, P., Wilcox, B.A., Kapan, D.D., Bennett, S.N., 2018. Mosquito vector-associated microbiota: metabarcoding bacteria and eukaryotic symbionts across habitat types in Thailand endemic for dengue and other arthropod-borne diseases. Ecol. Evol. 8, 1352–1368.

Trappeniers, K., Matetovici, I., Van Den Abbeele, J., De Vooght, L., 2019. The tsetse fly displays an attenuated immune response to its secondary symbiont, *Sodalis glossinidius*. Front. Microbiol. 10, 1650.

Trexler, J.D., Apperson, C.S., Zurek, L., Gemeno, C., Schal, C., Kaufman, M., Walker, E., Watson, D.W., Wallace, L., 2003. Role of bacteria in mediating the oviposition responses of *Aedes albopictus* (Diptera: Culicidae). J. Med. Entomol. 40, 841–848.

Turelli, M., 2010. Cytoplasmic incompatibility in populations with overlapping generations. Evolution 64, 232–241.

Turelli, M., Hoffmann, A.A., 1991. Rapid spread of an inherited incompatibility factor in California *Drosophila*. Nature 353, 440–442.

Valenzuela, J.G., Francischetti, I.M., Pham, V.M., Garfield, M.K., Ribeiro, J.M., 2003. Exploring the salivary gland transcriptome and proteome of the *Anopheles stephensi* mosquito. Insect Biochem. Mol. Biol. 33, 717–732.

Valzania, L., Coon, K.L., Vogel, K.J., Brown, M.R., Strand, M.R., 2018a. Hypoxia-induced transcription factor signaling is essential for larval growth of the mosquito *Aedes aegypti*. Proc. Natl. Acad. Sci. U. S. A. 115, 457–465.

Valzania, L., Martinson, V.G., Harrison, R.E., Boyd, B.M., Coon, K.L., Brown, M.R., Strand, M.R., 2018b. Both living bacteria and eukaryotes in the mosquito gut promote growth of larvae. PLoS Negl. Trop. Dis. 12, e0006638.

Van Tol, S.D., Dimopoulos, G., 2016. Influences of the mosquito microbiota on vector competence. Adv. Insect Physiol. 51, 243–291.

Verhulst, N.O., Beijleveld, H., Knols, B.G., Takken, W., Schraa, G., Bouwmeester, H.J., Smallegange, R.C., 2009. Cultured skin microbiota attracts malaria mosquitoes. Malar. J. 8, 302.

Verhulst, N.O., Andriessen, R., Groenhagen, U., Bukovinszkiné Kiss, G., Schulz, S., Takken, W., Van Loon, J.J., Schraa, G., Smallegange, R.C., 2010. Differential attraction of malaria mosquitoes to volatile blends produced by human skin bacteria. PLoS One 5, e15829.

Verhulst, N.O., Qiu, Y.T., Beijleveld, H., Maliepaard, C., Knights, D., Schulz, S., Berg-Lyons, D., Lauber, C.L., Verduijn, W., Haasnoot, G.W., Mumm, R., Bouwmeester, H.J., Claas, F.H., Dicke, M., Van Loon, J.J., Takken, W., Knight, R., Smallegange, R.C., 2011. Composition of human skin microbiota affects attractiveness to malaria mosquitoes. PLoS One 6, e28991.

Vieira, C.S., Waniek, P.J., Mattos, D.P., Castro, D.P., Mello, C.B., Ratcliffe, N.A., Garcia, E.S., Azambuja, P., 2014. Humoral responses in *Rhodnius prolixus*: bacterial feeding induces differential patterns of antibacterial activity and enhances mRNA levels of antimicrobial peptides in the midgut. Parasit. Vectors 7, 232.

Vieira, C.S., Moreira, O.C., Batista, K.K.S., Ratcliffe, N.A., Castro, D.P., Azambuja, P., 2018. The NF-kB inhibitor, IMD-0354, affects immune gene expression, bacterial microbiota and *Trypanosoma cruzi* infection in *Rhodnius prolixus* midgut. Front. Physiol. 9, 1189.

Vogel, K.J., Valzania, L., Coon, K.L., Brown, M.R., Strand, M.R., 2017. Transcriptome sequencing reveals large-scale changes in axenic *Aedes aegypti* larvae. PLoS Neglect. Trop. Dis. 11, e0005273.

Walker, E.D., Merritt, R.W., 1991. Behavior of larval *Aedes triseriatus* (Diptera: Culicidae). J. Med. Entomol. 28, 581–589.

Walker, E.D., Olds, E.J., Merritt, R.W., 1988. Gut content analysis of mosquito larvae (Diptera: Culicidae) using DAPI stain and epifluorescence microscopy. J. Med. Entomol. 25, 551–554.

Walker, T., Johnson, P.H., Moreira, L.A., Iturbe-Ormaetxe, I., Frentiu, F.D., McMeniman, C.J., Leong, Y.S., Dong, Y., Axford, J., Kriesner, P., Lloyd, A.L., Ritchie, S.A., O'Neill, S.L., Hoffmann, A.A., 2011. The wMel *Wolbachia* strain blocks dengue and invades caged *Aedes aegypti* populations. Nature 476, 450–453.

Waltmann, A., Willcox, A.C., Balasubramanian, S., Borrini Mayori, K., Mendoza Guerrero, S., Salazar Sanchez, R.S., Roach, J., Condori Pino, C., Gilman, R.H., Bern, C., Juliano, J.J., Levy, M.Z., Meshnick, S.R., Bowman, N.M., 2019. Hindgut microbiota in laboratory-reared and wild *Triatoma infestans*. PLoS Negl. Trop. Dis. 13, e0007383.

Wamwiri, F.N., Alam, U., Thande, P.C., Aksoy, E., Ngure, R.M., Aksoy, S., Ouma, J.O., Murilla, G.A., 2013. *Wolbachia, Sodalis* and trypanosome co-infections in natural populations of *Glossina austeni* and *Glossina pallidipes*. Parasit. Vectors 6, 232.

Wang, J., Aksoy, S., 2012. PGRP-LB is a maternally transmitted immune milk protein that influences symbiosis and parasitism in tsetse's offspring. Proc. Natl. Acad. Sci. U. S. A. 109, 10552–10557.

Wang, S., Jacobs-Lorena, M., 2013. Genetic approaches to interfere with malaria transmission by vector mosquitoes. Trends Biotechnol. 31, 185–193.

Wang, J., Wu, Y., Yang, G., Aksoy, S., 2009. Interactions between mutualist *Wigglesworthia* and tsetse peptidoglycan recognition protein (PGRP-LB) influence trypanosome transmission. Proc. Natl. Acad. Sci. U. S. A. 106, 12133–12138.

Wang, Y., Gilbreath 3rd, T.M., Kukutla, P., Yan, G., Xu, J., 2011. Dynamic gut microbiome across life history of the malaria mosquito *Anopheles gambiae* in Kenya. PLoS One 6, e24767.

Wang, X., Liu, T., Wu, Y., Zhong, D., Zhou, G., Su, X., Xu, J., Sotero, C.F., Sadruddin, A.A., Wu, K., Chen, X.G., Yan, G., 2018. Bacterial microbiota assemblage in *Aedes albopictus* mosquitoes and its impacts on larval development. Mol. Ecol. 27, 2972–2985.

Wei, G., Lai, Y., Wang, G., Chen, H., Li, F., Wang, S., 2017. Insect pathogenic fungus interacts with the gut microbiota to accelerate mosquito mortality. Proc. Natl. Acad. Sci. U. S. A. 114, 5994–5999.

Weiss, B.L., Wang, J.W., Aksoy, S., 2011. Tsetse immune system maturation requires the presence of obligate symbionts in larvae. PLoS Biol. 9, e1000619.

Weiss, B.L., Maltz, M., Aksoy, S., 2012. Obligate symbionts activate immune system development in the tsetse fly. J. Immunol. 188, 3395–3403.

Weiss, B.L., Wang, J.W., Maltz, M.A., Wu, Y.N., Aksoy, S., 2013. Trypanosome infection establishment in the tsetse fly gut is influenced by microbiome-regulated host immune barriers. PLoS Pathog. 9, e1003318.

Weiss, B.L., Maltz, M.A., Vigneron, A., Wu, Y., Walter, K.S., O'Neill, M.B., Wang, J., Aksoy, S., 2019. Colonization of the tsetse fly midgut with commensal *Kosakonia cowanii* Zambiae inhibits trypanosome infection establishment. PLoS Pathog. 15, e1007470.

Welburn, S.C., Arnold, K., Maudlin, I., Gooday, G.W., 1993. *Rickettsia*-like organisms and chitinase production in relation to transmission of trypanosomes by tsetse flies. Parasitology 107 (Pt. 2), 141–145.

Weurman, C., 1946. Investigations concerning the symbiosis of bacteria in *Triatoma Infestans*. Antonie Van Leeuwenhoek 11, 129–138.

WHO, 2015. Chagas disease in Latin America: an epidemiological update based on 2010 estimates. Wkly Epidemiol. Rec. 90, 33–43.

Wiegmann, B.M., Trautwein, M.D., Winkler, I.S., Barr, N.B., Kim, J.W., Lambkin, C., Bertone, M.A., Cassel, B.K., Bayless, K.M., Heimberg, A.M., Wheeler, B.M., Peterson, K.J., Pape, T., Sinclair, B.J., Skevington, J.H., Blagoderov, V., Caravas, J., Kutty, S.N., Schmidt-Ott, U., Kampmeier, G.E., Thompson, F.C., Grimaldi, D.A., Beckenbach, A.T., Courtney, G.W., Friedrich, M., Meier, R., Yeates, D.K., 2011. Episodic radiations in the fly tree of life. Proc. Natl. Acad. Sci. U. S. A. 108, 5690–5695.

Wigglesworth, V.B., 1926. Digestion in the tsetse-fly. Parasitology 21, 288–312.

Wigglesworth, V.B., 1936. Symbiotic bacteria in a blood-sucking insect, *Rhodnius prolixus* Stål (Hemiptera, Triatominae). Parasitology 28, 284–289.

Wilcox, J.L., Dunbar, H.E., Wolfinger, R.D., Moran, N.A., 2003. Consequences of reductive evolution for gene expression in an obligate endosymbiont. Mol. Microbiol. 48, 1491–1500.

Wood, S.F., 1951. Importance of feeding and defecation times of insect vectors in transmission of Chagas disease. J. Econ. Entomol. 44, 52–54.

Wotton, R.S., Chaloner, D.T., Yardley, C.A., Merritt, R.W., 1997. Growth of *Anopheles* mosquito larvae on dietary microbiota in aquatic surface microlayers. Med. Vet. Entomol. 11, 65–70.

Wu, P., Sun, P., Nie, K., Zhu, Y., Shi, M., Xiao, C., Liu, H., Liu, Q., Zhao, T., Chen, X., Zhou, H., Wang, P., Cheng, G., 2019. A gut commensal bacterium promotes mosquito permissiveness to arboviruses. Cell Host Microbe 25, 101–112e5.

Xi, Z., Khoo, C.C., Dobson, S.L., 2005. *Wolbachia* establishment and invasion in an *Aedes aegypti* laboratory population. Science 310, 326–328.

Xi, Z., Ramirez, J.L., Dimopoulos, G., 2008. The *Aedes aegypti* toll pathway controls dengue virus infection. PLoS Pathog. 4, e1000098.

Zaspel, J.M., Hoy, M.A., 2008. Microbial diversity associated with the fruit-piercing and blood-feeding moth *Calyptra thalictri* (Lepidoptera: Noctuidae). Ann. Entomol. Soc. Am. 101, 1050–1055.

Zouache, K., Raharimalala, F.N., Raquin, V., Tran-Van, V., Raveloson, L.H., Ravelonandro, P., Mavingui, P., 2011. Bacterial diversity of field-caught mosquitoes, *Aedes albopictus* and *Aedes aegypti*, from different geographic regions of Madagascar. FEMS Microbiol. Ecol. 75, 377–389.

Zumaya-Estrada, F.A., Martínez-Barnetche, J., Lavore, A., Rivera-Pomar, R., Rodríguez, M.H., 2018. Comparative genomics analysis of triatomines reveals common first line and inducible immunity-related genes and the absence of Imd canonical components among hemimetabolous arthropods. Parasit. Vectors 11, 48.

CHAPTER EIGHT

Toxin-mediated protection against natural enemies by insect defensive symbionts

Kerry M. Oliver[a], **Steve J. Perlman**[b]
[a]Department of Entomology, University of Georgia, Athens, GA, United States
[b]Department of Biology, University of Victoria, Victoria, BC, Canada

Contents

Abstract

All organisms have evolved a wide range of strategies to protect against natural enemies, from complex avoidance behaviours to sophisticated immune systems, and insects are no exception. A more recently recognized, and likely widespread, mode of protection for insects is through the use of microbial symbionts. Many of the best studied examples of defensive symbiosis in insects involve infections with maternally inherited symbionts. While inherited symbionts can protect their insect hosts by priming innate immune systems, or competing with enemies for scarce recourses, they have also been found to synthesize a wide diversity of toxins that effectively kill, weaken, or deter parasitic wasps, parasitic nematodes, pathogenic fungi, and predatory spiders. In this chapter, we provide an overview of defensive symbioses in insects, focusing in particular on toxin-based protection.

Advances in Insect Physiology, Volume 58
ISSN 0065-2806
https://doi.org/10.1016/bs.aiip.2020.03.005

1. Introduction

Insects are key components of terrestrial ecosystems, often functioning as the dominant herbivores, decomposers and natural enemies in a given community. These diverse functional roles are often aided by the presence of resident microbes, which can bring novel capabilities that complement existing host function or provide novel functions to the host (Douglas, 2015; Engel and Moran, 2013; Feldhaar, 2011; Frago et al., 2012; Hansen and Moran, 2014; Oliver and Russell, 2016). The basis for the establishment, and often long-term persistence of insect-microbe associations is simply that the latter encode far greater biosynthetic potential as a group relative to the former (Moran, 2007; Smith, 1989). Hence while partnering with microbes may incur both short- and long-term costs, the acquisition of new or improved functions via symbiont infection may also bring net fitness benefits that stabilize specific insect-microbe associations. While widespread roles for symbionts in degrading plant polymers and provisioning limiting nutrients have long been recognized (Baumann, 2005; Breznak and Brune, 1994; Brune, 2014; Buchner, 1965), the past 15 years have also seen the emergence of defensive symbiosis as a common phenomenon in insects (Ballinger and Perlman, 2019; Brownlie and Johnson, 2009; Flórez et al., 2015; Van Arnam et al., 2018; Vorburger, 2014). Perhaps this is not surprising given that microbes, especially bacteria, are well known for producing diverse, potent toxins that harm animal tissues or other microbes (Flórez et al., 2015; Van Arnam et al., 2018), combined with the shared threat of enemy attack, regardless of feeding niche. Of course, insects have evolved a range of behaviours and endogenous physiological defences, both cellular and humoral, to avoid or thwart attack from enemies (Gross, 1993; Lemaitre and Hoffmann, 2007; Strand, 2008). However, many insects also leverage microbial infections to aid in their defences, including the deployment of tactics that incorporate microbe-encoded toxins, which will be the focus of this review.

1.1 Insect-microbe symbioses

Insect tissues exposed to the environment, including the cuticle and gut, are routinely colonized by microorganisms. Cuticular symbionts are often involved in pathogen defence, while gut symbionts may facilitate nutrient acquisition, degrade harmful compounds or protect against ingested pathogens (Douglas, 2015; Engel and Moran, 2013). Gut associated-microbes persist extracellularly and are typically environmentally transmitted (e.g. acquired in diet and excreted) with microbial diversity and specificity, as well as any host specialization (e.g. symbiont crypts in midgut), being highly variable across taxa and depending on the environment. Other microbes are largely or entirely restricted to the tissues or cells of their insect hosts. These 'endosymbionts' often, but not always, persist intracellularly and are transmitted vertically from mothers to eggs or embryos (Moran et al., 2008). By definition, such heritable symbionts are specialized to the symbiotic lifestyle, although to differing degrees. Some associations are mutually obligate in which the symbiont(s) and host each require the other for survival and reproduction. These tend to be ancient, highly-specialized associations where specific bacteria are sequestered into host cells called bacteriocytes and involve sophisticated metabolic coordination (where studied) between partners (Douglas, 2014; Feng et al., 2019; Moran et al., 2008; Wilson and Duncan, 2015). For example, all plant sap-restricted (i.e. xylem or phloem) herbivores house bacteriocyte-associated symbionts that provision amino acids and co-factors that occur in insufficient quantities in their diets (Baumann, 2005; Buchner, 1965). Past symbiont infections enabled the use of plant-sap as a niche, which in turn facilitated diversification of aphids, cicadas and other sternorrhynchan and auchenorrhynchan hemipterans. The genomes of bacteriocyte-associated symbionts are highly reduced, often tiny, compared to free-living bacteria but tellingly retain genes for the biosynthesis of key nutrients (Moran et al., 2008; Shigenobu et al., 2000). This repeating pattern combined with the mere presence of these symbionts in all individuals of all species feeding on nutrient-restricted substrates, provides compelling correlational evidence of the generality of these nutritional mutualists, although few experimental studies confirm these roles (Buchner, 1965; Gunduz and Douglas, 2009). One nutritional symbiont, *Profftella armatura* (betaproteobacteria), also encodes a polyketide toxin called diaphorin that potentially functions in defence (Nakabachi et al., 2013) and is discussed in more detail below.

A second class of heritable symbionts are facultative from the host perspective, meaning they are not generally required for host survival, although insect survival may be contingent on their presence under some conditions (Oliver et al., 2014). Heritable facultative symbionts are very common, arguably infecting most insect species, although distributions both within and among taxa are highly variable (Moran et al., 2008; Weinert et al., 2015). While most facultative symbionts persist intracellularly, they can vary substantially in the tissues they inhabit, which are determined by host and symbiont genotypes. The maternal inheritance patterns of facultative symbionts provide two routes for symbiont spread and persistence within host populations: (1) they can spread by providing net fitness benefits to hosts, and/or (2) manipulate sexual reproduction in ways that favour the spread of infected females (Werren and O'Neill, 1997). In other words, if heritable facultative symbionts are present at moderate infection frequencies, then they are likely having important effects on host biology. Most of the early work on heritable facultative symbionts focused on reproductive manipulators as their symbiont-mediated phenotypes (e.g. biased sex ratios) were often conspicuous. Theoretical and experimental work identified four major types of reproductive manipulation: (a) male-killing, where the sons of symbiont-infected females die early in development, (b) feminization, where symbiont-infected genetic males are transformed into phenotypic females that can transmit symbionts, (c) parthenogenesis-induction, where symbiont-infected females reproduce clonally, and (d) cytoplasmic incompatibility, where matings between symbiont-infected males and uninfected females produce virtually no offspring, resulting in a relative benefit to symbiont-infected females (Werren and O'Neill, 1997). The bacterial lineages associated with these strategies, including *Wolbachia, Cardinium, Spiroplasma, Rickettsia,* and *Arsenophonus,* infect a broad range of insect taxa (Duron et al., 2008). Many of these same species were subsequently found to perform beneficial roles, especially in host defence (e.g. Hedges et al., 2008; Jaenike et al., 2010; Teixeira et al., 2008; Xie et al., 2014). In addition, other facultative symbiont species not previously associated with reproductive manipulation were also found to confer benefits to hosts. These include defence against thermal stress, pathogens, parasites, parasitoids and predators, as well as expansion of dietary breadth and other interactions with food plants (Hansen and Moran, 2014; Oliver and Moran, 2009). Most, but not all, defensive symbionts characterized in insects are heritable, facultative symbionts, so they will be the focus of this review.

1.2 Defensive symbionts

Insect are attacked by a range of microbial pathogens (viruses, bacteria, fungi and eukaryotic microbes), parasites (especially nematodes), parasitoids (especially hymenopteran wasps), and predators from diverse taxa. While variable across groups, common defences include avoidance and escape behaviours, physical barriers, as well as cellular and humoral responses when barriers are breached (Lemaitre and Hoffmann, 2007). Cellular responses include the phagocytosis of invading pathogens and the encapsulation and asphyxiation by haemocytes of larger invaders, including parasitoids (Strand, 2008). Humoral defences involve the production of inducible bioactive factors including antimicrobial peptides (AMPs) that may target specific groups of bacteria and fungi (Sheehan et al., 2018; Zasloff, 2002). In some groups, more specialized and sophisticated endogenous responses have evolved, including the chemical explosions of bombardier beetles and the cantharidin toxins of blister beetles (Nelsen et al., 2014).

To bolster endogenous defences, insects may benefit from symbiont infections in three, not mutually exclusive ways (Haine, 2008). First, infection with heritable symbionts may prime the innate immune system such that responses are stronger upon subsequent enemy challenge. For example, some strains of *Wolbachia* that have been artificially transfected from *Drosophila* into mosquitoes upregulate host immunity pathways, effectively suppressing a range of vectored human pathogens, including Dengue, Zika, Chikungunya, West Nile virus and *Plasmodium* (Caragata et al., 2019). A second means by which symbiont infection can provide host protection is via direct competition with natural enemies for limiting resources. In *Drosophila melanogaster*, *Wolbachia* is reported to compete with Drosophila C virus for cholesterol (Caragata et al., 2013), while the symbiont *Spiroplasma* competes with internally-developing parasitoids for lipids (Paredes et al., 2016). While some level of resource competition between facultative symbionts and eukaryotic enemies is inevitable, the relatively low abundance of the symbionts, together with often varying tissue tropism and phylogenetic distance between them suggests this mode may not generally produce large effects. However, small effects may tip the precarious balance in the host-parasite tug-of-war, and may serve as an adjuvant for other means of symbiont-protection. For example, limiting resources, such as cholesterol or lipids (Caragata et al., 2013; Paredes et al., 2016) may serve to trigger the immune system, or may be involved in packaging or delivery of toxic symbiont-derived factors.

Finally, as a third strategy, hosts may employ the arsenal of bacterial encoded toxins to their own benefit. Bacteria synthesize an astonishing range of secreted factors that harm other entities. In free-living bacteria, toxins mediate competition with other microbes and protect against predators, while in pathogenic bacteria, toxins may harm host tissues, subvert host immune responses, or harm competing microbes. Most are proteins that are synthesized within the cell and exported to targets (exotoxins), but others, lipopolysaccharides (LPS), are membrane associated compounds released after cell disruption (endotoxins) (Kumar et al., 2019). Among the most potent compounds characterized, bacterial toxins exhibit virulence via a wide range of activities, including inhibition of DNA/protein synthesis, activation or inhibition of adenylate cyclase, pore formation or cell lysis, and modification of signalling pathways (Kumar et al., 2019; Schmitt et al., 1999). In general, toxin production presumably benefits the bacterial pathogen as enhanced virulence is expected to increase opportunities for replication and transmission. It is also useful to distinguish between toxic factors that are employed to enter and affect eukaryotic cells, as opposed to prokaryotic ones.

Several symbionts from diverse insect taxa use microbial toxins to kill parasites, parasitoids and predators. Although there are still only a handful of documented cases, employment of toxic factors is probably the most common mechanism of symbiont-mediated protection. This may perhaps be because, similar to qualitative versus quantitative defences, concepts that are especially well-developed in the literature on plant defences against herbivores (Coley et al., 1985; Feeny, 1975), toxins may be less expensive for the host than global stimulation of the immune system or depletion of limiting resources.

For the remainder of this chapter, we will discuss in detail the known examples of protection involving production of toxic factors by inherited symbionts. We have attempted to restrict our discussion to cases where there is an especially strong connection between the symbiont, defence, and a specific toxic factor (see also Table 1 for an overview of the specific examples we review). In some cases, however, it has not yet been demonstrated whether protection is occurring in nature, and against whom. This highlights the challenges in studying defensive symbiosis in nature (Oliver et al., 2014) as it requires a good understanding of both an insect's symbionts and ecologically relevant natural enemies. We have also restricted our discussion to toxins where there has been some functional work directly linking them to protection. With the revolution in genome sequencing, we are finding that many symbiont genomes, especially those of facultative symbionts, harbour a great diversity of genes with sequence similarity to toxins that have been characterized in other microbes, but we have no idea whether the putative symbiont factor is indeed a toxin, much less if it is involved in protection.

Table 1 Summary of the best characterized inherited defensive symbionts in which toxins have been implicated in protection.

Symbiont	Insect host	Distribution and Prevalence	Mode of transmission	Toxin	Natural enemy targeted by toxin
Burkholderia	Lagriine beetles (Coleoptera: Tenebrionidae)	Found in 2 genera; 80–90% of adult females are infected	Extracellular, housed in female accessory glands, and deposited over the surface of the egg	Different strains produce diverse polyketides, aromatic glycosides, and lipopeptides	Fungi and bacteria
Hamiltonella	Aphids (Hemiptera: Aphididae)	Widely distributed in aphids with variable prevalence among species; protection has been demonstrated in five species	Intracellular, in embryos and eggs	Phage encoded YD repeat, CdtB, and Shiga-like toxins	Parasitic wasps
Profftella	*Diaphorina citri*—Asian citrus psyllid (Hemiptera: Liviidae)	Only known thus far from 1 species, where it infects all individuals (obligate)	Intracellular, bacteriome-restricted	Diaphorin polyketide	Unknown
Pseudomonas	Paederine beetles (Coleoptera: Staphylinidae)	Widely distributed across paederines, but within species, at least some individuals are uninfected	Females deposit bacteria over the surface of the egg; larvae can acquire and transmit symbionts by eating eggs	Pederin polyketide	Predatory spiders
Spiroplasma	*Drosophila* flies (Diptera: Drosophilidae)	~20 infected species, with prevalence ranging widely, from ~4–85%; protection has been demonstrated in three species	Inside egg	Ribosome-inactivating protein	Parasitic wasps, parasitic nematodes
Streptomyces	Philanthine beewolves (Hymenoptera: Crabronidae)	Found in three genera; virtually all females are infected (obligate)	Housed in adult female antennal glands and smeared in offspring's burrow, picked up by emerging adult female to continue the cycle	Cocktail of streptochlorin and piericidin derivatives	Fungi

2. Case studies

2.1 Toxins associated with obligate symbionts

While most insect inherited defensive symbionts that have thus far been described are facultative, two lineages, both of which employ toxins, stand out because they are obligate (i.e. the symbionts appear to be absolutely required for the host)—*Profftella* symbionts of psyllids, and *Streptomyces* symbionts of beewolves.

2.1.1 Profftella

The β-proteobacterial symbiont *Profftella armatura* infects the Asian citrus psyllid *Diaphorina citri* (Hemiptera: Liviidae), a devastating pest of citrus that vectors the citrus pathogen *Liberobacter asiaticus*, where it is housed in a symbiont organ (or bacteriome), along with another obligate symbiont, *Carsonella ruddi* (Nakabachi et al., 2013). As is common in many other sap-feeding Hemiptera, the two symbionts provide essential complementary vitamins and amino acids that are missing from the host's diet; *Profftella* synthesizes riboflavin and biotin and *Carsonella* synthesizes histidine, tryptophan and other amino acids. Despite having a highly reduced genome, typical of obligate nutritional symbionts, 15% of *Profftella*'s genome is composed of genes involved in the synthesis of a complex polyketide called diaphorin that is related to pederin made by symbionts of *Paederus* rove beetles. Phylogenetic analysis clearly indicates that the diaphorin synthesis genes were acquired via horizontal transfer. Although diaphorin is toxic to yeast and cultured mammalian and insect cells, it is not as toxic as pederin (Yamada et al., 2019). This may be because in diaphorin, the C10 methoxy group of pederin has been replaced by a hydroxyl group, which is thought to decrease hydrophobicity and binding to the ribosome (Mosey and Floreancig, 2012). Laboratory studies show that diaphorin is also toxic to aphids and ladybird beetles when injected. However, it is not yet known which specific natural enemies diaphorin is effective against in nature. The Asian citrus psyllid has many natural enemies, including generalist predators (Michaud, 2004) and more specialized parasitoid wasps (Chen and Stansly, 2014). Since the related pederin toxin from rove beetles (discussed below) protects only against some groups of generalist predators, we might expect a similar protection against generalists by diaphorin. It would also be useful to determine whether *Profftella* occurs in psyllids other than *D. citri*.

2.1.2 *Streptomyces and beewolves*

Solitary digger wasps in three genera in the subfamily Philanthinae (Hymenoptera: Crabronidae), commonly known as beewolves, are engaged in an amazing obligate defensive symbiosis with a lineage of *Streptomyces* bacteria. Adult females dig a burrow in the cell, where they lay a single egg and provision it with prey items. They also smear the burrow with *Streptomyces* that are housed inside specialized antennal glands. The wasp larva incorporates the bacteria into its cocoon, protecting it against generalist pathogenic fungi and bacteria in the soil. Experimental removal of bacteria resulted in over 90% larval mortality (Kaltenpoth et al., 2005), which explains why aposymbiotic wasps are rarely encountered. Upon emergence, the adult female wasp picks up bacteria from the cocoon, which colonize her antennal glands, continuing the cycle anew (Kaltenpoth et al., 2010). The symbiosis evolved once and has been stably maintained for ~70 million years. Although there is a strong degree of phylogenetic congruence between hosts and symbionts, wasps have exchanged symbiont strains on a number of occasions, possibly due to predation or reuse of nests (Kaltenpoth et al., 2014). Symbionts produce a complex cocktail of antibiotics, consisting primarily of streptochlorin and piericidin derivatives, and different compounds are effective against different pathogenic microbes (Engl et al., 2018; Kroiss et al., 2010). Interestingly, the complement of antibiotics has remained rather stable over the ~70 million years of the symbiosis.

2.2 Toxins associated with facultative heritable symbionts

2.2.1 *Pseudomonas and Paederus beetles*

The first report of symbiont-associated toxins being incorporated into host defence involves staphylinid (rove) beetles in the genus *Paederus* infected with the bacterium *Pseudomonas* (γ-proteobacteria). A factor contributing to the early discovery of this defensive symbiosis is that paederine beetles have long been of medical interest (possibly the cause of the biblical 'plague of boils') as painful lesions called Paederus dermatitis develop when the haemolymph of some paederine beetles contacts human skin. The blistering agent, named pederin, was identified as a polyketide amide in the 1960s (Frank and Kanamitsu, 1987; Gelmetti and Grimalt, 1993). Pederin was shown to be toxic to a range of cell lines and mice (Brega et al., 1968; Soldati et al., 1966). The presence of pederin in a diurnal and brightly coloured beetle (both atypical of rove beetles) suggested a potential role in defence against vertebrate or invertebrate predators (Dettner, 1987; Frank and Kanamitsu, 1987). Bioassays showed that pederin was not distasteful to vertebrate predators, nor did it reduce feeding of insect predators relative to pederin-free beetles. However, it was found to deter feeding by several species of wolf spiders and one jumping spider (Kellner and Dettner, 1996). While originally thought to be endogenously produced by the beetles, pederin was subsequently linked to infection by an undescribed *Pseudomonas* species and the discovery of a polyketide synthase (PKS) cluster in this bacterium confirmed the association (Kellner, 2002; Piel, 2002). Polyketide synthases comprise a large protein family synthesizing a range of bioactive compounds that are functionally diverse, including antifungal, antibacterial, anticancer, or immunosuppressive properties (Staunton and Weissman, 2001). Pederin belongs to the *trans*-AT PKS group. Members of this group lack acyltransferase (AT) domains, although ATs are encoded elsewhere in the polyketide synthase gene cluster and act *in trans* in an iterative fashion (Piel, 2009). The primary mode of action of pederin (and relatives -see below) is the inhibition of protein synthesis leading to cytotoxic effects (Brega et al., 1968; Burres and Clement, 1989; Mosey and Floreancig, 2012; Richter et al., 1997). Protein translation is disrupted by pederin's ability to bind to the *E*-site (final binding site for tRNA) of the large (60S) ribosome subunit resulting in the suppression of protein elongation (Dang et al., 2011; Nishimura et al., 2005). Other studies show that pederin also activates, possibly due to ribosome interference, protein kinase pathways, including p38 and JNK associated with stress-activated apoptosis (Lee et al., 2005).

In *Paederus*-associated *Pseudomonas*, the pederin synthase gene cluster is located on a horizontally transferred genomic island (Piel, 2002). Pederin is structurally very similar to the theopederins and onnamides isolated from bacteria inhabiting marine sponges (Fisch et al., 2009; Piel et al., 2004), diaphorin from the Asian citrus psyllid (Nakabachi et al., 2013—see above), and nosperin from lichens (Kampa et al., 2013). This structural similarity, combined with phylogenetic analyses (see below), suggests these compounds have been shared among distantly related organisms via lateral gene transfer among host-switching symbiotic bacteria. Recently, nonsymbiotic bacteria were also found to have pederin-like PKSs (e.g. Kust et al., 2018). Given that there are more than 600 species of *Paederus* beetles (and *Paederidus* beetles also bear this toxin), including 50 species known to cause contact dermatitis (Veraldi et al., 2013), PKS-based defences may be widespread in paederine beetles although the protective effects of infection have only been documented for two species. Moreover, that PKSs are widespread in the genus *Pseudomonas* (which is itself an incredibly widespread microbe) and that pederin-like toxins appear in diverse host-associated bacteria, suggests PKS-mediated defence may be among the most widespread symbiont defences in eukaryotes. While pederin-like toxins are associated with anti-predator phenotypes in terrestrial and marine systems, given the functional diversity of PKSs, as well as the taxonomic breadth of infected hosts, we may also expect to find symbiont-encoded polyketides that provide protection against parasites and pathogens.

2.2.2 *Spiroplasma*

Spiroplasma (Mollicutes) is an incredibly diverse, widespread, and likely nonmonophyletic lineage of host-associated microbes (Anbutsu and Fukatsu, 2011; Ballinger and Perlman, 2019; Regassa and Gasparich, 2006). *Spiroplasma* bacteria are found in a broad range of arthropods, including insects, arachnids, and crustaceans; a broad survey of western European terrestrial arthropods found that 7% (10/136 species) harboured *Spiroplasma* infections (Duron et al., 2008). Many *Spiroplasma* are gut commensals, others are pathogenic, including strains that cause disease in crabs, shrimp, and bees. Among the best-studied pathogenic *Spiroplasma* are leafhopper-vectored strains that cause disease in plants (Bove et al., 2003). Vertical transmission has evolved independently numerous times in this genus (Haselkorn et al., 2009), including male-killing strains infecting butterflies, lacewings, beetles, aphids, planthoppers, and *Drosophila* flies (Hayashi et al., 2016; Hurst et al., 1999; Jiggins et al., 2000; Poulson and Sakaguchi, 1961; Sanada-Morimura et al., 2013; Simon et al., 2011). Although inherited *Spiroplasma* are transmitted primarily from mothers to their offspring over ecological timescales, phylogenetic analyses clearly demonstrate extensive horizontal transmission over evolutionary timescales (Haselkorn et al., 2009). The mechanism of horizontal transmission is not known. In *Drosophila* and aphids, inherited *Spiroplasma* are abundant in haemolymph and haemolymph transfections are commonly used to establish infections in new host species (Haselkorn et al., 2013). Ectoparasitic mites have been shown to transfer *Spiroplasma* between fly hosts in the lab, and this may also be a common route of transfer in the wild (Jaenike et al., 2007).

Inherited *Spiroplasma* strains have been implicated in protection against diverse natural enemies in three different host lineages—*Drosophila* fruit flies, aphids, and tsetse flies. The mushroom-feeding woodland fly *Drosophila neotestacea* is commonly infected with a virulent generalist parasitic nematode, *Howardula aoronymphium*. Infection typically renders female *D. neotestacea* completely sterile. However, in flies that harbour *Spiroplasma* symbionts, fertility is fully restored, and nematodes appear sickly and small, and produce virtually no infective juveniles (Jaenike et al., 2010). Nematode infection is highly prevalent in the wild and the benefit provided by *Spiroplasma* to *D. neotestacea* is so great that *Spiroplasma*-infected flies are replacing their uninfected counterparts and rapidly spreading across N. America (Cockburn et al., 2013). In addition, *Spiroplasma* strains in *D. hydei*, *D. melanogaster*, and *D. neotestacea* have been shown to kill endoparasitic wasp larvae from two distantly related families—Braconidae and Figitidae (Haselkorn and Jaenike, 2015; Xie et al., 2010, 2014). Inherited *Spiroplasma* have been reported in at least 18 *Drosophila* species, suggesting that defence may be common.

Toxins called ribosome-inactivating proteins (RIPs) have been implicated in *Spiroplasma* protection against both nematodes and wasps. RIPs are *N*-glycosidases that target a highly conserved adenine residue in the α-sarcin/ricin loop of eukaryotic 28S ribosomal RNA. They are especially widely distributed in plants and bacteria, and include the well-studied and deadly poisons Shiga toxin and ricin. The genomes of the *Spiroplasma* symbionts of *D. neotestacea* (sNeo) and *D. melanogaster* (sMel) respectively encode four and three secreted and constitutively highly expressed RIPs (Ballinger et al., 2019). One of these RIPs was purified and shown to depurinate eukaryotic ribosomes (Hamilton et al., 2016). In addition, nematode and wasp ribosomes show characteristic signatures of RIP attack when parasitizing *Spiroplasma*-infected flies. Transfection experiments showed that sMel, unlike sNeo, does not provide protection (or show signatures of ribosome depurination) against nematodes (Ballinger et al., 2019; Haselkorn and Jaenike, 2015). This may be because sNeo contains highly divergent RIPs, and different RIPs may be effective against different enemies. However, it is still completely unknown how *Spiroplasma* RIPs enter target cells. RIPs are typically classified into two types. Type II RIPs consist of an A unit that has glycosidase activity once inside the cell and a B unit that binds to a surface cell receptor, whereas Type I RIPs lack the A unit and are therefore considered less toxic. While *Spiroplasma* RIPs do not have a recognizable A unit, they have a long N-terminal region of unknown function. Transgenically expressing *Spiroplasma* RIPs inside *Drosophila* cells results in depurination of fly ribosomes (Garcia-Arraez et al., 2019), demonstrating that the glycosidase is effective against host ribosomes; host cells might, thus be resistant to toxin entry.

Genome sequencing has found that a wide range of *Spiroplasma*, including cultivable gut commensals, encode a diverse array of RIPs whose function is unknown (Ballinger and Perlman, 2019; Hamilton et al., 2016).

RIP evolution appears be highly dynamic, and includes pseudogenized copies, as well as RIPs that are found on plasmids. The case of the European common red ant *Myrmica rubra* illustrates the dynamic nature of RIPs. Whereas ants in the native European and introduced North American range both harbour *Spiroplasma*, only symbionts of native ants appear to encode RIPs (Ballinger et al., 2018). The dynamic nature of RIPs suggest that they are gained and lost in conjunction with effectiveness against natural enemies.

Spiroplasma-mediated protection has also been demonstrated in aphids and tsetse flies. The mechanism of protection is not understood in either group, and aphid and tsetse *Spiroplasma* genomes have not yet been sequenced, so we do not know whether they harbour RIPs. A subset of *Spiroplasma* strains has been found to protect *A. pisum* pea aphids against the aphid fungal pathogen *Pandora neoaphidis* (Łukasik et al., 2013c) reducing overall aphid mortality as well as fungal sporulation. Transfection of pea aphid *Spiroplasma* into *Sitobion avenae* grain aphids also conferred protection (Łukasik et al., 2013b). A number of unrelated facultative symbionts, including *Regiella*, *Rickettsia* and *Rickettsiella*, also provided protection (Łukasik et al., 2013c), with greater efficacy than most *Spiroplasma*, suggesting that defence is quite general. On the other hand, *Spiroplasma* strains do not appear to protect aphids against parasitic wasps. Mathe-Hubert et al. (2019) recently transfected 12 different *Spiroplasma* strains into the same pea aphid clonal genotype and tested their ability to protect against three strains of the parasitic wasp *Aphidius ervi*. They found that only one symbiont strain protected against the most virulent wasp strain. This strain was quite costly, suggesting that wasps were less successful because aphids were in poor health. Another recent study found that *A. ervi* wasps were less attracted to volatiles emitted by plants that had been fed on by *Spiroplasma*-infected aphids than by uninfected ones (Frago et al., 2017), suggesting an indirect way by which *Spiroplasma* might benefit aphids. The mechanism underlying this change in preference is unknown, but this effect was also observed in aphid harbouring *Hamiltonella*, *Regiella*, *Serratia*, and *Rickettsia* facultative symbionts.

Finally, three species of *Glossina* tsetse fly were recently found to harbour *Spiroplasma* infections (Doudoumis et al., 2017). In controlled lab infections, *Spiroplasma*-positive flies were less likely to be infected with trypanosome gut parasites (Schneider et al., 2019); trypanosomatids were also more likely to be found in symbiont-free flies in the wild as well. The *Spiroplasma* in tsetse has very different dynamics and tissue tropism than in *Drosophila*, as it is found in the gut and not haemolymph, and at very high densities in larvae. More work is need to determine whether *Spiroplasma* infections result in reductions of gut parasites in tsetse, and if so, whether this effect is mediated by symbiont-encoded toxins.

2.2.3 *Hamiltonella*

Unlike the highly promiscuous *Spiroplasma,* another well-studied defensive symbiont, *Hamiltonella defensa,* appears restricted to hemipteran insects and is best studied in aphids (Moran et al., 2005b; Oliver and Higashi, 2019; Oliver et al., 2010). Residing in the Yersiniaceae (order Enterobacteriales of the γ-proteobacteria), *Hamiltonella* forms a clade with two other common aphid facultative symbionts, *Regiella insecticola* and *Fukatsuia symbiotica* (Patel et al., 2019). Examined strains from this symbiont clade are all host-restricted, maternally-transmitted aerobic heterotrophs; although phylogenies indicate that occasional host switching occurs within and among species (Chevignon et al., 2018; Henry et al., 2013; Russell et al., 2003). In the lab, *Hamiltonella* can be readily moved among distinct aphid clones and species via microinjection, although not all genotypes are equally amenable to establishing stable infections (Łukasik et al., 2015; Niepoth et al., 2018; Oliver et al., 2005; Parker et al., 2017; Vorburger et al., 2010). The most likely natural routes of lateral transfer are wasp ovipositors (Gehrer and Vorburger, 2012) and food plants (Li et al., 2018; Oliver et al., 2008), but the frequency of lateral transfer on ecological timescales (and hence its importance in symbiont maintenance in host populations) remains unknown. In aphids, *Hamiltonella* persists mostly in the haemolymph although it can establish in bacteriocytes and the surrounding sheath cells (Brandt et al., 2017; Moran et al., 2005b). Numerous published *Hamiltonella* genomes from pea aphids show that inventories of genes involved in housekeeping functions, nutrient acquisition, and metabolism are very similar among strains (and similar to other aphid facultative symbionts from the Yersiniaceae generally). In contrast, strains vary considerably in mobile DNA content, which comprises about ca. 25% of the typical *Hamiltonella* genome. This mobile DNA also results in large scale genome rearrangements among strains from distinct phylogenetic clades, while strains within the same clade exhibit similar architecture and mobile DNA content (Chevignon et al., 2018; Degnan et al., 2009).

Aphids have been developed as excellent models for studying phenotypic effects of infection with facultative symbionts, due to the ease of transfer among clonal lines, and the resulting abilities to ascribe symbiont-associated differences directly to impacts of the transferred symbionts (Brisson and

Stern, 2006; Oliver et al., 2010). Using such experimental approaches, *Hamiltonella* was first shown to protect the pea aphid, *Acyrthosiphon pisum,* from attack by its most common parasitoid, the aphidiine braconid wasp, *Aphidius ervi* (Oliver et al., 2003). While *Hamiltonella* infections were subsequently shown to modify aphid defensive and wasp oviposition behaviours (Dion et al., 2011a; Oliver et al., 2012), wasps nonetheless readily lay eggs in infected aphids, where *Hamiltonella* causes mortality to developing embyros or larvae, depending on symbiont strain; indicating a physiological basis of resistance (Martinez et al., 2014; Oliver et al., 2003). Later, *Hamiltonella* was shown to protect other aphid species from parasitoids, including *Aphis fabae* (black bean aphid), *Aphis craccivora* (cowpea aphid), *Rhopalosiphum padi* (bird cherry-oat aphid) and possibly *Metopolophium dirhodum* (rose-grain aphid) (Asplen et al., 2014; Leybourne et al., 2020; Monticelli et al., 2019; Schmid et al., 2012; Vorburger, 2014).

There is extensive strain variation in *Hamiltonella*-mediated defences (Oliver and Higashi, 2019). For example, a given strain often targets specific wasp species and different strains in an aphid species can confer variable levels of resistance against a single parasitoid species (Asplen et al., 2014; Cayetano and Vorburger, 2013, 2015; Hopper et al., 2018; Martinez et al., 2014, 2016; McLean and Godfray, 2015; Oliver et al., 2005; Schmid et al., 2012). Some studies have failed to find that particular *Hamiltonella* strains protect against one or more parasitoid species (Clarke et al., 2017; Łukasik et al., 2013a). However, the aforementioned variation in target specificity suggests that symbiont defences may simply be effective against other wasp species. Alternatively, in some systems *Hamiltonella* infections may be maintained in host populations by providing other benefits, including thermal tolerance, as suggested for pea aphids (Russell and Moran, 2006), or influencing interactions with food plants as shown for whiteflies and grain aphids (Li et al., 2018; Su et al., 2015). Given that about one-third of the 5000 or so aphid species are estimated to carry *Hamiltonella* (Zytynska and Weisser, 2016), aphid/*Hamiltonella* interactions may represent a wide-spread protective mutualism influencing interactions across a wide array of plant systems and their associated communities (McLean, 2019; Oliver and Higashi, 2019).

While *Hamiltonella*-based resistance to *A. ervi* was known to be physiological, based on experimental assays (Oliver et al., 2003), the first clues regarding the specific factors causing harm to wasps were derived from the first genomic studies. Most notably, partial sequencing of a moderately protective *Hamiltonella* strain identified a toxin-encoding bacteriophage, called APSE (for *Acyrthosiphon pisum* secondary endosymbiont), as a key factor of interest (Moran et al., 2005a; van der Wilk et al., 1999). In addition to the mere maintenance of an intact eukaryotic-toxin carrying temperate phage in a heritable symbiont, the case for APSE roles in defence were strengthened by variation in levels of protection correlating with specific APSE variants, in conjunction with APSEs occurring at high abundance relative to *Hamiltonella* and rt-qPCR assays showing that phage toxins were highly expressed (Martinez et al., 2014; Moran et al., 2005a; Oliver et al., 2005).

Subsequent experimental work confirmed that APSEs were required for the protection against parasitoids as phage loss resulted in the complete loss of the protective phenotype (Oliver et al., 2009), while reintroduction completely restored protection (Lynn-Bell et al., 2019). APSEs have other effects important in the regulation and maintenance of *Hamiltonella* (Weldon and Oliver, 2016), including lowering the within aphid abundance of *Hamiltonella,* which impacts infection costs (Weldon et al., 2013). The ability to culture *Hamiltonella*, APSE and parasitoids in vitro, established that *Hamiltonella* + APSE kills parasitoids without any aphid factors and confirmed in vivo findings that *Hamiltonella* alone is insufficient (Brandt et al., 2017). Mortality to wasps in conditioned media following the removal of *Hamiltonella*/APSE showed that a soluble factor is released from the symbionts that disrupts wasp development (Brandt et al., 2017). Furthermore, strains with published genomes that are virtually identical except for the presence of APSE confer either no protection (A2C, no phage) or nearly complete protection (AS3 with APSE) and the introduction of APSE in vivo or in vitro rapidly restores the protection (Brandt et al., 2017; Chevignon et al., 2018; Lynn-Bell et al., 2019).

APSEs are double-stranded DNA phages in the order Caudovirales with tail morphology similar to the Podoviridae and virion assembly genes showing homology to the P22 phage of *Salmonella enterica* (van der Wilk et al., 1999). Comparative genomics of APSEs indicate these phages exhibit mosaic genomes (33–41 kb) with those regions associated with regulatory and structural function showing highly similar gene content and sequence identity, while a module containing virulence factors, including the toxins hypothesized to function in defence, displays high variability among strains (Chevignon et al., 2018; Degnan and Moran, 2008a,b; Dennis et al., 2017; Martinez et al., 2014; Moran et al., 2005a; Rouïl et al., 2020). APSE variants with distinct phage backbones, and occurring in *Hamiltonella* strains from different clades, can encode the same toxins. For example, the most common APSEs in North American pea aphids associated with alfalfa (*Medicago sativa*) are APSE2 and APSE8 (Weldon, 2015). These each encode very similar homologs (99.3% nt similarity) of the active subunit (*cdtB*) of cytolethal distending toxin (CDT) (Chevignon et al., 2018; Moran et al., 2005a). Parasitism assays show that these *cdtB*-encoding APSE2 and APSE8 *Hamiltonella* confer similar and moderate levels of protection against *A. ervi* and no protection against the related aphidiine parasitoid *Praon pequodorum* despite the aforementioned variation in phage and bacterial strains (Doremus and Oliver, 2017; Martinez et al., 2014, 2016). Thus, along with phage loss and restoration studies, these correlations further support phage toxins as primary determinants of strength of resistance.

Cytolethal distending toxins are genotoxins associated with more than 30 proteobacterial and actinobacterial species, including the enteric pathogens *Yersinia, Escherichia* and *Shigella* that are related to *Hamiltonella* (Jinadasa et al., 2011; Scuron et al., 2016). In most studied mammalian pathogens CDTs are AB_2-type heterotrimeric toxins comprised of the active subunit CdtB and the binding subunits CdtA and CdtC. The latter mediates toxin binding to the cell surface and delivery of CdtB into the targeted cell where it ultimately localizes to the nucleus; all three subunits are required for full activity (Jinadasa et al., 2011; Scott and Kaper, 1994). The active subunit, CdtB, displays nuclease activity (homology to DNase 1), which arrests the cell cycle (usually in G2/M but in G1/S in some cell lines) leading to cell distention and death (Comayras et al., 1997; Lara-Tejero and Galan, 2000; Pérès et al., 1997). Not all cell types are equally susceptible to CDT intoxication, which potentially explains why wasps are often killed by *Hamiltonella* with APSE2 or 8 with no apparent harm to aphids. More perplexing is how CdtBs exit *Hamiltonella* cells, then bind and enter wasp cells given that CdtA and CdtB are not present on the APSE or *Hamiltonella* chromosomes (Chevignon et al., 2018; Degnan and Moran, 2008a; Degnan et al., 2009). Some cdtBs also occur in the *Hamiltonella* strains infecting black bean aphids (uncharacterized phage variant), *Chaitophorus* aphids (APSE6) and *Bemisia tabaci* whiteflies (APSE7) (Degnan and Moran, 2008a; Dennis et al., 2017). They also occur in some strains of *Arsenophonus*, a widely distributed symbiont that frequently carries APSEs or phage elements related to APSE (Duron, 2014). It is unclear whether any *Arsenophonus* strains are protective, but in the red gum lerp psyllid, *Glycaspis brimblecombei,* there is a positive correlation between the prevalence of *Arsenophonus* infection and prevalence of attack by parasitic wasps (Hansen et al., 2007). Some cdtBs also occur on the main chromosome of *Fukatsuia,* which lacks APSEs; these *cdtB* homologs are more similar to those occurring in *Yersinia* rather than those found in *Hamiltonella*-APSEs (Patel et al., 2019).

Recently, *cdtBs* were found in the genomes of several aphids and drosophilid flies (Verster et al., 2019). The presence in flies of *cdtBs* most closely related to those encoded by APSEs is especially surprising given that *Hamiltonella* has not been convincingly found outside of the Hemiptera. Perhaps some of these horizontal transfers were acquired from APSEs associated with *Arsenophonus* instead. It is not yet known whether the horizontally acquired *cdtB* genes function in defence as hypothesized (Verster et al., 2019), although their active sites are conserved, despite low overall protein sequence identity with their microbial homologs (Verster et al., 2019). In an interesting parallel, recent genome surveys have found that *Bemisia* whiteflies and *Aedes* and *Culex* mosquitoes have acquired RIP toxin genes via horizontal transfer (Lapadula et al., 2017, 2019) although the donor was not *Spiroplasma* (see above).

The most protective strains of *Hamiltonella* characterized to date are those associated with APSE3, which encode a YD (tyrosine/aspartate)-repeat protein, a member of the rearrangement hotspot (RHS) family of toxins, in the phage virulence cassette. In pea aphids, APSE3-*Hamiltonella* confers complete (or nearly so) protection against the parasitoid *A. ervi*, but as with APSE 2 and 8, these strains show no activity against the related wasp *P. pequordoum* (Martinez et al., 2014, 2016, 2018; Oliver et al., 2005, 2009). YDps also occur in the APSEs of protective *Hamiltonella* infecting black bean aphids (Dennis et al., 2017). Common in both Gram-negative and Gram-positive bacteria, RHS/YD-repeat proteins, including those in APSEs, are typically large, polymorphic exotoxins of 1400 or more AA residues with the N-terminal core exhibiting YD repeats and the C-termini being highly variable, often with no homology to known effectors (Chevignon et al., 2018; Degnan and Moran, 2008a; Jamet and Nassif, 2015). Hence the specific mechanism(s) of action are not known. Related RHS/YD-repeat toxins are found in the mammal pathogen *Pseudomonas aeruginosa* (Kung et al., 2012), multiple strains of the nematode symbiont *Xenorhabdus* (e.g. Kim et al., 2017), the sulphur-oxidizing *Bathymodiolus* symbionts of hydrothermal vent mussels (Sayavedra et al., 2015), and *Vibrio* from zebrafish (Stephens et al., 2015). They are also found in the entomopathogenic *Yersinia entomophaga,* which exhibit insecticidal activity (Busby et al., 2013). While many of these are thought to be defensive (Hillman and Goodrich-Blair, 2016), Rhs/YD proteins from phylogenetically diverse bacterial lineages may be involved in bacterial kin recognition and mediate intracellular competition among bacteria via contact-dependent inhibition of growth (Koskiniemi et al., 2013). Aphids are frequently infected with more than one facultative symbiont, but symbiont distributions are nonrandom, with some combinations highly enriched and others depleted (Rock et al., 2018; Smith et al., 2015). While several factors may contribute to structure the heritable microbiome, microbe-microbe competition mediated by toxins such as YD are also potentially important in this regard (Doremus and Oliver, 2017; Leclair et al., 2017; McLean et al., 2018; Oliver et al., 2006; Rock et al., 2018; Weldon et al., 2020).

The third toxin family associated with APSEs are Shiga toxins, a group of functionally related exotoxins that belong to the ribosome-inactivating protein (RIP) family discussed above with respect to *Spiroplasma*. Shiga toxins are also known from toxigenic *E. coli* strains (STEC, e.g. serotype 0157:H7) and *Shigella dysenteriae,* which cause haemorrhagic colitis and haemolytic uraemic syndrome in humans (Johannes and Römer, 2010). As with APSE-associated Shiga toxins, those from pathogenic bacteria are encoded by a range of bacteriophages, which through horizontal transmission contribute to genetic variation and confer novel traits that benefit the bacterial host (Herold et al., 2004). Lainhart et al. (2009) hypothesized that phage-encoded Shiga toxins in mammalian pathogens may have originally evolved in bacteria to protect against their eukaryotic predators. Their experiments show that the presence of the ciliate *Tetrahymena thermophila* induces toxin production, which kills the predator. The Shiga holotoxin has an AB_5 molecular configuration, in which the active subunit StxA binds to five identical fragments that form StxB. The StxB subunit binds to glycolipid receptors (globotriaosylceramides) resulting in tubular invaginations critical to uptake into host cells. As with other RIPs, once inside the cell, StxA inhibits protein synthesis by cleaving a specific adenine base of 28S rRNA, preventing amino-acyl tRNA binding and hence chain elongation (Endo et al., 1988; Johannes and Römer, 2010; Saxena et al., 1989). In several APSE variants (1, 4 and 5) it has been found that the P7 protein shows weak homology to StxB (Degnan and Moran, 2008a; Dykstra et al., 2014; van der Wilk et al., 1999), while P9 has been hypothesized to be the functional analog of StxA.

The protective phenotype of Stx-encoding APSE5 from *Hamiltonella* infecting *Uroleucon rudbeckiae* has not been characterized, but a *Hamiltonella* strain APSE1 from pea aphids confers high levels of protection against the braconid *A. ervi*, and low levels against the chalcid wasp *Aphelinus abdominalis* (McLean and Godfray, 2015). Also APSE4/*Hamiltonella* infecting the cowpea aphid, *Aphis craccivora* conferred high levels of protection against two *Binodoxys* species, but not two other braconid parasitoids (Asplen et al., 2014; Dykstra et al., 2014). In APSE4, the *stxB* homolog (P7) is pseudogenized by an inactivating insertion, raising questions as to the importance of the putative delivery submit in this protective mutualism. In the aforementioned study where Shiga toxins protect their bacterial hosts against protist predators, disruption of StxB substantially reduced StxA activity (Lainhart et al., 2009).

An emerging topic in symbiosis research is to ask how changing temperature affects the stability and outcome of symbiosis (Corbin et al., 2017); this is especially pertinent with global climate change. Interestingly, *Hamiltonella* protection fails at higher temperatures (Doremus et al., 2018; Higashi and Oliver, 2020) and this may be partially or wholly mediated by defensive toxin thermal sensitivity. CdtBs are well-known to be heat labile (Johnson and Lior, 1988) and *cdtB*-encoding APSE2 and APSE8 strains completely fail at only moderately warmer temperatures (Doremus et al., 2018). YD-encoding APSE3 strains only partially fail under exposure to warmer temperatures, but little is known about how YD proteins are affected by heat. Shiga toxins do not appear heat labile (Rasooly and Do, 2010), but APSE do not encode homologs of StxA and it is unclear whether the putative P9 homolog is heat labile; also no parasitism assays to date indicate whether these strains also fail at higher temperatures.

The strains of *Hamiltonella* infecting whiteflies form a distinct clade, are restricted to bacteriocytes, and the APSE phages appear inactivated (Chevignon et al., 2018; Gottlieb et al., 2008; Rollat-Farnier et al., 2015). It is likely then, that *Hamiltonella* is not generally serving as a defensive symbiont in whiteflies, although a recent study found a correlation between *Hamiltonella* frequency and parasitism (Qi et al., 2019). There are reports that whitefly *Hamiltonella* can influence the transmission of plant pathogenic viruses (Gottlieb et al., 2010), impact sex ratios (Shan et al., 2019), and affect plant signalling to suppress induced defences (Su et al., 2015), a finding recently reported in aphids as well (Li et al., 2019).

Aphids host two other defensive facultative symbionts that are close relatives of *Hamiltonella*—*Regiella insecticola*and *Fukatsuia symbiotica*. *Regiella insecticola* was the first aphid symbiont shown to provide defence against the specialized fungal pathogens *Pandora neoaphidis* (Order Entomophthorales) (Scarborough et al., 2005) and *Zoophthora occidentalis,* but not the generalist pathogen *Beauveria bassiana* (Parker et al., 2013). Subsequently, most other common aphid facultative symbionts, even distantly related species, were shown to protect against *Pandora*, including *Rickettsia* (α-proteobacteria), *Spiroplasma* (Mollicutes), and a range of γ-proteobacteria: *Fukatsuia, Rickettsiella* and possibly *Hamiltonella* (Heyworth and Ferrari, 2015; Łukasik et al., 2013c; Weldon et al., 2020). The mechanism of symbiont-based resistance to fungal pathogens is unknown. But given that almost all aphid facultative symbionts confer this trait, a more general response, such as immune priming, may underlie this form of symbiont-encoded protection. A strain of *Regiella* from *Myzus persicae* (green peach aphid) was also found to confer protection against the aphidiine parasitoid, *Aphidius colemani,* in its native host and when transferred to black bean aphids (Vorburger et al., 2010). The same strain also conferred resistance against *A. ervi* when transferred into pea aphids (Hansen et al., 2012; Vorburger et al., 2010), and a naturally occurring pea aphid strain conferred protection against this same parasitoid, but only in some aphid genotypes (Oliver et al., 2003; Weldon et al., 2020). *Regiella* does not contain APSE phages, and while the genome of the protective *Regiella* strain from *Myzus* was found to encode additional toxins and pathogenicity factors not found in a 'nonprotective' strain from pea aphids, it remains unclear how *Regiella* harms parasitoids (Degnan et al., 2010; Hansen et al., 2012). The 'nonprotective' *Regiella* strain from the Hansen et al. (2012) study appears similar to a strain that did provide protection against the same parasitoid depending on aphid genotypes (Weldon et al., 2020).

Forming a clade of facultative symbionts with *Regiella* and *Hamiltonella,* the symbiont *Fukatsuia symbiotica* has been reported to provide the full suite of defensive services seen across all pea aphid facultative symbionts (against parasitoids, fungal pathogens or thermal stress) (Heyworth and Ferrari, 2015). Another study found that the dominant North American strain infecting alfalfa-associated pea aphids provided none of these benefits. Instead, the authors hypothesized that *Fukatsuia* exploits the *Hamiltonella* protective mutualism and spreads within pea aphid populations via hitchhiking (Doremus and Oliver, 2017; Doremus et al., 2018). In Lachninae aphids, *Fukatsuia* is thought to provision B vitamins, as their *Buchnera* obligate symbionts have lost this ability (Meseguer et al., 2017; Russell et al., 2017). As with *Regiella, Fukatsuia* also does not contain APSEs but as noted above does have an intact *cdtB* homolog encoded on the main chromosome. This *cdtB* carrying strain, however, provided no protection against the wasp *A. ervi.* All three of the symbionts in this clade encode a range of overlapping pathogenicity factors, including a large number of repeats in toxin (RTX) (Patel et al., 2019). Named after the repeating glycine and aspartate repeats at the C-terminus, RTX toxins are associated with a variety of gram-negative bacteria with protein translocation across the bacterial membranes typically provided via Type-I secretion systems (T1SS) (Linhartová et al., 2010). While some RTXs are associated with cytotoxic activity, including pore-forming toxins, others are involved with a variety of biological activities (e.g. multifunctional autoprocessing RTX or MARTX toxins). It is unclear if RTX, or the other common pathogenicity factors encoded in these genomes, aid in protective phenotypes, either directly or indirectly.

2.2.4 *Burkholderia* and *Lagria* beetles

The most recent addition to the list of inherited toxin-producing defensive symbionts of insects is found in tenebrionid beetles in the subfamily Lagriinae. All species examined thus far (in the genera *Lagria* and *Ecnolagria*) harbour extracellular *Burkholderia* bacteria that are housed in adult female accessory glands in the reproductive tract, and that are vertically transmitted to offspring (Flórez and Kaltenpoth, 2017). In lab assays, the bacteria were found to protect eggs and larvae from pathogenic fungi (Flórez et al., 2017). Although both male and female larvae host the symbiont, adult males are not infected. Also, 80–90% of adult females are infected in the wild, suggesting that although the infection is highly prevalent, it is not obligate. Multiple strains of *Burkholderia* are involved in the symbiosis, and individual beetles can harbour more than one oligotype. Although these strains are all closely related to *B. gladioli*, there are interesting differences. For example, the most abundant strain in the field is relatively rare in beetles that have been maintained in the lab. It is as yet uncultivable and has a reduced genome that encodes toxic compounds that are not found in other lagriine *Burkholderia* symbionts, such as lagriamide, a novel polyketide that is similar to poisons found in ascidians (and suspected to be synthesized by ascidian bacterial symbionts) and whose biosynthetic gene cluster has been acquired by horizontal gene transfer (Flórez et al., 2018; Waterworth et al., 2019). On the other hand, the most abundant strain in lab-reared beetles is rare in the field. It can be cultured and produces a wide range of poisons, including polyketides, aromatic glycosides, and lipopeptides, some of which exhibit activity against bacteria or fungi in lab assays (Dose et al., 2018; Flórez et al., 2018). There is thus a rich pool of defensive *Burkholderia*, with a diverse repertoire of antimicrobial compounds. Although at least some lagriine *Burkholderia* are vertically transmitted, phylogenetic analysis demonstrates that there is much horizontal transmission. Interestingly, *B. gladioli* is a common plant pathogen and the most common cultivable lagriine *Burkholderia* was shown to be able to be transmitted from beetles to soybeans, where they induced an immune response and caused a reduction in seed production (Flórez and Kaltenpoth, 2017). The diversity of lagriine *Burkholderia* and the current inability to culture the most abundant field strain makes it challenging to understand whether and how *Burkholderia* strains mediate interactions between beetles and their host plants.

3. Three major questions in toxin-mediated defensive symbiosis

In this section, we highlight three major unresolved questions in the field.

1. *From where do endosymbiont-encoded toxins originate?*

As noted above, many of the toxins functioning in characterized defensive symbioses have wide distributions in bacteria, often facilitated by their localization on phages, plasmids and other mobile elements. For example, the toxins found in *Hamiltonella*, have homologs in diverse bacteria. In *Hamiltonella*, these toxins occur on temperate phages (APSEs), which possibly moved from the widespread facultative symbiont *Arsenophonus* into seemingly hemipteran-restricted symbionts (Duron, 2014). More broadly, ds-DNA tailed bacteriophages have mosaic genomes comprised of modules with distinct evolutionary histories. These modules typically contain sets of genes that function together, often encoding completely different proteins (Casjens and Thuman-Commike, 2011). APSEs, for example, contain a module with homology to a nonintegrative phage φSG1 infecting *Sodalis*, a symbiont of tsetse flies (Clark et al., 2007; Degnan and Moran, 2008a). Thus, toxins may be shuffled within and among bacteriophages, which in turn infect variable ranges of symbiont hosts. This modular evolution of phages may create novel toxins. As noted above, several APSEs encode homologs of StxB, important for toxin delivery into target cells, but not the active toxin StxA; instead, a hypothesized analog to StxA is present (Degnan and Moran, 2008a). Coinfections of facultative symbionts are common in insects (Ferrari et al., 2012; Jing et al., 2014; Rock et al., 2018; Russell et al., 2013; Smith et al., 2015; Toju and Fukatsu, 2011; Zchori-Fein et al., 2014; Zytynska and Weisser, 2016), providing ample ecological opportunities for mobile-element mediated toxin transfers. Recombinant phage with new toxins may not be beneficial to bacterial or animal hosts, or may impose costs, resulting in their inactivation and loss. Virtually all defensive symbionts arise from lineages with free-living or host-restricted pathogens as near ancestors. Hence many pathogenicity factors, including some toxins, were likely present on chromosomes at the initiation of the symbiosis. For example, while their function has not been characterized, multiple, highly similar copies of RTX occur in *Fukatsuia, Hamiltonella* and *Regiella* indicating they were likely present in their common ancestor with some diversification occurring after the split (Patel et al., 2019). However, it is important to note that all of the primary toxins thought to be involved in host defence occur on mobile elements, and some (e.g. PKS) are shared broadly across the animal world.

2. Do symbiont toxins harm hosts? & 3. How do natural enemies fight back/ protect themselves against symbiont-mediated toxins?

Defensive symbiont toxins also have the potential to harm the host of the defensive symbiont. While defensive symbionts clearly provide conditional benefits that allow for their spread and persistence in host populations, infections can also potentially incur both constitutive and induced costs (Cockburn et al., 2013; Jaenike et al., 2010; Kach et al., 2018; Martinez et al., 2018; Oliver et al., 2008; Smith et al., 2015; Vorburger and Gouskov, 2011). Many defensive symbionts lack genes for the biosynthesis of essential amino acids and other nutrients, yet retain transporter genes for their import. This indicates that they are nutritional parasites on some level contributing constitutive costs connected to maintaining infections (Ankrah et al., 2017; Chevignon et al., 2018; Patel et al., 2019). Of course, repeated interactions with the host immune system may also contribute to constitutive costs as antimicrobial peptides or other factors may be continuously deployed, in part to limit the distribution and/or proliferation of symbionts (Login et al., 2011; Ratzka et al., 2012). Costs of deploying toxins in host defence may also result in induced costs as the toxins that harm enemy tissues may also harm host tissues. While induced costs have not been reported in insect defensive symbionts (Vorburger et al., 2013) these costs are harder to measure as the enemy-challenge itself introduces costs that are difficult to disentangle from induced costs. Little is known about toxin delivery mechanisms in defensive symbionts, which are likely to impact induced costs. For example, in aphids, parasitism may induce the lysis of *H. defensa*, which may release toxins into the haemocoel and expose aphid tissues to these toxins (Martinez et al., 2014). However, if other factors deliver and insert toxins into target cells, or symbionts invade target cells and then lyse, then induced costs may be minimal. Transgenic *Drosophila melanogaster* engineered to express a *Spiroplasma* ribosome-inactivating protein showed reduced fitness, in terms of reduced adult survival and elevated embryonic mortality (Garcia-Arraez et al., 2019), but it is hard to determine how much this is due to toxins that remain inside the fly cell and damage fly ribosomes, as opposed to secretion and re-entry.

Some defensive toxins, including *cdtB,* lack the binding subunits required for delivery, but other factors may have replaced these functions due to the modular nature of toxin evolution. As a reminder, many defensive symbionts exhibit high target specificity. Pederin in rove beetles harmed spiders but not insect predators (Kellner and Dettner, 1996), and *H. defensa* strains are often effective only against particular parasitoid species, even closely related ones (Asplen et al., 2014; Cayetano and Vorburger, 2015; Martinez et al., 2016; McLean and Godfray, 2015). This could indicate that the defensive symbiont toxins or delivery mechanisms are highly specific, and so would be less likely to harm unrelated hosts.

Parasitoids can rapidly evolve resistance to defensive symbionts, and this is hypothesized to be mediated by toxins in *H. defensa*/APSE-infected aphids (Dennis et al., 2017; Dion et al., 2011b; Rouchet and Vorburger, 2014). More generally, coevolutionary interactions may maintain resistance-conferring symbiont and toxin diversity, and counter-resistance mechanisms in parasitoids (Hafer and Vorburger, 2019; Vorburger and Perlman, 2018). Target specificity may also mediate competition between rival parasitoids (Kraft et al., 2017; McLean and Godfray, 2017), which can select for or against specific defensive toxins.

Parasite counter-resistance is one of the least-studied aspects of defensive symbiosis, and also one of the most promising in terms of disentangling underlying mechanisms of protection. As mentioned above, the best examples of both fine-scale specificity and enemy counter-resistance come from aphid symbionts, with studies demonstrating effective protection against some genera but not others (e.g. Asplen et al., 2014, Martinez et al., 2016), all the way to specificity at the level of host, symbiont, and wasp genotype (Rouchet and Vorburger, 2012). Wasps can rapidly evolve resistance to symbiont-mediated protection (Rouchet and Vorburger, 2014), for example by superparasitism (Oliver et al., 2012) or avoiding symbiont-infected aphids (Łukasik et al., 2013a). Mateos and colleagues (Mateos et al., 2016) recently found three *Drosophila*-parasitic wasp species that were able to overcome *Spiroplasma* protection. These resistant wasps are all close relatives of susceptible species, which presents a promising opportunity to study mechanism of protection and counter-resistance in this system.

Another way to study enemy counter-resistance is to examine the evolutionary persistence of symbiosis and defence. For example, philanthine beewolves have hosted *Streptomyces* symbionts with a conserved toxin repertoire for ~70 million years, which shows that there has been little resistance evolution in this system. Resistance may be difficult to evolve here because there is a complex cocktail of poisons to detoxify and because the target microbes are opportunistic pathogens living in the soil, with therefore weaker selection for resistance than a developing parasitoid for example. Thus, we can contrast this stability with *Hamiltonella* and *Spiroplasma* defensive symbioses, which are highly dynamic, with symbionts and/or symbiont toxins being rapidly gained and lost, presumably due to the effectiveness of the toxin and the cost/benefit of carrying it.

For phage-encoded toxins, their production has the potential to mediate competition between bacterial cells with and without phage. This can have implications for protective function and the regulation of the defensive symbiosis. For example, within an individual aphid, not all *Hamiltonella* cells are infected with APSEs (Degnan and Moran, 2008a). If toxin production or other mechanisms result in phage infected cells suffering costs relative to uninfected cells, then within-host selection has the potential to reduce the frequency of APSE-carrying *Hamiltonella*, and phage losses may become fixed through vertical transmission bottlenecks. For APSE3 *Hamiltonella*, spontaneous phage losses occur repeatedly, while APSE2/8 associated *Hamiltonella* appear stable (Oliver et al., 2009). Phage loss results in complete loss of protection, but also increases in *Hamiltonella* abundance, which correlates with severe fitness costs to aphids where host-level selection is expected to remove them from the population (Oliver et al., 2009; Weldon et al., 2013). As with some pathogens (e.g. Sturm et al., 2011), toxin production may also be metabolically costly, which could also select for phage-free bacteria or add to constitutive infection costs.

In pathogens, in addition to performing 'offensive' roles in harming host tissue, toxins may also serve defensive functions by compromising the immunity of the bacterium's host, or even benefit bacteria in ways unrelated to these direct interactions, including roles in motility and biofilm formation (Rudkin et al., 2017). The defensive toxins identified in insect symbionts may similarly perform roles beyond direct toxicity. While such roles are unknown at present, and difficulty to study, some common putative toxins in defensive symbionts, including RTX toxins, are already associated with diverse functions in other bacteria (Linhartová et al., 2010). Defensive symbionts have reduced genomes compared to their free-living relatives, yet about a quarter of their genomes is comprised of mobile DNA (Chevignon et al., 2018; Degnan et al., 2009, 2010; Patel et al., 2019). Pathogenicity islands derived from acquired phages, plasmids or transposons, contain diverse toxins, adhesins, secretion systems, and iron uptake systems, which can have roles in pathogenicity and/or symbiosis (Hacker and Kaper, 2000; Patel et al., 2019). Hence, many facultative symbiont genomes are full of putative toxins and other pathogenicity factors. Yet it is entirely unclear what these do, and whether they play a role in protective symbiosis or provide other benefits to the bacterial or insect host. Long term host-restriction in heritable symbiosis combined with reduced opportunities for horizontal exchange with other bacteria, might presumably lead to selection for attenuated virulence and their removal—so why are they retained? Are they providing functions in service of the defensive symbiosis, such as roles required for persistence and vertical transmission in their hosts? This would involve navigating interactions with the insect immune system, reaching and occupying vital tissue for transmission, or resolving competition with other microbes. Alternatively, the bulk of these may be maintained by selfish genetic elements that prevent their removal. Toxin-antitoxin systems (aka 'addition modules') can enable plasmid persistence by eliminating cells that did not inherit a copy, but when present on the main chromosome they may 'stabilize' genomic islands by preventing large deletions (Van Melderen and Saavedra De Bast, 2009). Other factors, including toxin-antidote bacteriocins may perform similar functions (Inglis et al., 2013).

4. Conclusion

It is a truly exciting time to study defensive symbionts of insects. Genomic and proteomic approaches are allowing us to not only identify novel toxic factors, but to also examine population-level variation in protective ability. Advances in experimental approaches, such as the ability to culture and genetically transform symbionts (Brandt et al., 2017; Masson et al., 2018), or to express symbiont genes in model organisms amenable to transgenics (Beckmann et al., 2017; Harumoto and Lemaitre, 2018; LePage et al., 2017), will allow us to probe function in exciting detail. And of course, we have only reached the tip of the iceberg in discovering new defensive symbioses in the wild.

Acknowledgements

Funding support was provided by the National Science Foundation award 1754302 to KMO and grants from the Natural Sciences and Engineering Council of Canada and the Swiss National Science Foundation (Sinergia project CRSII3_154396) to SJP.

References

Anbutsu, H., Fukatsu, T., 2011. *Spiroplasma* as a model insect endosymbiont. Environ. Microbiol. Rep. 3 (2), 144–153.

Ankrah, N.Y.D., Luan, J., Douglas, A.E., 2017. Cooperative metabolism in a three-partner insect-bacterial symbiosis revealed by metabolic modeling. J. Bacteriol. 199 (15), e00872-16.

Asplen, M.K., Bano, N., Brady, C.M., Desneux, N., Hopper, K.R., Malouines, C., Oliver, K.M., White, J.A., Heimpel, G.E., 2014. Specialisation of bacterial endosymbionts that protect aphids from parasitoids. Ecol. Entomol. 39 (6), 736–739.

Ballinger, M.J., Perlman, S.J., 2019. The defensive *Spiroplasma*. Curr. Opin. Insect. Sci. 32, 36–41.

Ballinger, M.J., Moore, L.D., Perlman, S.J., 2018. Evolution and diversity of inherited *Spiroplasma* symbionts in *Myrmica* ants. Appl. Environ. Microbiol. 84 (4), e02299-17.

Ballinger, M.J., Gawryluk, R.M.R., Perlman, S.J., 2019. Toxin and genome evolution in a *Drosophila* defensive symbiosis. Genome Biol. Evol. 11 (1), 253–262.

Baumann, P., 2005. Biology of bacteriocyte-associated endosymbionts of plant sap-sucking insects. Annu. Rev. Microbiol. 59, 155–189.

Beckmann, J.F., Ronau, J.A., Hochstrasser, M., 2017. A *Wolbachia* deubiquitylating enzyme induces cytoplasmic incompatibility. Nat. Microbiol. 2 (5), 17007.

Bove, J.M., Renaudin, J., Saillard, C., Foissac, X., Garnier, M., 2003. *Spiroplasma citri,* a plant pathogenic mollicute: relationships with its two hosts, the plant and the leafhopper vector. Annu. Rev. Phytopathol. 41, 483–500.

Brandt, J.W., Chevignon, G., Oliver, K.M., Strand, M.R., 2017. Culture of an aphid heritable symbiont demonstrates its direct role in defence against parasitoids. Proc. R. Soc. B Biol. Sci. 284 (1866), 20171925.

Brega, A., Falaschi, A., Decarli, L., Pavan, M., 1968. Studies on mechanism of action of pederine. J. Cell Biol. 36 (3), 485–496.

Breznak, J.A., Brune, A., 1994. Role of microorganisms in the digestion of lignocellulose by termites. Annu. Rev. Entomol. 39, 453–487.

Brisson, J.A., Stern, D.L., 2006. The pea aphid, *Acyrthosiphon pisum*: an emerging genomic model system for ecological, developmental and evolutionary studies. Bioessays 28 (7), 747–755.

Brownlie, J.C., Johnson, K.N., 2009. Symbiont-mediated protection in insect hosts. Trends Microbiol. 17 (8), 348–354.

Brune, A., 2014. Symbiotic digestion of lignocellulose in termite guts. Nat. Rev. Microbiol. 12 (3), 168–180.

Buchner, P., 1965. Endosymbiosis of Animals With Plant Microorganisms. Rev. Eng. ed. [Translated by Bertha Mueller, with the collaboration of Francis H. Fockier]. ed. Interscience Publishers.

Burres, N.S., Clement, J.J., 1989. Antitumor-activity and mechanism of action of the novel marine natural-products mycalamide-a and mycalamide-b and onnamide. Cancer Res. 49 (11), 2935–2940.

Busby, J.N., Panjikar, S., Landsberg, M.J., Hurst, M.R.H., Lott, J.S., 2013. The BC component of ABC toxins is an RHS-repeat-containing protein encapsulation device. Nature 501 (7468), 547.

Caragata, E.P., Rances, E., Hedges, L.M., Gofton, A.W., Johnson, K.N., O'Neill, S.L., McGraw, E.A., 2013. Dietary cholesterol modulates pathogen blocking by *Wolbachia*. PLoS Pathog. 9 (6), e1003459.

Caragata, E.P., Tikhe, C.V., Dimopoulos, G., 2019. Curious entanglements: interactions between mosquitoes, their microbiota, and arboviruses. Curr. Opin. Virol. 37, 26–36.

Casjens, S.R., Thuman-Commike, P.A., 2011. Evolution of mosaically related tailed bacteriophage genomes seen through the lens of phage P22 virion assembly. Virology 411 (2), 393–415.

Cayetano, L., Vorburger, C., 2013. Genotype-by-genotype specificity remains robust to average temperature variation in an aphid/endosymbiont/parasitoid system. J. Evol. Biol. 26 (7), 1603–1610.

Cayetano, L., Vorburger, C., 2015. Symbiont-conferred protection against hymenopteran parasitoids in aphids: how general is it? Ecol. Entomol. 40 (1), 85–93.

Chen, X.L., Stansly, P.A., 2014. Biology of *Tamarixia radiata* (Hymenoptera: Eulophidae), parasitoid of the citrus greening disease vector *Diaphorina citri* (Hemiptera: Psylloidea): a mini review. Fla. Entomol. 97 (4), 1404–1413.

Chevignon, G., Boyd, B.M., Brandt, J.W., Oliver, K.M., Strand, M.R., 2018. Culture-facilitated comparative genomics of the facultative symbiont *Hamiltonella defensa*. Genome Biol. Evol. 10 (3), 786–802.

Clark, A.J., Pontes, M., Jones, T., Dale, C., 2007. A possible heterodimeric prophage-like element in the genome of the insect endosymbiont *Sodalis glossinidius*. J. Bacteriol. 189 (7), 2949–2951.

Clarke, H.V., Cullen, D., Hubbard, S.F., Karley, A.J., 2017. Susceptibility of *Macrosiphum euphorbiae* to the parasitoid *Aphidius ervi*: larval development depends on host aphid genotype. Entomol. Exp. Appl. 162 (2), 148–158.

Cockburn, S.N., Haselkorn, T.S., Hamilton, P.T., Landzberg, E., Jaenike, J., Perlman, S.J., 2013. Dynamics of the continent-wide spread of a *Drosophila* defensive symbiont. Ecol. Lett. 16 (5), 609–616.

Coley, P.D., Bryant, J.P., Chapin, F.S., 1985. Resource availability and plant antiherbivore defense. Science 230 (4728), 895–899.

Comayras, C., Tasca, C., Peres, S.Y., Ducommun, B., Oswald, E., De Rycke, J., 1997. *Escherichia coli* cytolethal distending toxin blocks the HeLa cell cycle at the G2/M transition by preventing cdc2 protein kinase dephosphorylation and activation. Infect. Immun. 65 (12), 5088–5095.

Corbin, C., Heyworth, E.R., Ferrari, J., Hurst, G.D.D., 2017. Heritable symbionts in a world of varying temperature. Heredity 118, 10–20.

Dang, Y.J., Schneider-Poetsch, T., Eyler, D.E., Jewett, J.C., Bhat, S., Rawal, V.H., Green, R., Liu, J.O., 2011. Inhibition of eukaryotic translation elongation by the antitumor natural product Mycalamide B. RNA 17 (8), 1578–1588.

Degnan, P.H., Moran, N.A., 2008a. Diverse phage-encoded toxins in a protective insect endosymbiont. Appl. Environ. Microbiol. 74 (21), 6782–6791.

Degnan, P.H., Moran, N.A., 2008b. Evolutionary genetics of a defensive facultative symbiont of insects: exchange of toxin-encoding bacteriophage. Mol. Ecol. 17 (3), 916–929.

Degnan, P.H., Yu, Y., Sisneros, N., Wing, R.A., Moran, N.A., 2009. *Hamiltonella defensa,* genome evolution of protective bacterial endosymbiont from pathogenic ancestors. Proc. Natl. Acad. Sci. U. S. A. 106 (22), 9063–9068.

Degnan, P.H., Leonardo, T.E., Cass, B.N., Hurwitz, B., Stern, D., Gibbs, R.A., Richards, S., Moran, N.A., 2010. Dynamics of genome evolution in facultative symbionts of aphids. Environ. Microbiol. 12 (8), 2060–2069.

Dennis, A.B., Patel, V., Oliver, K.M., Vorburger, C., 2017. Parasitoid gene expression changes after adaptation to symbiont-protected hosts. Evolution 71 (11), 2599–2617.

Dettner, K., 1987. Chemosystematics and evolution of beetle chemical defense. Annu. Rev. Entomol. 32, 17–48.

Dion, E., Polin, S.E., Simon, J.C., Outreman, Y., 2011a. Symbiont infection affects aphid defensive behaviours. Biol. Lett. 7 (5), 743–746.

Dion, E., Zele, F., Simon, J.C., Outreman, Y., 2011b. Rapid evolution of parasitoids when faced with the symbiont-mediated resistance of their hosts. J. Evol. Biol. 24 (4), 741–750.

Doremus, M.R., Oliver, K.M., 2017. Aphid heritable symbiont exploits defensive mutualism. Appl. Environ. Microbiol. 83 (8), 03276-16.

Doremus, M.R., Smith, A.H., Kim, K.L., Holder, A.J., Russell, J.A., Oliver, K.M., 2018. Breakdown of a defensive symbiosis, but not endogenous defences, at elevated temperatures. Mol. Ecol. 27 (8), 2138–2151.

Dose, B., Niehs, S.P., Scherlach, K., Florez, L.V., Kaltenpoth, M., Hertweck, C., 2018. Unexpected bacterial origin of the antibiotic icosalide: two-tailed depsipeptide assembly in multifarious *Burkholderia* symbionts. ACS Chem. Biol. 13 (9), 2414–2420.

Doudoumis, V., Blow, F., Saridaki, A., Augustinos, A., Dyer, N.A., Goodhead, I., Solano, P., Rayaisse, J.B., Takac, P., Mekonnen, S., Parker, A.G., Abd-Alla, A.M.M., Darby, A., Bourtzis, K., Tsiamis, G., 2017. Challenging the *Wigglesworthia, Sodalis, Wolbachia* symbiosis dogma in tsetse flies: *Spiroplasma* is present in both laboratory and natural populations. Sci. Rep. 7, 4699.

Douglas, A.E., 2014. Molecular dissection of nutrient exchange at the insect-microbial interface. Curr. Opin. Insect. Sci. 4, 23–28.

Douglas, A.E., 2015. Multiorganismal insects: diversity and function of resident microorganisms. Annu. Rev. Entomol. 60, 17–34.

Duron, O., 2014. *Arsenophonus* insect symbionts are commonly infected with APSE, a bacteriophage involved in protective symbiosis. FEMS Microbiol. Ecol. 90 (1), 184–194.

Duron, O., Bouchon, D., Boutin, S., Bellamy, L., Zhou, L.Q., Engelstadter, J., Hurst, G.D., 2008. The diversity of reproductive parasites among arthropods: *Wolbachia* do not walk alone. BMC Biol. 6, 27.

Dykstra, H.R., Weldon, S.R., Martinez, A.J., White, J.A., Hopper, K.R., Heimpel, G.E., Asplen, M.K., Oliver, K.M., 2014. Factors limiting the spread of the protective symbiont *Hamiltonella defensa* in *Aphis craccivora* aphids. Appl. Environ. Microbiol. 80 (18), 5818–5827.

Endo, Y., Tsurugi, K., Yutsudo, T., Takeda, Y., Ogasawara, T., Igarashi, K., 1988. Site of action of a vero toxin (vt2) from E*scherichia-coli* O157-H7 and of Shiga toxin on eukaryotic ribosomes—rna n-glycosidase activity of the toxins. Eur. J. Biochem. 171 (1–2), 45–50.

Engel, P., Moran, N.A., 2013. The gut microbiota of insects—diversity in structure and function. FEMS Microbiol. Rev. 37 (5), 699–735.

Engl, T., Kroiss, J., Kai, M., Nechitaylo, T.Y., Svatos, A., Kaltenpoth, M., 2018. Evolutionary stability of antibiotic protection in a defensive symbiosis. Proc. Natl. Acad. Sci. U. S. A. 115 (9), E2020–E2029.

Feeny, P., 1975. Biochemical coevolution between plants and their insect herbivores. In: Gilbert, L.E., Raven, P.H. (Eds.), Coevolution of Animals and Plants: Symposium V, First International Congress of Systematic and Evolutionary Biology, Boulder, Colorado, August, 1973. University of Texas Press, Austin, USA, pp. 3–19.

Feldhaar, H., 2011. Bacterial symbionts as mediators of ecologically important traits of insect hosts. Ecol. Entomol. 36 (5), 533–543.

Feng, H.L., Edwards, N., Anderson, C.M.H., Althaus, M., Duncan, R.P., Hsu, Y.C., Luetje, C.W., Price, D.R.G., Wilson, A.C.C., Thwaites, D.T., 2019. Trading amino acids at the aphid-*Buchnera* symbiotic interface. Proc. Natl. Acad. Sci. U. S. A. 116 (32), 16003–16011.

Ferrari, J., West, J.A., Via, S., Godfray, H.C.J., 2012. Population genetic structure and secondary symbionts in host-associated populations of the pea aphid complex. Evolution 66 (2), 375–390.

Fisch, K.M., Gurgui, C., Heycke, N., van der Sar, S.A., Anderson, S.A., Webb, V.L., Taudien, S., Platzer, M., Rubio, B.K., Robinson, S.J., Crews, P., Piel, J., 2009. Polyketide assembly lines of uncultivated sponge symbionts from structure-based gene targeting. Nat. Chem. Biol. 5 (7), 494–501.

Flórez, L.V., Kaltenpoth, M., 2017. Symbiont dynamics and strain diversity in the defensive mutualism between *Lagria* beetles and *Burkholderia*. Environ. Microbiol. 19 (9), 3674–3688.

Flórez, L.V., Biedermann, P.H.W., Engl, T., Kaltenpoth, M., 2015. Defensive symbioses of animals with prokaryotic and eukaryotic microorganisms. Nat. Prod. Rep. 32 (7), 904–936.

Flórez, L.V., Scherlach, K., Gaube, P., Ross, C., Sitte, E., Hermes, C., Rodrigues, A., Hertweck, C., Kaltenpoth, M., 2017. Antibiotic-producing symbionts dynamically transition between plant pathogenicity and insect-defensive mutualism. Nat. Commun. 8, 15172.

Flórez, L.V., Scherlach, K., Miller, I.J., Rodrigues, A., Kwan, J.C., Hertweck, C., Kaltenpoth, M., 2018. An antifungal polyketide associated with horizontally acquired genes supports symbiont-mediated defense in *Lagria villosa* beetles. Nat. Commun. 9, 2478.

Frago, E., Dicke, M., Godfray, H.C.J., 2012. Insect symbionts as hidden players in insect-plant interactions. Trends Ecol. Evol. 27 (12), 705–711.

Frago, E., Mala, M., Weldegergis, B.T., Yang, C.J., McLean, A., Godfray, H.C.J., Gols, R., Dicke, M., 2017. Symbionts protect aphids from parasitic wasps by attenuating herbivore-induced plant volatiles. Nat. Commun. 8, 1860.

Frank, J.H., Kanamitsu, K., 1987. *Paederus, sensu lato* (Coleoptera, Staphylinidae)—natural-history and medical importance. J. Med. Entomol. 24 (2), 155–191.

Garcia-Arraez, M.G., Masson, F., Escobar, J.C.P., Lemaitre, B., 2019. Functional analysis of RIP toxins from the *Drosophila* endosymbiont *Spiroplasma poulsonii*. BMC Microbiol. 19, 46.

Gehrer, L., Vorburger, C., 2012. Parasitoids as vectors of facultative bacterial endosymbionts in aphids. Biol. Lett. 8 (4), 613–615.

Gelmetti, C., Grimalt, R., 1993. *Paederus* dermatitis—an easy diagnosable but misdiagnosed eruption. Eur. J. Pediatr. 152 (1), 6–8.

Gottlieb, Y., Ghanim, M., Gueguen, G., Kontsedalov, S., Vavre, F., Fleury, F., Zchori-Fein, E., 2008. Inherited intracellular ecosystem: symbiotic bacteria share bacteriocytes in whiteflies. FASEB J. 22 (7), 2591–2599.

Gottlieb, Y., Zchori-Fein, E., Mozes-Daube, N., Kontsedalov, S., Skaljac, M., Brumin, M., Sobol, I., Czosnek, H., Vavre, F., Fleury, F., Ghanim, M., 2010. The transmission efficiency of tomato yellow leaf curl virus by the whitefly *Bemisia tabaci* is correlated with the presence of a specific symbiotic bacterium species. J. Virol. 84 (18), 9310–9317.

Gross, P., 1993. Insect behavioral and morphological defenses against parasitoids. Annu. Rev. Entomol. 38, 251–273.

Gunduz, E.A., Douglas, A.E., 2009. Symbiotic bacteria enable insect to use a nutritionally inadequate diet. Proc. R. Soc. B Biol. Sci. 276 (1658), 987–991.

Hacker, J., Kaper, J.B., 2000. Pathogenicity islands and the evolution of microbes. Annu. Rev. Microbiol. 54, 641–679.

Hafer, N., Vorburger, C., 2019. Diversity begets diversity: do parasites promote variation in protective symbionts? Curr. Opin. Insect Sci. 32, 8–14.

Haine, E.R., 2008. Symbiont-mediated protection. Proc. R. Soc. B Biol. Sci. 275 (1633), 353–361.

Hamilton, P.T., Peng, F.N., Boulanger, M.J., Perlman, S.J., 2016. A ribosome-inactivating protein in a *Drosophila* defensive symbiont. Proc. Natl. Acad. Sci. U. S. A. 113 (2), 350–355.

Hansen, A.K., Moran, N.A., 2014. The impact of microbial symbionts on host plant utilization by herbivorous insects. Mol. Ecol. 23 (6), 1473–1496.

Hansen, A.K., Jeong, G., Paine, T.D., Stouthamer, R., 2007. Frequency of secondary symbiont infection in an invasive psyllid relates to parasitism pressure on a geographic scale in California. Appl. Environ. Microbiol. 73 (23), 7531–7535.

Hansen, A.K., Vorburger, C., Moran, N.A., 2012. Genomic basis of endosymbiont-conferred protection against an insect parasitoid. Genome Res. 22 (1), 106–114.

Harumoto, T., Lemaitre, B., 2018. Male-killing toxin in a bacterial symbiont of *Drosophila*. Nature 557 (7704), 252.

Haselkorn, T.S., Jaenike, J., 2015. Macroevolutionary persistence of heritable endosymbionts: acquisition, retention and expression of adaptive phenotypes in *Spiroplasma*. Mol. Ecol. 24 (14), 3752–3765.

Haselkorn, T.S., Markow, T.A., Moran, N.A., 2009. Multiple introductions of the *Spiroplasma* bacterial endosymbiont into *Drosophila*. Mol. Ecol. 18 (6), 1294–1305.

Haselkorn, T.S., Cockburn, S.N., Hamilton, P.T., Perlman, S.J., Jaenike, J., 2013. Infectious adaptation: potential host range of a defensive endosymbiont in *Drosophila*. Evolution 67 (4), 934–945.

Hayashi, M., Watanabe, M., Yukuhiro, F., Nomura, M., Kageyama, D., 2016. A nightmare for males? A maternally transmitted male-killing bacterium and strong female bias in a green lacewing population. PLoS One 11 (6), e0155794.

Hedges, L.M., Brownlie, J.C., O'Neill, S.L., Johnson, K.N., 2008. *Wolbachia* and virus protection in insects. Science 322 (5902), 702.

Henry, L.M., Peccoud, J., Simon, J.C., Hadfield, J.D., Maiden, M.J.C., Ferrari, J., Godfray, H.C.J., 2013. Horizontally transmitted symbionts and host colonization of ecological niches. Curr. Biol. 23 (17), 1713–1717.

Herold, S., Karch, H., Schmidt, H., 2004. Shiga toxin-encoding bacteriophages—genomes in motion. Int. J. Med. Microbiol. 294 (2–3), 115–121.

Heyworth, E.R., Ferrari, J., 2015. A facultative endosymbiont in aphids can provide diverse ecological benefits. J. Evol. Biol. 28 (10), 1753–1760.

Higashi, C.H.V., Oliver, K.M., 2020. Warmer nights offer no respite for a defensive mutualism. J. Anim. Ecol (in press).

Hillman, K., Goodrich-Blair, H., 2016. Are you my symbiont? Microbial polymorphic toxins and antimicrobial compounds as honest signals of beneficial symbiotic defensive traits. Curr. Opin. Microbiol. 31, 184–190.

Hopper, K.R., Kuhn, K.L., Lanier, K., Rhoades, J.H., Oliver, K.M., White, J.A., Asplen, M.K., Heimpel, G.E., 2018. The defensive aphid symbiont *Hamiltonella defensa* affects host quality differently for *Aphelinus glycinis* versus *Aphelinus atriplicis*. Biol. Control 116, 3–9.

Hurst, G.D.D., von der Schulenburg, J.H.G., Majerus, T.M.O., Bertrand, D., Zakharov, I.A., Baungaard, J., Volkl, W., Stouthamer, R., Majerus, M.E.N., 1999. Invasion of one insect species, *Adalia bipunctata*, by two different male-killing bacteria. Insect Mol. Biol. 8 (1), 133–139.

Inglis, R.F., Bayramoglu, B., Gillor, O., Ackermann, M., 2013. The role of bacteriocins as selfish genetic elements. Biol. Lett. 9 (3), 20121173.

Jaenike, J., Polak, M., Fiskin, A., Helou, M., Minhas, M., 2007. Interspecific transmission of endosymbiotic *Spiroplasma* by mites. Biol. Lett. 3 (1), 23–25.

Jaenike, J., Unckless, R., Cockburn, S.N., Boelio, L.M., Perlman, S.J., 2010. Adaptation via symbiosis: recent spread of a *Drosophila* defensive symbiont. Science 329 (5988), 212–215.

Jamet, A., Nassif, X., 2015. New players in the toxin field: polymorphic toxin systems in bacteria. MBio 6 (3), 00285-15.

Jiggins, F.M., Hurst, G.D.D., Jiggins, C.D., Von der Schulenburg, J.H.G., Majerus, M.E.N., 2000. The butterfly *Danaus chrysippus* is infected by a male-killing *Spiroplasma* bacterium. Parasitology 120, 439–446.

Jinadasa, R.N., Bloom, S.E., Weiss, R.S., Duhamel, G.E., 2011. Cytolethal distending toxin: a conserved bacterial genotoxin that blocks cell cycle progression, leading to apoptosis of a broad range of mammalian cell lineages. Microbiology 157 (7), 1851–1875.

Jing, X.F., Wong, A.C.N., Chaston, J.M., Colvin, J., McKenzie, C.L., Douglas, A.E., 2014. The bacterial communities in plant phloem-sap-feeding insects. Mol. Ecol. 23 (6), 1433–1444.

Johannes, L., Römer, W., 2010. Shiga toxins—from cell biology to biomedical applications. Nat. Rev. Microbiol. 8 (2), 105–116.

Johnson, W.M., Lior, H., 1988. A new heat-labile cytolethal distending toxin (CLDT) produced by *Campylobacter* spp. Microb. Pathog. 4 (2), 115–126.

Kach, H., Mathe-Hubert, H., Dennis, A.B., Vorburger, C., 2018. Rapid evolution of symbiont-mediated resistance compromises biological control of aphids by parasitoids. Evol. Appl. 11 (2), 220–230.

Kaltenpoth, M., Gottler, W., Herzner, G., Strohm, E., 2005. Symbiotic bacteria protect wasp larvae from fungal infestation. Curr. Biol. 15 (5), 475–479.

Kaltenpoth, M., Goettler, W., Koehler, S., Strohm, E., 2010. Life cycle and population dynamics of a protective insect symbiont reveal severe bottlenecks during vertical transmission. Evol. Ecol. 24 (2), 463–477.

Kaltenpoth, M., Roeser-Mueller, K., Koehler, S., Peterson, A., Nechitaylo, T.Y., Stubblefield, J.W., Herzner, G., Seger, J., Strohm, E., 2014. Partner choice and fidelity stabilize coevolution in a cretaceous-age defensive symbiosis. Proc. Natl. Acad. Sci. U. S. A. 111 (17), 6359–6364.

Kampa, A., Gagunashvili, A.N., Gulder, T.A.M., Morinaka, B.I., Daolio, C., Godejohann, M., Miao, V.P.W., Piel, J., Andresson, O.S., 2013. Metagenomic natural product discovery in lichen provides evidence for a family of biosynthetic pathways in diverse symbioses. Proc. Natl. Acad. Sci. U. S. A. 110 (33), E3129–E3137.

Kellner, R.L.L., 2002. Molecular identification of an endosymbiotic bacterium associated with pederin biosynthesis in *Paederus sabaeus* (Coleoptera: Staphylinidae). Insect Biochem. Mol. Biol. 32 (4), 389–395.

Kellner, R.L.L., Dettner, K., 1996. Differential efficacy of toxic pederin in deterring potential arthropod predators of *Paederus* (Coleoptera: Staphylinidae) offspring. Oecologia 107 (3), 293–300.

Kim, I.H., Aryal, S.K., Aghai, D.T., Casanova-Torres, A.M., Hillman, K., Kozuch, M.P., Mans, E.J., Mauer, T.J., Ogier, J.C., Ensign, J.C., Gaudriault, S., Goodman, W.G., Goodrich-Blair, H., Dillman, A.R., 2017. The insect pathogenic bacterium *Xenorhabdus innexi* has attenuated virulence in multiple insect model hosts yet encodes a potent mosquitocidal toxin. BMC Genomics 18, 927.

Koskiniemi, S., Lamoureux, J.G., Nikolakakis, K.C., t'Kint de Roodenbeke, C., Kaplan, M.D., Low, D.A., Hayes, C.S., 2013. Rhs proteins from diverse bacteria mediate intercellular competition. Proc. Natl. Acad. Sci. U. S. A. 110 (17), 7032–7037.

Kraft, L.J., Kopco, J., Harmon, J.P., Oliver, K.M., 2017. Aphid symbionts and endogenous resistance traits mediate competition between rival parasitoids. PLoS One 12 (7), e0180729.

Kroiss, J., Kaltenpoth, M., Schneider, B., Schwinger, M.G., Hertweck, C., Maddula, R.K., Strohm, E., Svatos, A., 2010. Symbiotic streptomycetes provide antibiotic combination prophylaxis for wasp offspring. Nat. Chem. Biol. 6 (4), 261–263.

Kumar, R., Feltrup, T.M., Kukreja, R.V., Patel, K.B., Cai, S., Singh, B.R., 2019. Evolutionary features in the structure and function of bacterial toxins. Toxins 11 (1), 15.
Kung, V.L., Khare, S., Stehlik, C., Bacon, E.M., Hughes, A.J., Hauser, A.R., 2012. An rhs gene of *Pseudomonas aeruginosa* encodes a virulence protein that activates the inflammasome. Proc. Natl. Acad. Sci. U. S. A. 109 (4), 1275–1280.
Kust, A., Mares, J., Jokela, J., Urajova, P., Hajek, J., Saurav, K., Voracova, K., Fewer, D.P., Haapaniemi, E., Permi, P., Rehakova, K., Sivonen, K., Hrouzek, P., 2018. Discovery of a pederin family compound in a nonsymbiotic bloom-forming cyanobacterium. ACS Chem. Biol. 13 (5), 1123–1129.
Lainhart, W., Stolfa, G., Koudelka, G.B., 2009. Shiga toxin as a bacterial defense against a eukaryotic predator, *Tetrahymena thermophila*. J. Bacteriol. 191 (16), 5116–5122.
Lapadula, W.J., Marcet, P.L., Mascotti, M.L., Sanchez-Puerta, M.V., Ayub, M.J., 2017. Metazoan ribosome inactivating protein encoding genes acquired by horizontal gene transfer. Sci. Rep. 7, 1863.
Lapadula, W.J., Mascotti, M.L., Ayub, M.J., 2019. Recurrent acquisition of alien ribotoxin encoding genes by insect genomes. bioRxiv. https://doi.org/10.1101/2019.12.20.880922.
Lara-Tejero, M., Galan, J.E., 2000. A bacterial toxin that controls cell cycle progression as a deoxyribonuclease I-like protein. Science 290 (5490), 354–357.
Leclair, M., Polin, S., Jousseaume, T., Simon, J.C., Sugio, A., Morliere, S., Fukatsu, T., Tsuchida, T., Outreman, Y., 2017. Consequences of coinfection with protective symbionts on the host phenotype and symbiont titres in the pea aphid system. Insect Sci. 24 (5), 798–808.
Lee, K.-H., Nishimura, S., Matsunaga, S., Fusetani, N., Horinouchi, S., Yoshida, M., 2005. Inhibition of protein synthesis and activation of stress-activated protein kinases by onnamide a and theopederin B, antitumor marine natural products. Cancer Sci. 96 (6), 357–364.
Lemaitre, B., Hoffmann, J., 2007. The host defense of *Drosophila melanogaster*. Annu. Rev. Immunol. 25, 697–743.
LePage, D.P., Metcalf, J.A., Bordenstein, S.R., On, J.M., Perlmutter, J.I., Shropshire, J.D., Layton, E.M., Funkhouser-Jones, L.J., Beckmann, J.F., Bordenstein, S.R., 2017. Prophage WO genes recapitulate and enhance *Wolbachia*-induced cytoplasmic incompatibility. Nature 543 (7644), 243.
Leybourne, D.J., Bos, J.I.B., Valentine, T.A., Karley, A.J., 2020. The price of protection: a defensive endosymbiont impairs nymph growth in the bird cherry-oat aphid, *Rhopalosiphum padi*. Insect Sci. 27 (1), 69–85.
Li, Q., Fan, J., Sun, J.X., Wang, M.Q., Chen, J.L., 2018. Plant-mediatedhorizontal transmission of *Hamiltonella defensa* in the wheat aphid *Sitobion miscanthi*. J. Agric. Food Chem. 66 (51), 13367–13377.
Li, Q., Fan, J., Sun, J., Zhang, Y., Hou, M., Chen, J., 2019. Anti-plant defense response strategies mediated by the secondary symbiont *Hamiltonella defensa* in the wheat aphid *Sitobion miscanthi*. Front. Microbiol. 10, 2419.
Linhartová, I., Bumba, L., Mašín, J., Basler, M., Osička, R., Kamanová, J., Procházková, K., Adkins, I., Hejnová-Holubová, J., Sadílková, L., Morová, J., Šebo, P., 2010. RTX proteins: a highly diverse family secreted by a common mechanism. FEMS Microbiol. Rev. 34 (6), 1076–1112.
Login, F.H., Balmand, S., Vallier, A., Vincent-Monégat, C., Vigneron, A., Weiss-Gayet, M., Rochat, D., Heddi, A., 2011. Antimicrobial peptides keep insect endosymbionts under control. Science 334 (6054), 362–365.
Łukasik, P., Dawid, M.A., Ferrari, J., Godfray, H.C.J., 2013a. The diversity and fitness effects of infection with facultative endosymbionts in the grain aphid, *Sitobion avenae*. Oecologia 173 (3), 985–996.

Łukasik, P., Guo, H., Van Asch, M., Ferrari, J., Godfray, H.C.J., 2013b. Protection against a fungal pathogen conferred by the aphid facultative endosymbionts *Rickettsia* and *Spiroplasma* is expressed in multiple host genotypes and species and is not influenced by co-infection with another symbiont. J. Evol. Biol. 26 (12), 2654–2661.

Łukasik, P., van Asch, M., Guo, H.F., Ferrari, J., Godfray, H.C.J., 2013c. Unrelated facultative endosymbionts protect aphids against a fungal pathogen. Ecol. Lett. 16 (2), 214–218.

Łukasik, P., Guo, H.F., van Asch, M., Henry, L.M., Godfray, H.C.J., Ferrari, J., 2015. Horizontal transfer of facultative endosymbionts is limited by host relatedness. Evolution 69 (10), 2757–2766.

Lynn-Bell, N.L., Strand, M.R., Oliver, K.M., 2019. Bacteriophage acquisition restores protective mutualism. Microbiology 165 (9), 985–989.

Martinez, A.J., Weldon, S.R., Oliver, K.M., 2014. Effects of parasitism on aphid nutritional and protective symbioses. Mol. Ecol. 23 (6), 1594–1607.

Martinez, A.J., Kim, K.L., Harmon, J.P., Oliver, K.M., 2016. Specificity of multi-modal aphid defenses against two rival parasitoids. PLoS One 11 (5), e0154670.

Martinez, A.J., Doremus, M.R., Kraft, L.J., Kim, K.L., Oliver, K.M., 2018. Multi-modal defences in aphids offer redundant protection and increased costs likely impeding a protective mutualism. J. Anim. Ecol. 87 (2), 464–477.

Masson, F., Copete, S.C., Schupfer, F., Garcia-Arraez, G., Lemaitre, B., 2018. In vitro culture of the insect endosymbiont *Spiroplasma poulsonii* highlights bacterial genes involved in host-symbiont interaction. MBio. 9 (2), e00024-18.

Mateos, M., Winter, L., Winter, C., Higareda-Alvear, V.M., Martinez-Romero, E., Xie, J.L., 2016. Independent origins of resistance or susceptibility of parasitic wasps to a defensive symbiont. Ecol. Evol. 6 (9), 2679–2687.

Mathe-Hubert, H., Kaech, H., Ganesanandamoorthy, P., Vorburger, C., 2019. Evolutionary costs and benefits of infection with diverse strains of *Spiroplasma* in pea aphids. Evolution 73 (7), 1466–1481.

McLean, A.H.C., 2019. Cascading effects of defensive endosymbionts. Curr. Opin. Insect Sci. 32, 42–46.

McLean, A.H.C., Godfray, H.C.J., 2015. Evidence for specificity in symbiont-conferred protection against parasitoids. Proc. R. Soc. B Biol. Sci. 282 (1811), 20150977.

McLean, A.H.C., Godfray, H.C.J., 2017. The outcome of competition between two parasitoid species is influenced by a facultative symbiont of their aphid host. Funct. Ecol. 31 (4), 927–933.

McLean, A.H.C., Parker, B.J., Hrcek, J., Kavanagh, J.C., Wellham, P.A.D., Godfray, H.C.J., 2018. Consequences of symbiont co-infections for insect host phenotypes. J. Anim. Ecol. 87 (2), 478–488.

Meseguer, A.S., Manzano-Marin, A., Coeur D'Acier, A., Clamens, A.L., Godefroid, M., Jousselin, E., 2017. *Buchnera* has changed flatmate but the repeated replacement of co-obligate symbionts is not associated with the ecological expansions of their aphid hosts. Mol. Ecol. 26 (8), 2363–2378.

Michaud, J.P., 2004. Natural mortality of Asian citrus psyllid (Homoptera: Psyllidae) in Central Florida. Biol. Control 29 (2), 260–269.

Monticelli, L.S., Nguyen, L.T.H., Amiens-Desneux, E., Luo, C., Lavoir, A.V., Gatti, J.L., Desneux, N., 2019. The preference-performance relationship as a means of classifying parasitoids according to their specialization degree. Evol. Appl. 12 (8), 1626–1640.

Moran, N.A., 2007. Symbiosis as an adaptive process and source of phenotypic complexity. Proc. Natl. Acad. Sci. U. S. A. 104, 8627–8633.

Moran, N.A., Degnan, P.H., Santos, S.R., Dunbar, H.E., Ochman, H., 2005a. The players in a mutualistic symbiosis: insects, bacteria, viruses, and virulence genes. Proc. Natl. Acad. Sci. U. S. A. 102 (47), 16919–16926.

Moran, N.A., Russell, J.A., Koga, R., Fukatsu, T., 2005b. Evolutionary relationships of three new species of Enterobacteriaceae living as symbionts of aphids and other insects. Appl. Environ. Microbiol. 71 (6), 3302–3310.

Moran, N.A., McCutcheon, J.P., Nakabachi, A., 2008. Genomics and evolution of heritable bacterial symbionts. Annu. Rev. Genet. 42, 165–190.

Mosey, R.A., Floreancig, P.E., 2012. Isolation, biological activity, synthesis, and medicinal chemistry of the pederin/mycalamide family of natural products. Nat. Prod. Rep. 29 (9), 980–995.

Nakabachi, A., Ueoka, R., Oshima, K., Teta, R., Mangoni, A., Gurgui, M., Oldham, N.J., van Echten-Deckert, G., Okamura, K., Yamamoto, K., Inoue, H., Ohkuma, M., Hongoh, Y., Miyagishima, S., Hattori, M., Piel, J., Fukatsu, T., 2013. Defensive bacteriome symbiont with a drastically reduced genome. Curr. Biol. 23 (15), 1478–1484.

Nelsen, D.R., Nisani, Z., Cooper, A.M., Fox, G.A., Gren, E.C.K., Corbit, A.G., Hayes, W.K., 2014. Poisons, toxungens, and venoms: redefining and classifying toxic biological secretions and the organisms that employ them. Biol. Rev. 89 (2), 450–465.

Niepoth, N., Ellers, J., Henry, L.M., 2018. Symbiont interactions with non-native hosts limit the formation of new symbioses. BMC Evol. Biol. 18, 27.

Nishimura, S., Matsunaga, S., Yoshida, M., Hirota, H., Yokoyama, S., Fusetani, N., 2005. 13-Deoxytedanolide, a marine sponge-derived antitumor macrolide, binds to the 60S large ribosomal subunit. Bioorg. Med. Chem. 13 (2), 449–454.

Oliver, K.M., Higashi, C.H.V., 2019. Variations on a protective theme: *Hamiltonella defensa* infections in aphids variably impact parasitoid success. Curr. Opin. Insect Sci. 32, 1–7.

Oliver, K.M., Moran, N.A., 2009. Defensive symbionts in aphids and other insects. In: White, J.F., Torres, M.S. (Eds.), Defensive Mutualism in Microbial Symbiosis. CRC Press, pp. 129–147.

Oliver, K.M., Russell, J.A., 2016. Symbiosis, introduction to. In: Kliman, R.M. (Ed.), Encyclopedia of Evolutionary Biology. Academic Press, Oxford, pp. 282–290.

Oliver, K.M., Russell, J.A., Moran, N.A., Hunter, M.S., 2003. Facultative bacterial symbionts in aphids confer resistance to parasitic wasps. Proc. Natl. Acad. Sci. U. S. A. 100 (4), 1803–1807.

Oliver, K.M., Moran, N.A., Hunter, M.S., 2005. Variation in resistance to parasitism in aphids is due to symbionts not host genotype. Proc. Natl. Acad. Sci. U. S. A. 102 (36), 12795–12800.

Oliver, K.M., Moran, N.A., Hunter, M.S., 2006. Costs and benefits of a superinfection of facultative symbionts in aphids. Proc. R. Soc. B Biol. Sci. 273 (1591), 1273–1280.

Oliver, K.M., Campos, J., Moran, N.A., Hunter, M.S., 2008. Population dynamics of defensive symbionts in aphids. Proc. R. Soc. B Biol. Sci. 275 (1632), 293–299.

Oliver, K.M., Degnan, P.H., Hunter, M.S., Moran, N.A., 2009. Bacteriophages encode factors required for protection in a symbiotic mutualism. Science 325 (5943), 992–994.

Oliver, K.M., Degnan, P.H., Burke, G.R., Moran, N.A., 2010. Facultative symbionts in aphids and the horizontal transfer of ecologically important traits. Annu. Rev. Entomol. 55, 247–266.

Oliver, K.M., Noge, K., Huang, E.M., Campos, J.M., Becerra, J.X., Hunter, M.S., 2012. Parasitic wasp responses to symbiont-based defense in aphids. BMC Biol. 10, 11.

Oliver, K.M., Smith, A.H., Russell, J.A., 2014. Defensive symbiosis in the real world - advancing ecological studies of heritable, protective bacteria in aphids and beyond. Funct. Ecol. 28 (2), 341–355.

Paredes, J.C., Herren, J.K., Schupfer, F., Lemaitre, B., 2016. The role of lipid competition for endosymbiont-mediated protection against parasitoid wasps in *Drosophila*. MBio 7 (4), e01006-16.

Parker, B.J., Spragg, C.J., Altincicek, B., Gerardo, N.M., 2013. Symbiont-mediated protection against fungal pathogens in pea aphids: a role for pathogen specificity? Appl. Environ. Microbiol. 79 (7), 2455–2458.
Parker, B.J., McLean, A.H.C., Hrcek, J., Gerardo, N.M., Godfray, H.C.J., 2017. Establishment and maintenance of aphid endosymbionts after horizontal transfer is dependent on host genotype. Biol. Lett. 13 (5), 20170016.
Patel, V., Chevignon, G., Manzano-Marín, A., Brandt, J.W., Strand, M.R., Russell, J.A., Oliver, K.M., 2019. Cultivation-assisted genome of *Candidatus* Fukatsuia symbiotica; the enigmatic "X-type" symbiont of aphids. Genome Biol. Evol. 11 (12), 3510–3522.
Pérès, S.Y., Marchès, O., Daigle, F., Nougayrède, J.-P., Hérault, F., Tasca, C., De Rycke, J., Oswald, E., 1997. A new cytolethal distending toxin (CDT) from *Escherichia coli* producing CNF2 blocks HeLa cell division in G2/M phase. Mol. Microbiol. 24 (5), 1095–1107.
Piel, J., 2002. A polyketide synthase-peptide synthetase gene cluster from an uncultured bacterial symbiont of *Paederus* beetles. Proc. Natl. Acad. Sci. U. S. A. 99 (22), 14002–14007.
Piel, J., 2009. Metabolites from symbiotic bacteria. Nat. Prod. Rep. 26 (3), 338–362.
Piel, J., Hui, D.Q., Wen, G.P., Butzke, D., Platzer, M., Fusetani, N., Matsunaga, S., 2004. Antitumor polyketide biosynthesis by an uncultivated bacterial symbiont of the marine sponge *Theonella swinhoei*. Proc. Natl. Acad. Sci. U. S. A. 101 (46), 16222–16227.
Poulson, D.F., Sakaguchi, B., 1961. Nature of sex-ratio agent in *Drosophila*. Science 133 (346), 1489.
Qi, L.D., Sun, J.T., Hong, X.Y., Li, Y.X., 2019. Diversity and phylogenetic analyses reveal horizontal transmission of endosymbionts between whiteflies and their parasitoids. J. Econ. Entomol. 112 (2), 894–905.
Rasooly, R., Do, P.M., 2010. Shiga toxin Stx2 is heat-stable and not inactivated by pasteurization. Int. J. Food Microbiol. 136 (3), 290–294.
Ratzka, C., Gross, R., Feldhaar, H., 2012. Endosymbiont tolerance and control within insect hosts. Insects 3 (2), 553–572.
Regassa, L.B., Gasparich, G.E., 2006. *Spiroplasmas:* evolutionary relationships and biodiversity. Front. Biosci. 11, 2983–3002.
Richter, A., Kocienski, P., Raubo, P., Davies, D.E., 1997. The in vitro biological activities of synthetic 18-O-methyl mycalamide B, 10-epi-18-O-methyl mycalamide B and pederin. Anticancer Drug Des. 12 (3), 217–227.
Rock, D.I., Smith, A.H., Joffe, J., Albertus, A., Wong, N., O'Connor, M., Oliver, K.M., Russell, J.A., 2018. Context-dependent vertical transmission shapes strong endosymbiont community structure in the pea aphid, *Acyrthosiphon pisum*. Mol. Ecol. 27 (8), 2039–2056.
Rollat-Farnier, P.A., Santos-Garcia, D., Rao, Q., Sagot, M.F., Silva, F.J., Henri, H., Zchori-Fein, E., Latorre, A., Moya, A., Barbe, V., Liu, S.S., Wang, X.W., Vavre, F., Mouton, L., 2015. Two host clades, two bacterial arsenals: evolution through gene losses in facultative endosymbionts. Genome Biol. Evol. 7 (3), 839–855.
Rouchet, R., Vorburger, C., 2012. Strong specificity in the interaction between parasitoids and symbiont-protected hosts. J. Evol. Biol. 25 (11), 2369–2375.
Rouchet, R., Vorburger, C., 2014. Experimental evolution of parasitoid infectivity on symbiont-protected hosts leads to the emergence of genotype specificity. Evolution 68 (6), 1607–1616.
Rouïl, J., Jousselin, E., Coeur d'acier, A., Cruaud, C., Manzano-Marín, A., 2020. The protector within: comparative genomics of APSE phages across aphids reveals rampant recombination and diverse toxin arsenals. bioRxiv. https://doi.org/10.1101/2020.03.17.996009 2020.03.17.996009.
Rudkin, J.K., McLoughlin, R.M., Preston, A., Massey, R.C., 2017. Bacterial toxins: offensive, defensive, or something else altogether? PLoS Pathog. 13 (9), e1006452.

Russell, J.A., Moran, N.A., 2006. Costs and benefits of symbiont infection in aphids: variation among symbionts and across temperatures. Proc. R. Soc. B Biol. Sci. 273 (1586), 603–610.

Russell, J.A., Latorre, A., Sabater-Munoz, B., Moya, A., Moran, N.A., 2003. Side-stepping secondary symbionts: widespread horizontal transfer across and beyond the Aphidoidea. Mol. Ecol. 12 (4), 1061–1075.

Russell, J.A., Weldon, S., Smith, A.H., Kim, K.L., Hu, Y., Lukasik, P., Doll, S., Anastopoulos, I., Novin, M., Oliver, K.M., 2013. Uncovering symbiont-driven genetic diversity across north American pea aphids. Mol. Ecol. 22 (7), 2045–2059.

Russell, J.A., Oliver, K.M., Hansen, A.K., 2017. Band-aids for *Buchnera* and B vitamins for all. Mol. Ecol. 26 (8), 2199–2203.

Sanada-Morimura, S., Matsumura, M., Noda, H., 2013. Male killing caused by a *Spiroplasma* symbiont in the small brown planthopper, *Laodelphax striatellus*. J. Hered. 104 (6), 821–829.

Saxena, S.K., Obrien, A.D., Ackerman, E.J., 1989. Shiga toxin, Shiga-like toxin-ii variant, and ricin are all single-site rna n-glycosidases of 28-s rna when microinjected into xenopus oocytes. J. Biol. Chem. 264 (1), 596–601.

Sayavedra, L., Kleiner, M., Ponnudurai, R., Wetzel, S., Pelletier, E., Barbe, V., Satoh, N., Shoguchi, E., Fink, D., Breusing, C., Reusch, T.B.H., Rosenstiel, P., Schilhabel, M.B., Becher, D., Schweder, T., Markert, S., Dubilier, N., Petersen, J.M., 2015. Abundant toxin-related genes in the genomes of beneficial symbionts from deep-sea hydrothermal vent mussels. Elife 4, e07966.

Scarborough, C.L., Ferrari, J., Godfray, H.C., 2005. Aphid protected from pathogen by endosymbiont. Science 310 (5755), 1781.

Schmid, M., Sieber, R., Zimmermann, Y.S., Vorburger, C., 2012. Development, specificity and sublethal effects of symbiont-conferred resistance to parasitoids in aphids. Funct. Ecol. 26 (1), 207–215.

Schmitt, C.K., Meysick, K.C., O'Brien, A.D., 1999. Bacterial toxins: friends or foes? Emerg. Infect. Dis. 5 (2), 224–234.

Schneider, D.I., Saarman, N., Onyango, M.G., Hyseni, C., Opiro, R., Echodu, R., O'Neill, M., Bloch, D., Vigneron, A., Johnson, T.J., Dion, K., Weiss, B.L., Opiyo, E., Caccone, A., Aksoy, S., 2019. Spatio-temporal distribution of *Spiroplasma* infections in the tsetse fly (*Glossina fuscipes fuscipes*) in northern Uganda. PLoS Negl. Trop. Dis. 13 (8), e0007340.

Scott, D.A., Kaper, J.B., 1994. Cloning and sequencing of the genes encoding *Escherichia-coli* cytolethal distending toxin. Infect. Immun. 62 (1), 244–251.

Scuron, M.D., Boesze-Battaglia, K., Dlakić, M., Shenker, B.J., 2016. The cytolethal distending toxin contributes to microbial virulence and disease pathogenesis by acting as a tri-perditious toxin. Front. Cell. Infect. Microbiol. 6, 168.

Shan, H.W., Luan, J.B., Liu, Y.Q., Douglas, A.E., Liu, S.S., 2019. The inherited bacterial symbiont *Hamiltonella* influences the sex ratio of an insect host. Proc. R. Soc. B Biol. Sci. 286 (1915), 20191677.

Sheehan, G., Garvey, A., Croke, M., Kavanagh, K., 2018. Innate humoral immune defences in mammals and insects: the same, with differences ? Virulence 9 (1), 1625–1639.

Shigenobu, S., Watanabe, H., Hattori, M., Sakaki, Y., Ishikawa, H., 2000. Genome sequence of the endocellular bacterial symbiont of aphids *Buchnera* sp. APS. Nature 407 (6800), 81–86.

Simon, J.C., Boutin, S., Tsuchida, T., Koga, R., Le Gallic, J.F., Frantz, A., Outreman, Y., Fukatsu, T., 2011. Facultative symbiont infections affect aphid reproduction. PLoS One 6 (7), e21831.

Smith, J.M., 1989. Evolution—generating novelty by symbiosis. Nature 341 (6240), 284–285.

Smith, A.H., Lukasik, P., O'Connor, M.P., Lee, A., Mayo, G., Drott, M.T., Doll, S., Tuttle, R., Disciullo, R.A., Messina, A., Oliver, K.M., Russell, J.A., 2015. Patterns, causes and consequences of defensive microbiome dynamics across multiple scales. Mol. Ecol. 24 (5), 1135–1149.

Soldati, M., Fioretti, A., Ghione, M., 1966. Cytotoxicity of pederin and some of its derivatives on cultured mammalian cells. Experientia 22 (3), 176.

Staunton, J., Weissman, K.J., 2001. Polyketide biosynthesis: a millennium review. Nat. Prod. Rep. 18 (4), 380–416.

Stephens, W.Z., Wiles, T.J., Martinez, E.S., Jemielita, M., Burns, A.R., Parthasarathy, R., Bohannan, B.J.M., Guillemin, K., 2015. Identification of population bottlenecks and colonization factors during assembly of bacterial communities within the zebrafish intestine. MBio 6 (6), (e01163-15).

Strand, M.R., 2008. The insect cellular immune response. Insect Sci. 15 (1), 1–14.

Sturm, A., Heinemann, M., Arnoldini, M., Benecke, A., Ackermann, M., Benz, M., Dormann, J., Hardt, W.-D., 2011. The cost of virulence: retarded growth of *Salmonella typhimurium* cells expressing type iii secretion system 1. PLoS Pathog. 7 (7), e1002143.

Su, Q., Oliver, K.M., Xie, W., Wu, Q.J., Wang, S.L., Zhang, Y.J., 2015. The whitefly-associated facultative symbiont *Hamiltonella defensa* suppresses induced plant defences in tomato. Funct. Ecol. 29 (8), 1007–1018.

Teixeira, L., Ferreira, A., Ashburner, M., 2008. The bacterial symbiont *Wolbachia* induces resistance to rna viral infections in *Drosophila melanogaster.* PLoS Biol. 6 (12), 2753–2763.

Toju, H., Fukatsu, T., 2011. Diversity and infection prevalence of endosymbionts in natural populations of the chestnut weevil: relevance of local climate and host plants. Mol. Ecol. 20 (4), 853–868.

Van Arnam, E.B., Currie, C.R., Clardy, J., 2018. Defense contracts: molecular protection in insect-microbe symbioses. Chem. Soc. Rev. 47 (5), 1638–1651.

van der Wilk, F., Dullemans, A.M., Verbeek, M., van den Heuvel, J., 1999. Isolation and characterization of APSE-1, a bacteriophage infecting the secondary endosymbiont of *Acyrthosiphon pisum*. Virology 262 (1), 104–113.

Van Melderen, L., Saavedra De Bast, M., 2009. Bacterial toxin–antitoxin systems: more than selfish entities? PLoS Genet. 5 (3), e1000437.

Veraldi, S., Cuka, E., Chiaratti, A., Nazzaro, G., Gianotti, R., Suss, L., 2013. *Paederus fuscipes* dermatitis: a report of nine cases observed in Italy and review of the literature. Eur. J. Dermatol. 23 (3), 387–391.

Verster, K.I., Wisecaver, J.H., Karageorgi, M., Duncan, R.P., Gloss, A.D., Armstrong, E.E., Price, D.K., Menon, A.R., Ali, Z.M., Whiteman, N.K., 2019. Horizontal transfer of bacterial cytolethal distending toxin B genes to insects. Mol. Biol. Evol. 36 (10), 2105–2110.

Vorburger, C., 2014. The evolutionary ecology of symbiont-conferred resistance to parasitoids in aphids. Insect Sci. 21 (3), 251–264.

Vorburger, C., Gouskov, A., 2011. Only helpful when required: a longevity cost of harbouring defensive symbionts. J. Evol. Biol. 24 (7), 1611–1617.

Vorburger, C., Perlman, S.J., 2018. The role of defensive symbionts in host-parasite coevolution. Biol. Rev. Camb. Philos. Soc. 93 (4), 1747–1764.

Vorburger, C., Gehrer, L., Rodriguez, P., 2010. A strain of the bacterial symbiont *Regiella insecticola* protects aphids against parasitoids. Biol. Lett. 6 (1), 109–111.

Vorburger, C., Ganesanandamoorthy, P., Kwiatkowski, M., 2013. Comparing constitutive and induced costs of symbiont-conferred resistance to parasitoids in aphids. Ecol. Evol. 3 (3), 706–713.

Waterworth, S.C., Flórez, L.V., Rees, E.R., Hertweck, C., Kaltenpoth, M., Kwan, J.C., 2019. Horizontal gene transfer to a defensive symbiont with a reduced genome amongst a multipartite beetle microbiome. Biorxiv. https://doi.org/10.1101/780619.

Weinert, L.A., Araujo-Jnr, E.V., Ahmed, M.Z., Welch, J.J., 2015. The incidence of bacterial endosymbionts in terrestrial arthropods. Proc. R. Soc. B Biol. Sci. 282 (1807), 20150249.

Weldon, S.R., 2015. Matryoshka Mutualisms: Developing the Bacteriophage APSE-*Hamiltonella defensa*-*Acyrthosiphon pisum* System as a Model for Tripartite Symbioses. Unpublished Thesis, University of Georgia.

Weldon, S.R., Oliver, K.M., 2016. Diverse bacteriophage roles in an aphid-bacterial defensive mutualism. In: Hurst, C.J. (Ed.), Mechanistic Benefits of Microbial Symbionts. Springer, pp. 173–206.

Weldon, S.R., Strand, M.R., Oliver, K.M., 2013. Phage loss and the breakdown of a defensive symbiosis in aphids. Proc. R. Soc. B Biol. Sci. 280 (1751), 20122103.

Weldon, S.R., Russell, J.A., Oliver, K.M., 2020. More is not always better: coinfections with defensive symbionts generate highly variable outcomes. Appl. Environ. Microbiol. 86, e02537-19.

Werren, J.H., O'Neill, S.L., 1997. The evolution of heritable symbionts. In: Werren, J.H., O'Neill, S.L. (Eds.), Influential Passengers: Inherited Microorganisms and Arthoropod Reproduction. Oxford University Press, New York, USA, pp. 1–41.

Wilson, A.C.C., Duncan, R.P., 2015. Signatures of host/symbiont genome coevolution in insect nutritional endosymbioses. Proc. Natl. Acad. Sci. U. S. A. 112 (33), 10255–10261.

Xie, J.L., Vilchez, I., Mateos, M., 2010. *Spiroplasma* bacteria enhance survival of *Drosophila hydei* attacked by the parasitic wasp *Leptopilina heterotoma*. PLoS One 5 (8), e12149.

Xie, J., Butler, S., Sanchez, G., Mateos, M., 2014. Male killing *Spiroplasma* protects *Drosophila melanogaster* against two parasitoid wasps. Heredity 112 (4), 399–408.

Yamada, T., Hamada, M., Floreancig, P., Nakabachi, A., 2019. Diaphorin, a polyketide synthesized by an intracellular symbiont of the Asian citrus psyllid, is potentially harmful for biological control agents. PLoS One 14 (5), e0216319.

Zasloff, M., 2002. Antimicrobial peptides of multicellular organisms. Nature 415 (6870), 389–395.

Zchori-Fein, E., Lahav, T., Freilich, S., 2014. Variations in the identity and complexity of endosymbiont combinations in whitefly hosts. Front. Microbiol. 5, 310.

Zytynska, S.E., Weisser, W.W., 2016. The natural occurrence of secondary bacterial symbionts in aphids. Ecol. Entomol. 41 (1), 13–26.

CHAPTER NINE

The saboteur's tools: Common mechanistic themes across manipulative symbioses

Matthew R. Doremus[a], Martha S. Hunter[b]
[a]Graduate Interdisciplinary Program in Entomology & Insect Science, The University of Arizona, Tucson, AZ, United States
[b]Department of Entomology, The University of Arizona, Tucson, AZ, United States

Contents

Abstract

Insects and other arthropods often harbour intracellular bacterial associates. These bacterial symbionts cannot survive outside their host and rely on vertical transmission from infected mothers to their progeny. Thus, symbiont success is tied directly to reproductive success of female hosts. As a result of this intimate relationship, these heritable symbionts have evolved numerous strategies to increase the likelihood of their own transmission, some of which involve the direct manipulation of host reproduction to increase production or fitness of female progeny. These manipulations often come at the expense of male hosts or uninfected individuals, and include strategies such as inducing crossing incompatibilities that increase infected female fitness relative to uninfected females (cytoplasmic incompatibility), killing or feminizing male hosts, and inducing asexual production of female progeny (parthenogenesis). In the past decade, there has been substantial interest in the applied use of these manipulative bacteria in control of pest species, including suppressing disease carried by mosquito vectors of human pathogens. This, along with developments in molecular tools and techniques, has spurred major advances in understanding the mechanisms by which symbiont lineages manipulate their insect hosts, culminating with the identification

Advances in Insect Physiology, Volume 58
ISSN 0065-2806
https://doi.org/10.1016/bs.aiip.2020.03.003

of the bacterial genes responsible for key manipulations. Here, we review the major research advances in the mechanisms of symbiont-induced reproductive manipulation and compare the mechanisms of a number of common reproductive manipulators.

1. Introduction

Symbioses between insects and bacteria that live within them intracellularly or within bodily fluids are a "pandemic" (Duron et al., 2008; Werren et al., 2008); most insects bear these intimate associates. The bacteria in these partnerships are almost exclusively transmitted vertically from infected mothers to their progeny and are unable to survive outside of their host. To propagate, these heritable bacteria rely on the survival and successful reproduction of infected females. The intertwined fitness of host and symbiont has spurred the development of numerous strategies used by symbionts that increase the likelihood of their transmission. These include direct manipulation of the host's reproduction in a way that increases the number or fitness of female hosts, usually at the expense of male or uninfected hosts. These "reproductive manipulators" infect an estimated 40%–50% of all insects, as well as other arthropod lineages like crustaceans and chelicerates (Weinert et al., 2015; Zug and Hammerstein, 2012).

To date, four main types of reproductive manipulations have been identified from a diverse array of symbionts, representing many of the major bacterial phyla and even some microbial eukaryotes. The most common manipulation, cytoplasmic incompatibility (CI), is a crossing incompatibility that, at its simplest, results in offspring mortality in crosses between infected males and uninfected females. Bidirectional CI can also occur when both partners are infected with different symbiont strains (Werren et al., 2008). Mortality is averted when the female harbours the same symbiont strain, with bacteria or bacterial products in the infected cytoplasm "rescuing" the embryo. Another manipulative strategy is male-killing, in which infected males die at some point during development. A third is feminization, which occurs when genetically male hosts develop as functional females, and a fourth is parthenogenesis induction (PI), in which infected, mostly haplodiploid, hosts asexually produce infected daughters. While these phenotypes have historically been considered separately, there is increasing evidence that suggests that symbionts inducing different phenotypes may target the same host processes, or that symbionts causing one phenotype may contain the genetic capability to induce other manipulations (Fig. 1). For example, the successful transinfection of novel host species with well-characterized symbiont strains has sometimes revealed unexpected switches in manipulative phenotypes, and recent analyses of the genetic repertoire of these bacterial manipulators have revealed that many harbour the genes to induce multiple types of manipulations.

Despite their importance, the dependence of heritable symbionts on the host environment and our inability to culture them has made their study challenging. Typical microbiological methods, including direct manipulation of gene expression are not available. Although identified in numerous insect lineages as early as the 1920s, research on these bacteria was a daunting task until the development of molecular techniques like PCR. The difficulty of manipulating these symbionts has made the mechanisms by which control of host reproduction is achieved a long-standing mystery. However, by largely observational methods that take advantage of well-studied genetic host model systems, as well as various "omics" approaches, recent research on manipulative symbionts has made large strides towards elucidating the mechanisms by which these bacteria modify their arthropod hosts.

Here, we first review the current knowledge of the mechanisms controlling a variety of phenotypes conferred by heritable symbionts. In so doing, we hope to draw attention to those symbionts and mechanisms that are comparatively less well-characterized and compare potentially similar mechanisms of distantly related symbionts. We also discuss the mechanistic implications of the shifting phenotypes observed in novel manipulative infections, and the similarities seen across the four manipulative strategies. In doing so, we highlight common mechanistic themes of various arthropod-associated manipulative symbionts.

2. Cytoplasmic incompatibility (CI)

Cytoplasmic incompatibility is the most widely studied of the four reproductive manipulations, in part because of its ubiquity, and its occurrence in the model insect *Drosophila melanogaster*. CI also has long been of interest because of its potential to suppress pest populations via release of infected males (Laven, 1967). More recently, the discovery that some strains of the most common CI-inducing symbiont, *Wolbachia*, increase mosquito resistance to human vector-borne pathogens (Frentiu et al., 2014; Moreira et al., 2009; Walker et al., 2011) has refocused the community on understanding the mechanisms behind the CI and pathogen-blocking phenotypes

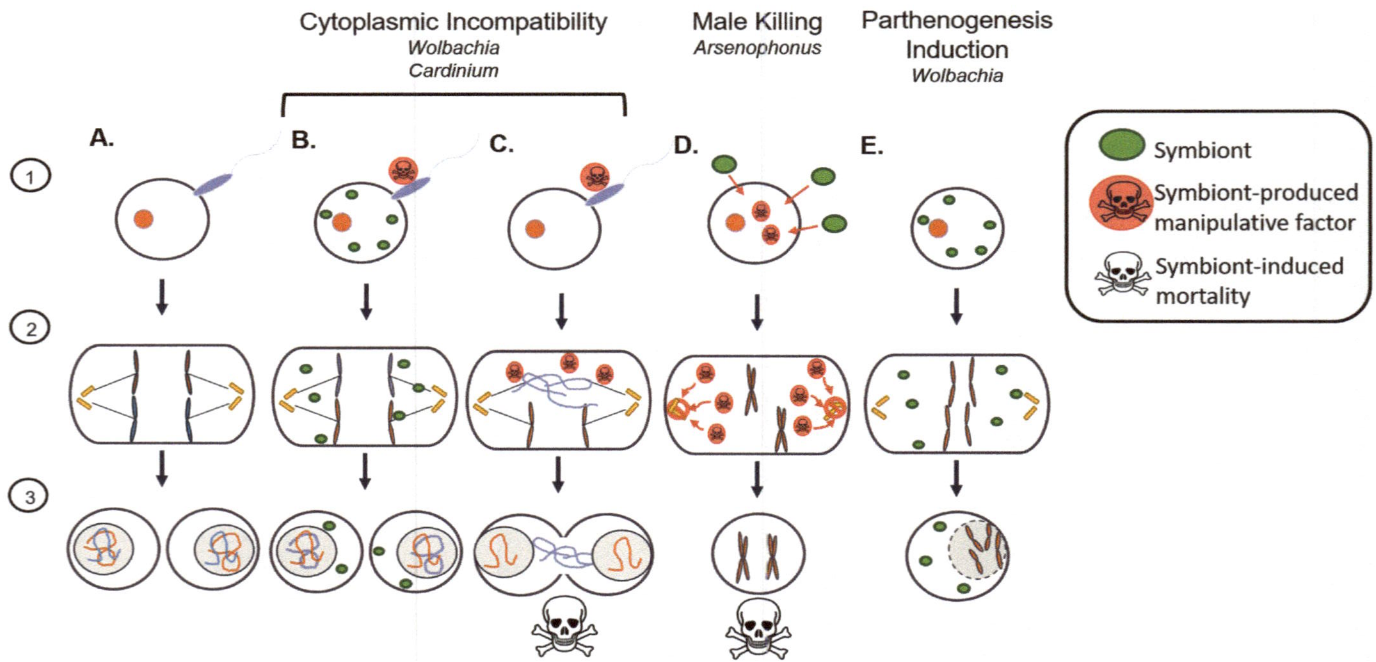

Fig. 1 See figure legend on opposite page.

(Caragata et al., 2013; Caragata et al., 2019; Rances et al., 2012; Thomas et al., 2018). This conferred resistance to pathogens, combined with the genetic drive capabilities of CI has spurred substantial additional interest in the study of CI-causing symbionts and has led to releases of *Wolbachia*-infected mosquitos in regions with dengue virus activity (Hoffmann et al., 2011).

Early crossing experiments of mosquitoes infected with different *Wolbachia* strains revealed the potential for complex crossing types between different strains, with some strains either rescuing or failing to rescue the CI of other strains, and even *Wolbachia* strains that could only rescue, but not induce CI (Atyame et al., 2011; Bourtzis et al., 1998; Charlat et al., 2004; Laven, 1967). The complexity of these crossing types conformed to a verbal model of "modification" of sperm in males and "rescue" by cytoplasm of infected females (Werren, 1997), and suggested that the CI modification/rescue process of CI must involve at least two genes: a modification and a rescue factor. The modification and rescue system is conceptually similar to bacterial toxin/antitoxin systems; in these systems, the short-lived antidote and long-lived toxin typically form an operon and are expressed together. The antidote prevents toxicity when present, often by directly

Fig. 1 Symbiont-induced reproductive manipulations that directly target the eukaryotic cell cycle. (A) Early stages of normal embryonic development including the first round of mitosis. (1) Fertilization of uninfected egg, (2) normal anaphase, with sister chromatids both condensed and fully separated, (3) completion of mitosis and the cell cycle culminating in two daughter cells. (B) Rescue of symbiont-induced cytoplasmic incompatibility beginning with (1) fertilization of CI-symbiont-infected egg cell with CI-modified sperm, (2) presence of CI-symbiont throughout mitosis promotes successful chromatid condensation and separation, (3) successful CI rescue and cell replication. (C) Cellular defects associated with CI mortality. (1) Fertilization of uninfected egg cell with CI-modified sperm, (2) Modification factor affects paternal chromatin condensation, (3) cell attempts to complete cell replication, resulting in paternal chromatin "bridging" between the two daughter cells prior to embryo mortality. (D) *Arsenophonus*-associated male-killing. (1) *Arsenophonus* male-killing factor diffuses into egg cell, (2) male-killing factor targets maternally-derived centrosomes that form in unfertilized male eggs, (3) centrosomes do not produce a mitotic spindle and the cell fails to divide, leading to male-specific embryonic mortality. (E) Gamete duplication variant of parthenogenesis induction. (1) Unfertilized egg infected with PI symbiont, (2) after DNA replication, sister chromatids do not fully separate during early embryonic mitosis, (3) daughter nuclei fuse back together, resulting in a diploid cell. Green circles represent the symbiont manipulator, circle skull represents a symbiont factor involved in manipulation. *Wolbachia* and *Spiroplasma* use a different mechanism(s) to kill males by targeting the dosage compensation system. For brevity, only *Arsenophonus* that specifically targets cell cycle-associated organelles is shown. *Cardinium* and *Rickettsia* PI target meiosis, where the reduction division is aborted, not mitosis as depicted here.

binding the toxin. However, the antidote must be continuously produced in order for continuous blocking of the toxin. Failure to produce the antidote, typically through the loss of the extrachromosomal element harbouring the toxin-antidote operon, results in exposure to the longer-lived toxin and results in the mortality of the host organism (Van Melderen and Saavedra De Bast, 2009).

After nearly a century since the first discovery of *Wolbachia* in mosquitoes, bacterial genes responsible for CI have been identified. A protein associated with the *Wolbachia*-infecting bacteriophage, WO, was identified in the spermathecae of uninfected female mosquitos after mating with infected males harbouring the *w*Pip strain (Beckmann and Fallon, 2013; Bordenstein and Bordenstein, 2016). This protein was named *cidA* and was found to be directly upstream of a second gene that harbours a deubiquitinase domain, *cidB*, suggesting that these two genes could form a CI factor operon, with *cidA* acting as the hypothetical rescue (or antidote) factor and *cidB* acting as the modification (toxin) factor (Beckmann et al., 2017; Beckmann et al., 2019). Based on other toxin-antidote systems found in free-living bacteria, it has been hypothesized that the genes for the CI factors also form an operon (Beckmann et al., 2017; Beckmann et al., 2019; Van Melderen and Saavedra De Bast, 2009). In further support of their status as a toxin-antidote system, *cidA* and *cidB* proteins show high binding affinity for one another (Beckmann et al., 2017). However, the operon status of these loci has been debated, given that the *cid* genes exhibit quite different expression levels (Lindsey et al., 2018). While operon-associated genes typically show similar expression levels, this is not always the case (Güell et al., 2011; Rocha, 2008). Identification of the *cid* promotor region(s) is required to confirm their status as an operon (Shropshire et al., 2019). Additionally, while the hypothetical rescue factor, *cidA*, is delivered along with the modified sperm of infected male mosquitoes, the proposed modification factor itself, *cidB*, has not been found in these sperm (Beckmann and Fallon, 2013; Shropshire et al., 2019). This suggests that the modification factor may sabotage sperm in a specific manner prior to the jettison of *Wolbachia* during sperm maturation (Clark et al., 2002), and the rescue factor restores sperm function but may not directly interact with the modification factor (Shropshire et al., 2019). Alternatively, given that *cidB* shows a six-fold lower expression level than *cidA*, the methods used for protein detection in sperm may not have captured a low-concentration protein like *cidB* (Beckmann and Fallon, 2013; Lindsey et al., 2018).

While their status as a true toxin-antidote system remains up for debate, the *w*Pip *cid* genes and their homologues in the *w*Mel *Wolbachia* strain infecting *D. melanogaster* were confirmed in separate studies to induce CI-like mortality when co-expressed in the testes of uninfected flies (Beckmann et al., 2017; LePage et al., 2017). The CI-like mortality caused by *cidA-cidB* expression was also rescued in crosses between transgenic males and *Wolbachia*-infected females (LePage et al., 2017). Lastly, the hypothesized role of *cidA* as the rescue factor was functionally confirmed via transgenic rescue crosses between *Wolbachia*-infected males and uninfected females expressing *cidA* (Shropshire et al., 2018). Throughout the rest of this section, we will refer to these the CI genes as "*cif*" (CI factor) genes when discussing general features shared across the different variants, and their specific names, *cid* (harbouring a deubiquitinase domain) or *cin* (harbouring PD-(D/E)xK nuclease domains) when discussing specific variants. Despite major advances in the genetics of CI, the mechanism(s) of modification and rescue is/are still unknown. For example, *cidA*-bound *cidB* retains the deubiquitinase activity required for CI-induced mortality, suggesting that, despite the high binding affinity for homologous *cif* protein pairs, this binding is not directly responsible for CI rescue (Beckmann et al., 2017). One model of rescue proposes that, rather than directly inhibiting the activity of the *cifB* toxin, the rescue factor, *cifA*, changes its cellular localization, or blocks its ability to interact with host proteins (Beckmann et al., 2017; Chen et al., 2019).

While studies using heterologous expression of *Wolbachia* genes confirmed the identity of CI factors, these studies also raised new mechanistic questions, including why the recapitulation of the modification step of CI in uninfected hosts requires co-expression of both *cifB* and *cifA*, when *cifB* is the hypothetical modification factor. The requirement for both *cifB* and *cifA* suggests an added level of complexity to the hypothesized CI mechanism, whereby both genes may be involved in modification, while only *cifA* is necessary for rescue (Shropshire et al., 2018). Alternatively, this apparent requirement for both factors may be an artefact of heterologous expression of the *cifB* toxin. These expression systems can result in much higher expression levels compared to native bacterial expression (Harumoto and Lemaitre, 2018). Heterologous *cifB* expression in the absence of *cifA* could provide abnormally high amounts of the modification factor in sperm that results in inviable sperm (Beckmann et al., 2019). Those sperm capable of fertilization may not have been exposed to the *cifB* factor, potentially masking this gene's role as the sole modification factor. Studies investigating sperm maturation and viability in *cifB*-expressing *D. melanogaster* could clarify the role of *cifA* and *cifB* in CI modification (Beckmann et al., 2019).

Genome comparisons of various CI-inducing *Wolbachia* have revealed tremendous diversity and copy number of *cif* loci (LePage et al., 2017; Lindsey et al., 2018). Generally, the identity and number of *cifA* and *cifB* copies present in a CI-*Wolbachia* genome relates to the crossing dominance of certain strains, with dominant CI-inducing strains harbouring more copies of the CI genes (Bonneau et al., 2018; LePage et al., 2017; Lindsey et al., 2018). This variability extends to the protein encoding domains within these *cif* genes, which can contain numerous domains of different functions. For example, while the original *cidB* gene identified by Beckmann et al. (2017, 2013) encodes a deubiquitinylating protease essential for CI crossing mortality, other *cifB* homologues encode modules that lack any identifiable protease and instead harbour other sets of protein domains including nucleases, restriction enzymes, and/or DNA repair proteins (Lindsey et al., 2018). Recently, functional tests of PD-(D/E)xK nuclease encoding *cif* variants from the *Culex pipiens*-infecting *w*Pip strain (which also harbours the *cid* variants; Beckmann et al., 2017) demonstrated that the nuclease variants are also capable of causing CI (Chen et al., 2019). Consistent with previous studies testing CI induced by the deubiquitinase-encoding *cid* pair (Beckmann et al., 2017; LePage et al., 2017), the so-called *cinA* and *cinB*, which contains two PD-(D/E)xK nucleases domains, induce CI-like mortality when co-expressed in the germline cells of uninfected male *D. melanogaster* (Chen et al., 2019). Additionally, expression of *cinA* alone in the germline of uninfected female flies mated with *Wolbachia*-infected males results in rescue of natural CI, similar to the *cid* system, in which *cidA* alone rescues CI (Chen et al., 2019; Shropshire et al., 2018). In further agreement with the *cid* system, *cinA* and *cinB* show high binding affinity for each other, but this binding does not inhibit the nuclease activity of *cinB* responsible for the mortality effect, suggesting that the antitoxin operates via some mechanism distinct from disruptive binding (Beckmann et al., 2017; Chen et al., 2019).

Despite the variability of homologues of *cifA* and *cifB*, the cytological effects of *Wolbachia*-CI appear consistent across strains, with CI-induced mortality resulting from the delayed condensation of paternal chromatin at the first mitotic division of the zygote, followed by fatal mitotic defects and chromatin bridging (Fig. 1; Tram et al., 2003; Beckmann et al., 2017;Chen et al., 2019; LePage et al., 2017). Specifically, these defects appear to arise from a delay in the recruitment of the histone complex to DNA of the male pronucleus prior to the onset of the first embryonic mitosis

(Landmann et al., 2009). Given that two variants of the *cif* genes cause similar cytological defects despite encoding different protein domains, it seems that there may be multiple ways in which CI-*Wolbachia* can modify sperm chromatin (Chen et al., 2019; LePage et al., 2017). Further research on the downstream mechanics of these deubiquitinase- and nuclease-based systems will reveal at what point these mechanisms converge to cause similar defects in male chromatin.

Wolbachia is not the only symbiont capable of causing CI. To date, three other intracellular bacteria have been implicated in causing CI in arthropod hosts, although two of these three have been discovered only recently (Hunter et al., 2003; König et al., 2019; Takano et al., 2017). The best-studied of these, *Cardinium hertigii* (Phylum Bacteroidetes), has been found to cause CI in a number of arthropod hosts (Gebiola et al., 2016; Hunter et al., 2003; Nakamura et al., 2012; Nguyen et al., 2017; Wu and Hoy, 2012). Genomic comparisons between CI-*Cardinium* and CI-*Wolbachia* revealed very little gene homology (Mann et al., 2017; Penz et al., 2012), and *Cardinium* lacks the *cif* genes responsible for *Wolbachia*-CI (Lindsey et al., 2018). Given the lack of homology between these two bacteria, it's likely that CI arose independently in these two lineages and operates via a different molecular mechanism (Penz et al., 2012). However, a cytological study of *Cardinium*-CI mortality revealed mitotic defects strikingly similar to those seen in *Wolbachia*-CI (Fig. 1; Gebiola et al., 2017). This is perhaps unsurprising, given the conservation of CI-induced cytological defects across *Wolbachia* strains harbouring different *cif* genes (Tram et al., 2003; Beckmann et al., 2017; LePage et al., 2017; Lindsey et al., 2018). That CI generated by two different lineages of bacteria via different upstream mechanisms results in similar cytological defects also suggests that there are multiple avenues for intracellular symbionts to sabotage the host cell cycle in a CI-like manner.

3. Male-killing

Some strains of manipulative symbionts induce sex-specific offspring mortality, only killing infected male hosts either early in embryonic development or later as juveniles. The elimination of male offspring may reduce resource competition for infected female hosts and, in cases involving cannibalistic hosts, even provide developing females with additional and easily accessible nutrition via their undeveloped brothers (Hurst and Majerus, 1993). Like CI, male-killing is a common manipulative phenotype, with at least four distinct symbiont lineages containing male-killers (Gherna et al., 1991; Hurst et al., 1992; Hurst et al., 1999a: Hurst et al., 1999b). Male-killers are especially prevalent and well-studied in several lepidopteran (Fukui et al., 2015; Hornett et al., 2006) and coleopteran hosts (Hurst et al., 1992, 1996, 1999a, 1999b; Lawson et al., 2001; Schulenburg et al., 2001; Werren et al., 1994) species, although male-killing strains of *Wolbachia* and another prevalent manipulative symbiont, *Spiroplasma*, occur naturally in several *Drosophila* fly species as well (Haselkorn, 2010). The natural occurrence of these two male-killing symbionts in the *Drosophila* model system, as well as the presence of a third male-killer, *Arsenophonus*, in *Nasonia vitripennis*, a parasitic wasp model system, has allowed for more intensive research into the genetic mechanisms controlling symbiont-induced male mortality.

The male-killing phenotype of *Drosophila*-infecting *Spiroplasma* is perhaps the best understood mechanistically of any reproductive manipulation. In the context of the XX/XY sex chromosome system of *Drosophila*, male-specific mortality arises from widespread apoptosis caused by improper X chromosome segregation (Bentley et al., 2007; Harumoto et al., 2016), genome-wide gene mis-expression (Cheng et al., 2016), and widespread neural defects in the central nervous system of male embryos (Martin et al., 2013; Tsuchiyama-Omura et al., 1988). The male-killing phenotype involves the manipulation of the dosage compensation machinery required for proper X chromosome gene expression in XY males, particularly a protein-RNA complex called male-specific lethal (MSL; Veneti et al., 2005; Harumoto et al., 2016). Normally, in the process of dosage compensation, the MSL complex is specifically recruited to the X chromosome in males, where it plays a role in overexpression of X chromosome genes (Penalva and Sánchez, 2003). *Spiroplasma*-infected males show improper localization of the MSL complex; rather than associating only with the single X chromosome of male hosts, the MSL complex also associates with autosomal chromosomes, resulting in altered levels of histone acetylation and overexpression of autosomal genes (Cheng et al., 2016). That *Spiroplasma* causes female mortality in transgenic *Drosophila* that express MSL genes and does not cause mortality in males with non-functional MSL, offers additional strong support for the MSL as the host target for the male-killing mechanism (Cheng et al., 2016; Harumoto et al., 2016; Veneti et al., 2005).

Most recently, a gene for *Spiroplasma* male-killing was identified and functionally confirmed to induce male-specific mortality via expression in uninfected *D. melanogaster* (Harumoto and Lemaitre, 2018). The expression of this plasmid encoded gene, *Spaid* (*S. poulsonii* androcidin), resulted in male-specific embryonic mortality, phenotypically consistent with male mortality due to *Spiroplasma* infection (Harumoto and Lemaitre, 2018). The *Spaid* gene contains an ankyrin and deubiquitinase domain, and expression of *Spaid*-constructs lacking either domain showed altered male-killing phenotypes with reduced efficacy (Harumoto and Lemaitre, 2018). Specifically, expression of the construct lacking the ankyrin domain did not result in male mortality and the construct did not localize with the MSL complex as observed for wild-type *Spaid*. Constructs without the deubiquitinase domain also showed reduced male mortality, and a lack of association between these constructs and host nuclei. Harumoto and Lemaitre (2018) conclude that the ankyrin domain of *Spaid* is required for its interaction with the host MSL complex, while the deubiquitinase domain is involved in the initial nuclear localization of *Spaid*.

The genome of the *D. melanogaster*-associated strain of male-killing *Spiroplasma* contains a single gene with a recognizable ankyrin domain: *Spaid* (Harumoto and Lemaitre, 2018). In contrast, two other widespread manipulators, *Wolbachia* and *Cardinium*, contain numerous genes with ankyrin domains, which are typically involved in interactions with eukaryotic proteins. Due to their role in eukaryotic protein interactions, ankyrin domain genes were the first genes suspected to be involved in reproductive manipulation (Fenn and Blaxter, 2006; Penz et al., 2012; Sinkins et al., 2005; Yamada et al., 2011). Interestingly, *Wolbachia* also contains a gene similar to *Spaid* containing an ankyrin and deubiquitinase domain. However, expression of this gene does not induce male-killing in uninfected flies (Yamada et al., 2011). More recently, a phage-associated gene has been linked to the male-killing phenotype of *Wolbachia* (Perlmutter et al., 2019). Expression of this gene, *wmk* (*WO-mediated killing*), induces a male-killing phenotype in uninfected *D. melanogaster* consistent with the phenotypes observed in natural male-killing *Wolbachia–Drosophila* symbioses (Perlmutter et al., 2019).

The candidate factors responsible for male-killing in *Wolbachia* and *Spiroplasma*, *wmk* and *Spaid*, respectively, while both found on extrachromosomal elements (a phage and a plasmid) are not homologous (Harumoto and Lemaitre, 2018; Perlmutter et al., 2019). While *Spaid* encodes an ankyrin and deubiquitase domain, *wmk* is predicted to encode two helix-turn-helix domains and an XRE DNA-binding domain (Harumoto and Lemaitre, 2018; Perlmutter et al., 2019). Given the lack of homology of these two factors, *Wolbachia* and *Spiroplasma* likely utilize different mechanisms to induce male-killing. However, these mechanisms seem to converge at their eukaryotic target: the dosage compensation system required for proper

X chromosome gene expression in male hosts (Fukui et al., 2015; Harumoto et al., 2018; Perlmutter et al., 2019). In a 2015 study of male-killing *Wolbachia* in the Lepidopteran host, *Ostrinia furnaculis*, Fukui et al. (2015) found that *Wolbachia* reduced expression of *masculinizer* (*masc*), a dosage compensation factor required for successful male development. Notably, Lepidoptera harbour a ZZ/ZW sex determination system, in which males are the homozygous sex (Kiuchi et al., 2014). In healthy ZZ males, the *masc* protein is involved in the repression of Z-encoded gene expression, acting as a reverse type of dosage compensation compared to XX/XY species (Kiuchi et al., 2014). Experimental reduction of *masc* expression in *Bombyx mori* results in male-specific mortality with a similar phenotype as *Wolbachia*-induced mortality of *O. furnaculis* (Fukui et al., 2015). In the *masc*-depleted, doomed males of either species, mortality is associated with overexpression of Z chromosome genes, consistent with the dosage compensation role of *masc* (Fukui et al., 2015). Furthermore, *Wolbachia*-induced male *O. furnaculis* mortality is rescued when male embryos are injected with additional *masc* transcripts, offering further support for *masc* being the target for *Wolbachia* male-killing in *O. furnaculis* (Fukui et al., 2015). Perlmutter et al. (2019) found evidence that *Wolbachia*'s male-killing factor, *wmk*, also targets the dosage compensation system in *Drosophila*. Given the relative conservation of *wmk* across male-killing *Wolbachia* strains, it's quite possible that the *wmk* gene is also responsible for *masc* suppression in *O. furnaculis* (Fukui et al., 2015; Perlmutter et al., 2019).

While *Spiroplasma* and *Wolbachia* use different genes to initiate the male-killing phenotype, their shared target in the dosage compensation system of males seems to result in similar cytological defects responsible for embryonic mortality. Both result in segregation defects during cell division and widespread embryonic apoptosis, although *Wolbachia*-killed males do not exhibit the neural defects found in *Spiroplasma*-killed males (Harumoto et al., 2018).

A third male-killing bacterium, *Arsenophonus*, utilizes a unique strategy to kill male hosts of the parasitic wasp, *Nasonia vitripennis* (Ferree et al., 2008). In this haplodiploid system, *Arsenophonus* selectively kills haploid male embryos by targeting and preventing the formation of the maternally-derived centrosomes that are required for male development in haplodiploid species (Fig. 1; Tram and Sullivan, 2000; Ferree et al., 2008). Loss of these centrosomes results in disorganization of spindle required for functional mitosis and leads to asynchronous mitosis and chromosome condensation, resulting in male-specific embryonic mortality (Ferree et al., 2008). This maternally-derived centrosome inhibition is also observed in *Arsenophonus*-infected female hosts, however female haplodiploid organisms preferentially use the paternal set of centrosomes for mitosis and so are unharmed (Ferree et al., 2008; Tram and Sullivan, 2000). In an elegant experiment, Ferree et al. (2008) tested the centrosome hypothesis by using a second selfish genetic element, the selfish host chromosome, Paternal Sex Ratio (PSR). PSR causes a male-skewed sex ratio by destroying the paternal genome in diploid eggs, resulting in haploid, incipient male, embryos. PSR also bypasses the requirement for maternal centrosomes in male development; instead, males inherit the paternal centrosomes which remain after the paternal genome is destroyed (Nur et al., 1988). Crosses between *Arsenophonus*-infected females and PSR-bearing males resulted in normal broods of males and females, providing evidence that this symbiont specifically targets maternal centrosomes to kill males (Ferree et al., 2008). The genes responsible for this phenotype are not yet characterized, although assembly of the *Arsenophonus nasoniae* genome has revealed an abundance of suspected toxins, a Type III secretion system, and a potentially functional polyketide synthase system which produces small peptides capable of transmembrane travel, any of which may play a role in its son-killing phenotype (Wilkes et al., 2010).

Finally, male-killing symbionts may target sex determination systems, and in this way be related to feminizers, or even parthenogenesis-inducers. For example, a *Wolbachia* strain, initially described as a feminizer, was found to cause female-skewed sex ratios via late-stage larval male mortality in the lepidopteran host, *Ostrinia scapulalis* (Kageyama et al., 2002; Kageyama and Traut, 2004). This mortality seems to arise from male expression of female isoforms of *doublesex* (*dsx*), a major gene controlling the sex determination system of many insects (Sugimoto and Ishikawa, 2012). Furthermore, this *Wolbachia* is now required for female *dsx* expression, and female hosts cured of *Wolbachia* die as larvae (Sugimoto and Ishikawa, 2012). Given that the sex determination and dosage compensation systems are distinct host targets, this *Wolbachia* strain in *O. scapulalis* seems to be capable of yet another means of male-killing.

While the mechanisms by which *Spiroplasma*, *Wolbachia*, and *Arsenophonus* induce male-specific mortality in their insect hosts have received the bulk of research attention, there are more male-killing bacterial lineages for which the male-killing mechanism is almost entirely unknown. Notable among these lesser studied male-killers are *Rickettsia* and *Flavobacterium*, both of which have been found in various beetle hosts, including many ladybird species (Hurst et al., 1992; Werren et al., 1994; Hurst et al., 1996; Hurst et al., 1999a, 1999b; Lawson et al., 2001; Schulenburg et al., 2001). These symbionts may target the dosage compensation system, like *Wolbachia* or *Spiroplasma*, or utilize a different strategy like the parasitoid-infecting *Arsenophonus*. Further research on these less understood symbioses and mechanisms will likely reveal a diverse array of strategies utilized by manipulative symbionts to kill male hosts.

4. Feminization

Symbiont-induced reproductive sabotage does not always result in male mortality. Some manipulative symbionts can modify the host phenotype to produce genetic males that are phenotypically female, and thus fully capable of spreading the symbiont (Werren et al., 2008). These feminizing microbes are not restricted to insects and chelicerates, and are also described in crustaceans (Bouchon et al., 1998; Dunn et al., 1993). Nor are these manipulative saboteurs strictly bacteria; several instances of feminizing microsporidia and other parasitic eukaryotes have been described in shrimp species like *Gammarus duebeni* (Dunn et al., 1993; Ironside et al., 2003; Terry et al., 1998; Pickup and Ironside, 2018). Crustaceans and insects have evolved distinct mechanisms for phenotypic sex determination, suggesting that, like those causing CI and male-killing, feminization symbionts can potentially interact with multiple distinct host sex determination pathways (Legrand et al., 1987; Martin et al., 1999; Verhulst et al., 2010; Verhulst and van de Zande, 2015).

In many crustaceans, male development is controlled by a specialized androgenic gland (Cordaux and Gilbert, 2017). This male-associated gland appears to be a common target for multiple crustacean-associated feminizing microbes, including the microsporidian, *Nosema,* which inhibits androgenic gland development (Rodgers-Gray et al., 2004). Another feminizing crustacean symbiont, *Wolbachia*, targets the androgenic gland in a similar manner, inhibiting its development and resulting in expression of the female phenotype in genetic male hosts of the pillbug, *Armadillidium vulgare* (reviewed in Cordaux and Gilbert, 2017). *Armadillidium vulgare* has a ZZ/ZW sex chromosome system, so that feminized males (ZZ) produce more feminized males via *Wolbachia* transmission (Cordaux and Gilbert, 2017). Over time, this has led to the loss of the W chromosome responsible for female development and complete dependence on *Wolbachia* for female production and population persistence (Cordaux and Gilbert, 2017).

Interestingly, studies investigating *A. vulgare* feminization found that certain ZZ individuals that were not infected with *Wolbachia* also exhibited a feminized phenotype ("f" females; Cordaux and Gilbert, 2017). These feminized but *Wolbachia*-free females were found to harbour a large nuclear insert (~3 Mb) thought to be responsible for the phenotypic switch to female development (Cordaux and Gilbert, 2017; Leclercq et al., 2016). In a striking example of horizontal gene transfer between a symbiont and its host, genomic analysis of the "f element" found homology with the feminizing *Wolbachia* strain found in *A. vulgare*, *w*VulC (Leclercq et al., 2016).

While it is still unclear whether a gene (or genes) within this insert is responsible for female development, or whether the insertion disrupted a gene required for male development, the gene contents of the f element provided an initial list of 3000 potential feminizing gene candidates (Leclercq et al., 2016). By comparing the genome of the feminizing *Wolbachia* strain *w*VulC, the source of the f element, with a closely related CI *Wolbachia* strain, Badawi et al. (2018) have narrowed this list of candidates to 35 genes. Expression profiles of these candidates throughout isopod development revealed two genes of particular interest: a gene lacking any known domain, *w*Vul_1821, and one containing an ankyrin and a protease domain, *w*Vul_1408 (Badawi et al., 2018). Curiously, *w*Vul_1408 is absent in the f element, suggesting this gene is not involved in feminization in f element-bearing individuals (Badawi et al., 2018).

Much of the work on feminizing symbionts has focused on crustacean-infecting microbes, culminating in a short list of feminizing candidate genes. However, the sex determination systems of crustaceans and insects are fundamentally different (Legrand et al., 1987; Verhulst and van de Zande, 2015). Crustacean sex determination is hormonally based and involves the androgenic gland; insect sex determination occurs on a cell by cell basis, involving a cascade dependent on sex-specific splicing variants of one or more transcripts (Verhulst and van de Zande, 2015). Notable among these genes are *doublesex*, the final gene in the cascade responsible for regulating sex-specific gene expression, and immediately above it, *transformer* (*tra*), a paralog of *dsx*, which controls *dsx* expression (Verhulst et al., 2010; Verhulst and van de Zande., 2015). Further upstream genes vary depending on insect species; however, the basal elements of the sex determination cascade, *dsx* and *tra*, are found throughout insects (Verhulst et al., 2010; Verhulst and van de Zande, 2015; Geuverink and Beukeboom, 2014). Given the massive differences between sex determination of these two major arthropods groups, it is likely that feminization arises from different mechanisms in crustacean- and insect-infecting feminizers.

In insects, *Wolbachia* is the most common feminizing symbiont, with feminizing strains identified in both lepidopteran hosts and a planthopper species (Hiroki et al., 2002; Narita et al., 2007; Narita et al., 2011; Negri et al., 2006). Notably, a feminizing *Wolbachia* found in the butterfly host, *Eurema mandarina*, is required for female development (Kageyama et al., 2017). Infected

E. mandarina have a ZZ/ZO sex chromosome system, rather than the typical ZZ/ZW system of lepidopterans; this host lacks the W chromosome responsible for female sex determination in many butterflies and moths. Like the *Wolbachia* in *A. vulgare*, the feminizing *Wolbachia* strain in *E. mandarina* seems to have replaced the nuclear sex determination system of its host (Kageyama et al., 2017). Antibiotic treatment of *E. mandarina* larvae induces expression of the male *dsx* isoform and results in intersex individuals. A similar result is seen when *Wolbachia*-infected larvae of *E. hecabe* (sister species to *E. mandarina*) are fed antibiotics (Narita et al., 2007). These results suggest that feminizing *Wolbachia* may target the sex determination cascade in insect hosts, given that the feminizing influence of *Wolbachia* is required throughout host development and similarly, insect sex determination occurs constantly on a cell-by-cell basis throughout insect development (Narita et al., 2007; Kageyama et al., 2017; Verhulst and van de Zande, 2015). In this set of experiments, antibiotic treatments also resulted in high rates of late-stage (pupal) host mortality, consistent with defects caused by disruption of the sex determination system (Kageyama et al., 2017; Narita et al., 2007).

A different strain of *Wolbachia*, now considered to be a late-stage male-killer, is also required for female development of the moth, *Ostrinia scapulalis* (Kageyama et al., 2002; Kageyama and Traut, 2004). While eventual loss of the W chromosome via the production of ZZ female offspring of feminized males is expected, loss of the female W chromosome leading to an obligatory requirement for a symbiont in a male-killing system is a surprising result. Intriguingly, this male-killer of *O. scapulalis* utilizes a seemly different technique from other *Wolbachia* to kill males by manipulating the sex determination cascade, resulting in a mismatch between host genotype and the splicing variant *dsx* it expresses (Sugimoto and Ishikawa, 2012). It is possible that the male-killing *Wolbachia* in *O. scapulalis* is a feminizing strain "gone bad." This could explain the manipulation of the sex determination system that results in male mortality, and the unlikely requirement of *Wolbachia* presence for female development. Genomic comparisons between the *O. scapulalis* strain and both feminizing and male-killing *Wolbachia* strains of other hosts may provide some insight into this strange male-killer, as well as the into the genetic basis of feminization in insect hosts.

Aside from *Wolbachia*, only one other symbiont has been documented to cause feminization of an arthropod host. One strain of *Cardinium* was shown to feminize a species of the haplodiploid *Brevipalpus* mite (Groot and Breeuwer, 2006; Weeks et al., 2001). *Cardinium* infection results in entirely haploid female mites, indicating that this *Cardinium* feminizes haploid males, and as the authors point out, is perhaps one of the few if not only examples of an entirely haploid metazoan (Groot and Breeuwer, 2006). Unfortunately, little else is known concerning the mechanism of this second bacterial symbiont's feminization. Homologues of *dsx*- and *tra*- have been identified in another mite species, *Metaseiulus occidentalis*, suggesting that the sex determination system of mites, and a potential host target for feminizing symbionts, may be similar for chelicerates and insects (Pomerantz et al., 2015).

5. Parthenogenesis induction

A fourth reproductive manipulation by heritable bacteria is the induction of parthenogenesis, in which the infected female host asexually produces genetically female offspring. Parthenogenesis-inducing (PI) symbionts have been described in numerous haplodiploid arthropod lineages, including parasitic wasps (Giorgini et al., 2010; Gottlieb et al., 2002; Pannebakker et al., 2004; Stouthamer et al., 1993; Zchori-Fein et al., 1995; Zchori-Fein et al., 2001), thrips (Arakaki et al., 2001), and some mite species (Weeks and Breeuwer, 2001). There are also instances of PI-associated bacteria occurring in diplodiploid hosts such as Collembola, booklice, armoured scales, and many beetle species (Pike and Kingcombe, 2009; reviewed in Ma and Schwander, 2017). Unfortunately, experimental confirmation of symbiont-induced parthenogenesis is difficult in these diplodiploid systems, as the symbiont is often required for host reproduction and is not amenable to typical manipulation strategies using antibiotics or temperature (Graber and Fallon, 2019; Ma and Schwander, 2017). The symbionts in these instances often remain phenotypically uncharacterized, and most of our present mechanistic understanding of PI results from studies of PI symbionts in haplodiploid hosts, typically parasitic wasp species.

To date, three symbiont lineages have been confirmed to cause PI in arthropod hosts (Ma and Schwander, 2017). The most widespread of these, *Wolbachia*, causes PI in numerous arthropod species, including minute parasitoid wasps in the genus *Trichogramma*, in which symbiont-induced PI was first confirmed (Stouthamer et al., 1990; Stouthamer and Werren, 1993). Parthenogenesis induction has also been shown to be caused by *Cardinium* in the parasitic wasp genus *Encarsia* (Zchori-Fein et al., 2004) and is associated with parthenogenesis in some armoured scale species (Gruwell et al., 2009; Provencher et al., 2005). Lastly, two lineages of

Rickettsia have been shown to cause PI in eulophid (Chalcidoidea) parasitoid species (Giorgini et al., 2010; Hagimori et al., 2006).

The mechanism of PI has been particularly challenging to study and is consequently less well understood compared to other reproductive manipulations. One challenge is that hosts infected with a PI symbiont have a tendency to become reliant on these symbionts for reproduction, such that sexual reproduction cannot be restored, which would allow for rigorous comparisons between infected and uninfected individuals. Attempts to cure the host of the symbionts often leads to reproductive failure either through production of infertile male offspring, sexual fertilization failure, and/or egg development failure (Giorgini et al., 2007; Graber and Fallon, 2019; Russell and Stouthamer, 2011; Timmermans and Ellers, 2009; Zchori-Fein et al., 1995). Secondly, PI symbionts are not known from a model insect such as *Drosophila,* or mosquitoes, or even the model hymenopteran parasitoid *Nasonia vitripennis*. Instead, many PI symbionts occur in especially minute species of endoparasitoid wasps (e.g., *Trichogramma* and *Encarsia),* further restricting the ability of researchers to manipulate these systems (Adachi-Hagimori et al., 2008; Giorgini et al., 2007; Gottlieb et al., 2002; Ma et al., 2015; Pannebakker et al., 2004; Stouthamer and Kazmer, 1994; Zchori-Fein et al., 1995). Finally, PI symbionts manipulate reproduction primarily in the hymenopteran superfamily Chalcidoidea, in which the normal mechanism of sex determination is not well understood (Van Wilgenburg et al., 2006). Thus, researchers have a doubly difficult challenge of understanding what pieces of the (unknown) sex determination system of chalcidoids are targeted by PI symbionts.

In spite of these difficulties, cytological analysis of egg and early embryonic development and microsatellite marker analysis (Ma and Schwander, 2017) have provided some insights into mechanism. These studies of symbiont-infected parthenogenetic eggs have revealed that PI symbionts cause parthenogenetic development via the restoration of haploid eggs to diploidy. While diploidy is unlikely to be sufficient for female development (in chalcidoid species), it is likely to be necessary in most instances (Giorgini et al., 2009; Ma and Schwander, 2017). There appear to be two main mechanistic strategies to restore diploidy: (1) disruption of early mitosis in newly oviposited embryos that causes mitotic products to coalesce again, known as gamete duplication (Gottlieb et al., 2002; Pannebakker et al., 2004; Stouthamer and Kazmer, 1994), and (2) disruption of the reductive division of meiosis, resulting in a diploid zygote prior to the beginning of mitosis (Adachi-Hagimori et al., 2008; Giorgini et al., 2007). These two types of diploidy restoration can have different effects on host heterozygosity, depending on when they occur within oogenesis and whether they allow for recombination (reviewed in Ma and Schwander, 2017).

The first cytological study of *Wolbachia*-PI, in *Trichogramma* wasps, revealed that diploidy restoration was the result of gamete duplication (Stouthamer and Kazmer, 1994). In this system, the egg cells undergo typical meiosis following oviposition, with three of the four chromatid products condensing into polar bodies, and the fourth becoming a haploid maternal pronucleus (chromosome $n=5$). Upon oviposition, this unfertilized, haploid egg, which would develop as a male in a sexual species, prepares for the first mitotic division. However, during mitotic anaphase, no spindle forms, the chromatid pairs segregate only slightly, and the nucleus does not divide, resulting in a diploid daughter cell ($2n=10$ chromosomes) (Fig. 1; Stouthamer and Kazmer, 1994). This type of diploidy restoration results in offspring that are homozygous at every locus. This finding appears to explain why PI *Wolbachia* has rarely been found in the superfamily of wasps Ichneumonoidea. In this superfamily, the dominant sex determination mechanism is complementary sex determination (CSD), whereby heterozygosity at any one of a number of sex determination alleles triggers female production, while hemizygosity results in males, as does high levels of homozygosity in diploid individuals (resulting in diploid, generally sterile males) (Zhishan et al., 2003). CSD does not appear to occur in all ichneumonoids, however, including the braconid *Asobara tabida* (Beukeboom et al., 2000), which also harbours a PI-inducing strain of *Wolbachia* (Ma et al., 2015). The absence of CSD may explain the presence of PI *Wolbachia* in this host (Ma et al., 2015).

Gamete duplication has been described in other instances of *Wolbachia*-PI via either cytology (Giorgini et al., 2007; Gottlieb et al., 2002; Pannebakker et al., 2004) or has been inferred by measuring heterozygosity via microsatellite markers (Plantard et al., 1998). However, there is variation in the timing of the aberrant mitosis step within gamete duplication, as seen in the PI-*Wolbachia*-infected parasitoid, *Muscidifurax uniraptor*. In this host, the initial mitosis cycle occurs normally, and instead the two resulting nuclei from the initial mitotic event fuse together during the early stages of the subsequent mitosis cycle, resulting in diploidy restoration (Gottlieb et al., 2002). This contrasts with the typical gamete duplication at the first mitotic division of *Trichogramma* wasps (Stouthamer and Kazmer, 1994) as well as in other parasitoid hosts, *Leptopilina clavipes* (Pannebakker et al., 2004) and *Encarsia formosa* (Giorgini et al., 2007). A similar process of nuclear fusion after the first mitotic division was observed in the parthenogenetic gall wasp, *Diplolepis rosae* (Stille and Dävring, 1980). However, while *D. rosae* does harbour *Wolbachia*, it has not been confirmed as the causative agent for PI in this host (Van Meer et al., 1999). Unfortunately, given that there are only a few detailed cytogenetic studies of *Wolbachia*-PI, is it not clear at this point whether these variations of gamete duplication are a result of genetic differences in PI symbiont strain, titre effects, or are due to expression of the PI phenotype in different host environments.

The two lineages of *Rickettsia* that cause PI occur in chalcidoid wasp species in the family Eulophidae, including *Neochrysocharis formosa* and *Pnigalio soemius* (Giorgini et al., 2010; Hagimori et al., 2006). Cytogenetic studies of *Rickettsia*-infected *N. formosa* found that this symbiont restores diploidy prior to mitosis, such that *N. formosa* eggs are functionally diploid because the reduction division of meiosis does not occur. Instead the eggs undergo just one round of divisional meiosis prior to the first mitotic division (Adachi-Hagimori et al., 2008). The final PI symbiont, *Cardinium*, appears widespread in another chalcidoid wasp lineage, the Aphelinidae, and was shown to cause PI in *Encarsia hispida* (Giorgini et al., 2007; Giorgini et al., 2009; Zchori-Fein et al., 2001; Zchori-Fein et al., 2004). Cytogenetic analysis of *Encarsia* eggs revealed a third, unique, mechanism of diploidy restoration that acts pre-mitotically (Giorgini et al., 2007). In this system, the first meiotic division is functionally aborted entirely, resulting in the egg cells undergoing only a single round of meiotic division resulting in diploid, rather than haploid eggs (Giorgini et al., 2007). Similar to the temporal variation observed in *Wolbachia*-induced gamete duplication, certain *Cardinium* strains seem to induce the aborting of meiotic division at different times. In *E. hispida* and *E. guadeloupe*, *Cardinium* causes the first meiotic division to fail, while in *E. tabacivora (= E. pergandiella)*, the first meiotic division occurs normally, however the two resulting nuclei fuse after this first round of meiosis (Giorgini et al., 2007). Both strategies, named central fusion, result in diploid egg cells. Given that all three lineages of PI symbionts restore diploidy through different mechanisms, it's likely that this manipulative strategy has convergently evolved multiple times in symbiotic bacteria.

These various cytological studies have at times revealed a fascinating secondary result: the production of diploid male wasps from PI symbiont-infected mothers exposed to antibiotics (Giorgini et al., 2009; Ma et al., 2015). These PI symbionts typically are found in haplodiploid systems, in which normal sexually produced males are haploid and diploid male production is aberrant (Giorgini et al., 2009). Examples of diploid male production have been observed in symbioses involving both *Wolbachia* and *Cardinium*, although it appears male diploidy does not occur in every instance of these PI symbioses (Giorgini et al., 2009; Ma et al., 2015). For example, while *Cardinium*-infected *E. hispida* females will produce diploid male progeny after exposure to antibiotics, other closely related host species like *E. guadeloupe* and *E. tabacivora* do not produce males at all and are rendered effectively sterile by antibiotics (Giorgini et al., 2007, 2009). It has been hypothesized that the diploid males of *E. hispida* result from relaxed selection on the host's production of normal haploid males via long-term infection with the PI-*Cardinium*, rather than *Cardinium* itself, however this has not yet been tested (Giorgini et al., 2009). However, diploid males are produced when their adult mothers are fed antibiotics. Hence, it is possible that bacterial products, whether live or dead, that remain in the egg are sufficient to induce diploidy, as seen in another parasitoid, *Telenomus nawai* (Arakaki et al., 2000). Alternatively, *Cardinium* could set in motion a developmental programme in the host pupal stage that eventually leads to diploidy of eggs, but performs a second, feminizing step within the oviposited egg. Antibiotic treatment then interrupts the second process, but not the first. Indeed, evidence from the mite-infecting feminizing strain of *Cardinium* shows that this symbiont feminizes haploid individuals independent of diploidy restoration (Groot and Breeuwer, 2006). Experiments designed to reduce *Cardinium* density prior to adult emergence, perhaps by exposing larval or pupal *E. hispida* to stressful temperatures (Doremus et al., 2019) or microinjected antibiotics, may help elucidate whether this symbiont feminizes host-diploid males or causes PI via a two-step diploidy-inducing and feminizing mechanism.

Wolbachia-infected *Asobara japonica* (Braconidae: Ichneumonoidea) females fed antibiotics also produce diploid males, although in this case, diploidy seems to result from a low-density *Wolbachia* infection (Ma et al., 2015). When *A. japonica* eggs retain a high-density *Wolbachia* infection they develop as diploid female parthenogens, while a low-level infection results in diploid males, and the absence of *Wolbachia* results in typical haploid male development (Ma et al., 2015). Another study found that antibiotic treatment of *Wolbachia* in *Trichogramma* leads to diploid male production (Tulgetske, 2010). These studies suggest that, like *Cardinium*, *Wolbachia*-induced PI may also proceed in a two-step fashion, in which *Wolbachia* first restores host diploidy, then performs a second, feminizing step to induce female development (Ma et al., 2015; Tulgetske, 2010). This could mean that PI and feminizing *Wolbachia* and *Cardinium* perform similar feminizing actions via the sex determination system, but PI symbionts also perform the additional diploidization step (Kageyama et al., 2017; Ma et al., 2015; Sugimoto and Ishikawa, 2012; Weeks et al., 2001). Unfortunately, we do not presently know the bacterial factors responsible for either mechanism, and this connection between feminization and PI remains speculative.

Outside of these cytogenetic studies, little is known concerning the mechanism of symbiont-induced parthenogenesis. Complete genomes are only available for a handful of PI-*Wolbachia* strains (Faddeeva-Vakhrusheva et al., 2017; Lindsey et al., 2016), and no genomes for PI-*Rickettsia* or PI-*Cardinium* have been published. A genomic comparison between a PI strain and other *Wolbachia* genomes did not reveal clear candidates for the bacterial factors responsible for PI (Lindsey et al., 2016). Notably, the PI strain used in this comparative analysis, a *Wolbachia* from *Trichogramma pretiosum*, (wTpre) seems to lack the WO bacteriophage that encodes both the CI and male-killing factors found in other *Wolbachia* strains (Gavotte et al., 2006; Lindsey et al., 2016). While other PI strains of *Wolbachia* do harbour the WO phage, it's absence in the genomes of several PI strains suggests that the phage does not play a direct role in parthenogenesis induction (Gavotte et al., 2006). The *w*Tpre genome also shows a widespread abundance of truncated versions of *Wolbachia* genes (Lindsey et al., 2016). A second *Wolbachia* strain, thought to be responsible for PI in the diploid collembolan host, *Folsomia candida*, harbours the largest recorded *Wolbachia* genome and also harbours many genes encoding various ankyrins (Pike and Kingcombe, 2009; Faddeeva-Vakhrusheva et al., 2017). Given the scarcity of PI genomes available, and the variation in overall genomic decay observed in PI-*Wolbachia* genomes, it is unclear whether the generalized abundance of truncated genes observed in *w*Tpre is unique to this *Wolbachia* strain or related strains from *Trichogramma*, or a more general hallmark of PI strains. The genome, like those of other *Wolbachia*, also contains many genes encoding products with ankyrin domains, which may play a role in host interaction (Lindsey et al., 2016). However, until more PI genomes are available, the bacterial factor(s) responsible for *Wolbachia*-PI is likely to continue to elude researchers.

6. Manipulative symbionts can induce multiple phenotypes

While the symbionts detailed in this review are usually denoted by the phenotypes they exhibit in their hosts (e.g., male-killers), symbiont lineages may not be restricted to performing a single type of manipulation. *Wolbachia*, the "master manipulator" has strains causing all four phenotypes (Werren et al., 2008). *Cardinium* can induce three distinct phenotypes (Hunter et al., 2003; Weeks et al., 2001; Zchori-Fein et al., 2001), and *Rickettsia* perform both male-killing and parthenogenesis induction (Hagimori et al., 2006; Hurst et al., 1992). However, any one strain of these bacteria is typically associated with a single phenotype. There have been notable exceptions to this rule, with certain strains of *Wolbachia* spontaneously causing novel phenotypes under different conditions. These instances of novel phenotypes suggest that at least some symbionts contain the genetic infrastructure to induce multiple manipulations.

The most frequent examples of phenotype switching come from male-killing *Wolbachia* that spontaneously show CI phenotypes, or vice versa (Hornett et al., 2008; Jaenike, 2007;Sasaki et al., 2002; Sasaki et al., 2005). In one case, this phenotype shift is the result of experimental transinfections of *Wolbachia* from one host species to another (Sasaki et al., 2002; Sasaki et al., 2005). Upon successful transinfection to the novel host, the *Wolbachia* strain, which originally caused CI, induced male-killing in its novel host (Sasaki et al., 2002; Sasaki et al., 2005). A study by Jaenike (2007) found similar results in *Drosophila* through introgression of CI-*Wolbachia*-infected *Drosophila recens* into the genetic background of a different host species, *D. subquinaria*. While the *Wolbachia* strain caused CI in *D. recens*, the phenotype switched to male-killing in *D. subquinaria* (Jaenike, 2007). Another study using *Drosophila* found that the male-killing phenotype of *Wolbachia* was not only reduced after exposure to high temperatures, but surviving males were infected with *Wolbachia* and exhibited CI (Hurst et al., 2000). There are even examples of spontaneous phenotype switching in natural host populations infected with male-killing *Wolbachia* (Hornett et al., 2008). Host resistance to the male-killing phenotype has been shown to arise in populations, likely due to the intense conflict that arises between the nuclear genome and the male-killing symbiont (Hornett et al., 2006; Hurst et al., 1992; Majerus and Majerus, 2010). In theory, the resistance mutation should spread quickly through the host population, rendering the male-killing strategy ineffective (Hurst et al., 1992). This was documented in *Hypolimnas* butterflies in the South Pacific, but interestingly, when the *Hypolimnas* host evolved resistance to male-killing, the *Wolbachia* that was formerly male-killing was then shown to cause CI (Hornett et al., 2008). Interestingly, these transitions seem to show comparable trends. When a CI strain is moved to a novel host, it may now induce a male-killing phenotype; when male-killing fails, the surviving infected males undergo CI modification.

These studies suggest that *Wolbachia* can contain the genetic tools to induce multiple phenotypes. Indeed, another study investigating *Drosophila pandora* found that females infected with male-killing *Wolbachia* could mate compatibly with males infected with a CI-*Wolbachia*, indicating that expression of one phenotype does not prevent a symbiont from expressing other aspects of different manipulative phenotypes (Richardson et al., 2016). The discovery of the *Wolbachia* genes responsible for both male-killing and CI, and the presence of both gene sets in *Wolbachia* strains, provides direct evidence that a single *Wolbachia* strain may harbour genes responsible for at least these two manipulations (Beckmann et al., 2017; LePage et al., 2017;Lindsey et al., 2018; Perlmutter et al., 2019). The genes responsible for CI, *cifA* and *cifB*, and male-killing, *wmk*, are both associated with the WO phage that is widespread throughout the *Wolbachia* (Lindsey et al., 2018; Perlmutter et al., 2019). For example, the CI-inducing strain from *D. melanogaster*, *w*Mel, also has the *wmk* gene and shares high overall genome similarity with the male-killing strain, *w*Rec from *D. recens*, raising the possibility that CI inducing strains like *w*Mel could potentially cause male-killing in novel host environments (Jaenike, 2007; Perlmutter et al., 2019).

Why certain *Wolbachia* impose a male-killing phenotype upon their host rather than a CI phenotype is not clear, although phenotype expression is not under obvious bacterial transcriptional control, as *Wolbachia* appears to express both gene sets at the same relative levels ($cifA > cifB = wmk$), regardless of the resulting phenotype (Perlmutter et al., 2019). Instead the resulting phenotype may be influenced by bacterial post-transcriptional or post-translational control, or be attributed to host genotype, as hosts tend to undergo rapid evolution of male-killing suppression, which has been repeatedly observed in male-killing systems (Hayashi et al., 2018; Hornett et al., 2006; Majerus and Majerus, 2010). For example, *Wolbachia* could express genes for both phenotypes, but in a male host that is resistant to male-killing, male-killing would fail, but if the *cif* genes are expressed, the surviving male would have CI-modified sperm. Such a scenario has occurred in populations of *Hypolimnas* butterflies (Hornett et al., 2008). This two-pronged modification approach, in which male-killing has priority over CI, would better ensure symbiont maintenance, despite host efforts to suppress these often-harmful manipulations. However, it seems strange that these *Wolbachia* strains retain, and even express, a functional *wmk* gene in resistant hosts, when other unnecessary genes are often lost in the genomes of heritable symbionts (Moran et al., 2008; Perlmutter et al., 2019). Phenotype switching has only been observed thus far in *Wolbachia*, but this may have more to do with the greater number of studied *Wolbachia*-host systems than an inability of other symbionts to induce multiple phenotypes.

7. The eukaryotic cell cycle is a common target of bacterial manipulators

Generally, symbiont-induced manipulations converge upon a common target: the eukaryotic cell cycle (Fig. 1). The only exception to this appears to be feminization, which may target the insect sex determination system in insect hosts and prevent differentiation of the androgenic gland responsible for male development in crustaceans (Narita et al., 2007; Sugimoto and Ishikawa, 2012; Kageyama et al., 2017; Cordaux and Gilbert, 2017). The other mechanisms seem to cause defects in chromatin that results in either malfunctional or modified cell cycle processes. In male-killing, the symbiont targets the dosage compensation system, a protein complex responsible for modifying male gene expression on sex chromosomes (Fukui et al., 2015; Harumoto et al., 2016; Perlmutter et al., 2019; Veneti et al., 2005). This leads to genome-wide altered gene expression, which results in widespread cellular defects that become apparent during embryonic mitosis, in which chromatin bridging in males results in embryonic death (Bentley et al., 2007; Harumoto et al., 2016; Harumoto et al., 2018). The CI strategy also targets host chromatin, however in CI the symbiont-attributed DNA modifications are not conveyed via a sex-specific protein complex (Gebiola et al., 2017; Tram et al., 2003). The actual targets of symbiont-induced CI are not yet identified, although histones and other proteins involved in chromatin wrapping prior to the first mitotic division may be involved (Landmann et al., 2009). These CI modifications also result in chromatin bridging during early embryonic mitosis (Gebiola et al., 2017; Tram et al., 2003). Finally, PI symbionts seem to modify either meiosis or mitosis, although these modifications do not kill the host; instead, PI symbionts either suppress entire cycles of meiosis or cause secondary merging of nuclear products from the cell cycle (Giorgini et al., 2007; Gottlieb et al., 2002; Pannebakker et al., 2004; Stouthamer and Kazmer, 1994). The PI phenotype also does not cause chromatin bridging like CI or male-killing; however, all three mechanisms require that the altered eukaryotic cell proceeds through the cycle cell. The cell cycle, one of the most conserved elements across eukaryotic species, has internal checkpoints that respond to various forms of DNA and spindle damage (Elledge, 1996). That three of the four symbiont-induced reproductive manipulations are able to both modify the host DNA and prevent detection of these modifications by these checkpoint mechanisms suggests that there may be conserved elements in these manipulations that bypass cell cycle checkpoints.

8. The importance of a manipulator's extrachromosomal parasites

Like their free-living cousins, many heritable symbionts harbour extrachromosomal elements like bacteriophages and plasmids (Bordenstein and Reznikoff, 2005). In several cases, these elements contain genes required for a symbiont-induced phenotype (Ballinger et al., 2019; Degnan and Moran, 2008; Oliver et al., 2009; Beckmann et al., 2017; LePage et al., 2017; Harumoto and Lemaitre, 2018; Perlmutter et al., 2019). These are not limited to reproductive manipulators, indeed there are notable examples of extrachromosomal elements playing direct roles in defensive symbioses (Ballinger and Perlman, 2019; Oliver et al., 2009). For example, certain *Spiroplasma* strains infecting *Drosophila* confer protection against fly natural enemies like nematodes and parasitoids (Ballinger and Perlman, 2017; Jaenike et al., 2010). The ribosome-inactivating toxins (RIPs) responsible for the invading parasite's mortality are often abundant within a *Spiroplasma* genome, with some RIPs located on *Spiroplasma*-associated plasmids (Ballinger and Perlman, 2017, 2019; Hamilton et al., 2016). In another case, an aphid defensive symbiont, *Hamiltonella defensa*, confers protection against aphid parasitoids (Oliver et al., 2005). However, this protective phenotype requires that *H. defensa* harbour the APSE (*Acyrthosiphon pisum* secondary endosymbiont) phage, which encodes toxins that target the developing wasp larva (Brandt et al., 2017; Oliver et al., 2009). In this system, phage loss results in a substantial loss of protection (Oliver et al., 2009), while restoration of phage infection restores the protective phenotype (Brandt et al., 2017; Lynn-Bell et al., 2019).

Extrachromosomal elements also play a major role in several reproductive manipulations. The same *Spiroplasma*-associated plasmid that harbours copies of the ribosome-inactivating toxin also contains *Spaid*, the factor responsible for *Spiroplasma*-induced male-killing in *Drosophila* (Harumoto and Lemaitre, 2018). *Wolbachia* is frequently infected with the WO phage that contains a region of genes thought to be involved in eukaryote interactions (Bordenstein and Bordenstein, 2016). Among these genes are *wmk*, the *Wolbachia* male-killing candidate factor, and the *cif* CI factors (Beckmann et al., 2017; LePage et al., 2017; Perlmutter et al., 2019).

Given that the extrachromosomal elements of two unrelated symbionts are responsible for at least some of their reproductive manipulations, research into the uncharacterized extrachromosomal elements of other manipulators is enticing. For example, a CI-inducing strain of *Cardinium*, harbours a plasmid that contains numerous genes potentially involved in insect host interaction (Penz et al., 2012). Interestingly, another strain of *Cardinium*, *c*Sfur, infecting the planthopper *Sogatella furcifera*, does not seem to contain an intact plasmid, but does have certain plasmid genes incorporated into its genome (Zeng et al., 2018). While *Cardinium* is known to cause CI in *S. furcifera* (Nakamura et al., 2012), it is not yet clear whether the strain in this study is a CI-inducing strain; furthermore, a whitefly-infecting *Cardinium* strain also harbours the plasmid but does not induce CI, leaving the nature of this plasmid's role in its host bacterium's biology a mystery (Santos-Garcia et al., 2014).

9. Conclusion

The research community investigating bacterial reproductive manipulators has made major strides in the past few decades, culminating with work identifying the bacterial genes responsible for several symbiont-induced manipulations in arthropod hosts (Beckmann et al., 2017; Harumoto and Lemaitre, 2018; LePage et al., 2017; Perlmutter et al., 2019; Shropshire et al., 2018). These and other studies have revealed a fascinatingly diverse array of bacterial lineages capable of manipulating host reproduction, many of which use entirely unique pathways to cause their respective manipulations. However, these studies have also revealed mechanistic themes consistent across the different phenotypes and symbiont lineages. The host cell cycle especially seems to be a consistent target for heritable symbionts prone to manipulation, which is perhaps not surprising given that this fundamental process is regulated by conserved genes across the hosts of these symbionts. These symbionts may target the cell cycle directly, as in the case of male-killing *Arsenophonus* that targets mitotic organelles (Ferree et al., 2008), or indirectly, like another male-killing symbiont, *Spiroplasma*, which sabotages an unrelated protein complex in a manner that results in an aberrant cell cycle (Harumoto and Lemaitre, 2018). These symbionts also themselves often harbour intracellular passengers like phages and plasmids that are essential for host manipulation. As research on these manipulative cell tenants progresses into the next decade, work on identifying host targets for manipulation will reveal more common themes and distinctions in these mechanisms. As these symbionts, most notably *Wolbachia*, see increased application towards control of arthropod pests, knowledge of these reproductive manipulations is of paramount importance.

References

Adachi-Hagimori, T., Miura, K., Stouthamer, R., 2008. A new cytogenetic mechanism for bacterial endosymbiont-induced parthenogenesis in Hymenoptera. Proc. R Soc. B Biol. Sci. 275 (1652), 2667–2673. https://doi.org/10.1098/rspb.2008.0792.

Arakaki, N., Noda, H., Yamagishi, K., 2000. *Wolbachia*-induced parthenogenesis in the egg parasitoid *Telenomus nawai*. Entomol. Exp. Appl. 96 (2), 177–184.

Arakaki, N., Miyoshi, T., Noda, H., 2001. *Wolbachia*-mediated parthenogenesis in the predatory thrips *Franklinothrips vespiformis* (Thysanoptera: Insecta). Proc. R Soc. B Biol. Sci. 268 (1471), 1011–1016. https://doi.org/10.1098/rspb.2001.1628.

Atyame, C.M., Duron, O., Tortosa, P., Pasteur, N., Fort, P., Weill, M., 2011. Multiple *Wolbachia* determinants control the evolution of cytoplasmic incompatibilities in *Culex pipiens* mosquito populations. Mol. Ecol. 20 (2), 286–298.

Badawi, M., Moumen, B., Giraud, I., Grève, P., Cordaux, R., 2018. Investigating the molecular genetic basis of cytoplasmic sex determination caused by *Wolbachia* endosymbionts in terrestrial isopods. Genes 9 (6), 290.

Ballinger, M.J., Perlman, S.J., 2017. Generality of toxins in defensive symbiosis: ribosome-inactivating proteins and defense against parasitic wasps in *Drosophila*. PLoS Pathog. 13 (7), e1006431.

Ballinger, M.J., Perlman, S.J., 2019. The defensive *Spiroplasma*. Curr. Opin. Insect Sci. 32, 36–41.

Ballinger, M.J., Gawryluk, R.M.R., Perlman, S.J., 2019. Toxin and genome evolution in a Drosophila defensive symbiosis. Genome Biol. Evol. 11 (1), 253–262.

Beckmann, J.F., Fallon, A.M., 2013. Detection of the *Wolbachia* protein WPIP0282 in mosquito spermathecae: implications for cytoplasmic incompatibility. Insect Biochem. Mol. Biol. 43 (9), 867–878.

Beckmann, J.F., Markowski, T.W., Witthuhn, B.A., Fallon, A.M., 2013. Detection of the Wolbachia-encoded DNA binding protein, HU beta, in mosquito gonads. Insect Biochem. Mol. Biol. 43 (3), 272–279.

Beckmann, J.F., Ronau, J.A., Hochstrasser, M.A., 2017. *Wolbachia* deubiquitylating enzyme induces cytoplasmic incompatibility. Nat. Microbiol. 2, 17007. https://doi.org/10.1038/nmicrobiol.2017.7.

Beckmann, J.F., Bonneau, M., Chen, H., Hochstrasser, M., Poinsot, D., Merçot, H., et al., 2019. The toxin–antidote model of cytoplasmic incompatibility: genetics and evolutionary implications. Trends Genet. 35 (3), 175–185.

Bentley, J.K., Veneti, Z., Heraty, J., Hurst, G.D.D., 2007. The pathology of embryo death caused by the male-killing *Spiroplasma* bacterium in *Drosophila nebulosa*. BMC Biol. 5 (1), 9.

Beukeboom, L.W., Ellers, J., Van Alphen, J.J.M., 2000. Absence of single-locus complementary sex determination in the braconid wasps *Asobara tabida* and *Alysia manducator*. Heredity 84 (1), 29.

Bonneau, M., Atyame, C., Beji, M., Justy, F., Cohen-Gonsaud, M., Sicard, M., et al., 2018. *Culex pipiens* crossing type diversity is governed by an amplified and polymorphic operon of *Wolbachia*. Nat. Commun. 9 (1), 319. https://doi.org/10.1038/s41467-017-02749-w.

Bordenstein, S.R., Bordenstein, S.R., 2016. Eukaryotic association module in phage WO genomes from *Wolbachia*. Nat. Commun. 7, 13155.

Bordenstein, S.R., Reznikoff, W.S., 2005. Mobile DNA in obligate intracellular bacteria. Nat. Rev. Microbiol. 3 (9), 688–699.

Bouchon, D., Thierry, R., Pierre, J., 1998. Evidence for widespread *Wolbachia* infection in isopod crustaceans: molecular identification and host feminization. Proc. R Soc. B Biol. Sci. 265 (1401), 1081–1090. https://doi.org/10.1098/rspb.1998.0402.

Bourtzis, K., Dobson, S.L., Braig, H.R., O'Neill, S.L., 1998. Rescuing *Wolbachia* have been overlooked. Nature 391 (6670), 852–853.

Brandt, J.W., Chevignon, G., Oliver, K.M., Strand, M.R., 2017. Culture of an aphid heritable symbiont demonstrates its direct role in defence against parasitoids. Proc. R Soc. B Biol. Sci. 284 (1866), e20171925.

Caragata, E.P., Rancès, E., Hedges, L.M., Gofton, A.W., Johnson, K.N., O'Neill, S.L., et al., 2013. Dietary cholesterol modulates pathogen blocking by *Wolbachia*. PLoS Pathog. 9 (6), e1003459.

Caragata, E.P., Rocha, M.N., Pereira, T.N., Mansur, S.B., Dutra, H.L.C., Moreira, L.A., 2019. Pathogen blocking in *Wolbachia*-infected *Aedes aegypti* is not affected by Zika and dengue virus co-infection. PLoS Negl. Trop. Dis. 13 (5), e0007443.

Charlat, S., Riegler, M., Baures, I., Poinsot, D., Stauffer, C., Mercot, H., 2004. Incipient evolution of *Wolbachia* compatibility types. Evolution 58 (9), 1901–1908.

Chen, H., Ronau, J.A., Beckmann, J.F., Hochstrasser, M., 2019. A *Wolbachia* nuclease and its binding partner provide a distinct mechanism for cytoplasmic incompatibility. Proc. Natl. Acad. Sci. 116 (44), 22314–22321.

Cheng, B., Kuppanda, N., Aldrich, J.C., Akbari, O.S., Ferree, P.M., 2016. Male-killing *Spiroplasma* alters behavior of the dosage compensation complex during *Drosophila melanogaster* embryogenesis. Curr. Biol. 26 (10), 1339–1345.

Clark, M.E., Veneti, Z., Bourtzis, K., Karr, T.L., 2002. The distribution and proliferation of the intracellular bacteria *Wolbachia* during spermatogenesis in *Drosophila*. Mech. Dev. 111 (1–2), 3–15.

Cordaux, R., Gilbert, C., 2017. Evolutionary significance of *Wolbachia*-to-animal horizontal gene transfer: female sex determination and the f element in the isopod *Armadillidium vulgare*. Genes 8 (7), 186.

Degnan, P.H., Moran, N.A., 2008. Diverse phage-encoded toxins in a protective insect endosymbiont. Appl. Environ. Microbiol. 74 (21), 6782–6791.

Doremus, M.R., Kelly, S.E., Hunter, M.S., 2019. Exposure to opposing temperature extremes causes comparable effects on *Cardinium* density but contrasting effects on *Cardinium*-induced cytoplasmic incompatibility. PLoS Pathog. 15 (8), e1008022. https://doi.org/10.1371/journal.ppat.1008022.

Dunn, A.M., Adams, J., Smith, J.E., 1993. Transovarial transmission and sex ratio distortion by a microsporidian parasite in a shrimp. J. Invertebr. Pathol. 61 (3), 248–252.

Duron, O., Bouchon, D., Boutin, S., Bellamy, L., Zhou, L., Engelstädter, J., et al., 2008. The diversity of reproductive parasites among arthropods: *Wolbachia* do not walk alone. BMC Biol. 6 (1), 1.

Elledge, S.J., 1996. Cell cycle checkpoints: preventing an identity crisis. Science 274 (5293), 1664–1672.

Faddeeva-Vakhrusheva, A., Kraaijeveld, K., Derks, M.F.L., Anvar, S.Y., Agamennone, V., Suring, W., et al., 2017. Coping with living in the soil: the genome of the parthenogenetic springtail *Folsomia candida*. BMC Genomics 18 (1), 493. https://doi.org/10.1186/s12864-017-3852-x.

Fenn, K., Blaxter, M., 2006. *Wolbachia* genomes: revealing the biology of parasitism and mutualism. Trends Parasitol. 22 (2), 60–65.

Ferree, P.M., Avery, A., Azpurua, J., Wilkes, T., Werren, J.H., 2008. A bacterium targets maternally inherited centrosomes to kill males in *Nasonia*. Curr. Biol. 18 (18), 1409–1414.

Frentiu, F.D., Zakir, T., Walker, T., Popovici, J., Pyke, A.T., van den Hurk, A., et al., 2014. Limited dengue virus replication in field-collected *Aedes aegypti* mosquitoes infected with *Wolbachia*. PLoS Negl. Trop. Dis. 8 (2), e2688.

Fukui, T., Kawamoto, M., Shoji, K., Kiuchi, T., Sugano, S., Shimada, T., et al., 2015. The Endosymbiotic bacterium *Wolbachia* selectively kills male hosts by targeting the masculinizing gene. PLoS Pathog. 11 (7), e1005048. https://doi.org/10.1371/journal.ppat.1005048.

Gavotte, L., Henri, H., Stouthamer, R., Charif, D., Charlat, S., Boulétreau, M., et al., 2006. A survey of the bacteriophage WO in the endosymbiotic bacteria *Wolbachia*. Mol. Biol. Evol. 24 (2), 427–435.

Gebiola, M., White, J.A., Cass, B.N., Kozuch, A., Harris, L.R., Kelly, S.E., et al., 2016. Cryptic diversity, reproductive isolation and cytoplasmic incompatibility in a classic biological control success story. Biol. J. Linn. Soc. 117 (2), 217–230. https://doi.org/10.1111/bij.12648.

Gebiola, M., Giorgini, M., Kelly, S.E., Doremus, M.R., Ferree, P.M., Hunter, M.S., 2017. Cytological analysis of cytoplasmic incompatibility induced by *Cardinium* suggests convergent evolution with its distant cousin *Wolbachia*. Proc. R Soc. B Biol. Sci. 284 (1862).

Geuverink, E., Beukeboom, L.W., 2014. Phylogenetic distribution and evolutionary dynamics of the sex determination genes *doublesex* and *transformer* in insects. Sex. Dev. 8 (1–3), 38–49.

Gherna, R.L., Werren, J.H., Weisburg, W., Cote, R., Woese, C.R., Mandelco, L., et al., 1991. *Arsenophonus nasoniae* gen. Nov., sp. nov., the causative agent of the son-killer trait in the parasitic wasp *Nasonia vitripennis*. Int. J. Syst. Evol. Microbiol. 41 (4), 563–565.

Giorgini, M., Hunter, M.S., Mancini, D., Pedata, P.A., 2007. Cytological evidence for two different mechanisms of thelytokous parthenogenesis in *Encarsia* parasitoids harbouring *Wolbachia* or *Cardinium* bacteria. In: Baster X Eur Workshop Insect Parasit, pp. 17–21.

Giorgini, M., Monti, M.M., Caprio, E., Stouthamer, R., Hunter, M.S., 2009. Feminization and the collapse of haplodiploidy in an asexual parasitoid wasp harboring the bacterial symbiont *Cardinium*. Heredity 102 (4), 365–371.

Giorgini, M., Bernardo, U., Monti, M.M., Nappo, A.G., Gebiola, M., 2010. *Rickettsia* symbionts cause parthenogenetic reproduction in the parasitoid wasp *Pnigalio soemius* (Hymenoptera: Eulophidae). Appl. Environ. Microbiol. 76 (8), 2589–2599.

Gottlieb, Y., Zchori-Fein, E., Werren, J.H., Karr, T.L., 2002. Diploidy restoration in *Wolbachia*-infected *Muscidifurax uniraptor* (Hymenoptera: pteromalidae). J. Invertebr. Pathol. 81 (3), 166–174.

Graber, L.C., Fallon, A.M., 2019. Tetracycline reduces feeding and reproduction of the parthenogenetic springtail, *Folsomia candida*. Symbiosis 77 (3), 257–264. https://doi.org/10.1007/s13199-018-00593-0.

Groot, T.V.M., Breeuwer, J.A.J., 2006. *Cardinium* symbionts induce haploid thelytoky in most clones of three closely related *Brevipalpus* species. Exp. Appl. Acarol. 39 (3–4), 257–271.

Gruwell, M.E., Wu, J., Normark, B.B., 2009. Diversity and phylogeny of *Cardinium* (Bacteroidetes) in armored scale insects (Hemiptera: Diaspididae). Ann. Entomol. Soc. Am. 102 (6), 1050–1061.

Güell, M., Yus, E., Lluch-Senar, M., Serrano, L., 2011. Bacterial transcriptomics: what is beyond the RNA horizome? Nat. Rev. Microbiol. 9 (9), 658–669.
Hagimori, T., Abe, Y., Date, S., Miura, K., 2006. The first finding of a *Rickettsia* bacterium associated with parthenogenesis induction among insects. Curr. Microbiol. 52 (2), 97–101. https://doi.org/10.1007/s00284-005-0092-0.
Hamilton, P.T., Peng, F., Boulanger, M.J., Perlman, S.J., 2016. A ribosome-inactivating protein in a *Drosophila* defensive symbiont. Proc. Natl. Acad. Sci. 113 (2), 350–355.
Harumoto, T., Lemaitre, B., 2018. Male-killing toxin in a bacterial symbiont of *Drosophila*. Nature 557 (7704), 252–255. https://doi.org/10.1038/s41586-018-0086-2.
Harumoto, T., Anbutsu, H., Lemaitre, B., Fukatsu, T., 2016. Male-killing symbiont damages host's dosage-compensated sex chromosome to induce embryonic apoptosis. Nat. Commun. 7, 12781.
Harumoto, T., Fukatsu, T., Lemaitre, B., 2018. Two common and unique strategies of male killing evolved in two distinct *Drosophila* symbionts. Proc. R Soc. B Biol. Sci. 285 (1875), 20172167. https://doi.org/10.1098/rspb.2017.2167.
Haselkorn, T.S., 2010. The *Spiroplasma* heritable bacterial endosymbiont of *Drosophila*. Fly 4 (1), 80–87.
Hayashi, M., Nomura, M., Kageyama, D., 2018. Rapid comeback of males: evolution of male-killer suppression in a green lacewing population. Proc. R Soc. B Biol. Sci. 285 (1877), 20180369.
Hiroki, M., Kato, Y., Kamito, T., Miura, K., 2002. Feminization of genetic males by a symbiotic bacterium in a butterfly, *Eurema hecabe* (Lepidoptera: Pieridae). Naturwissenschaften 89 (4), 167–170. https://doi.org/10.1007/s00114-002-0303-5.
Hoffmann, A.A., Montgomery, B.L., Popovici, J., Iturbe-Ormaetxe, I., Johnson, P.H., Muzzi, F., et al., 2011. Successful establishment of *Wolbachia* in *Aedes* populations to suppress dengue transmission. Nature 476, 454. https://doi.org/10.1038/nature10356.
Hornett, E.A., Charlat, S., Duplouy, A.M.R., Davies, N., Roderick, G.K., Wedell, N., et al., 2006. Evolution of male-killer suppression in a natural population. PLoS Biol. 4 (9), e283. https://doi.org/10.1371/journal.pbio.0040283.
Hornett, E.A., Duplouy, A.M.R., Davies, N., Roderick, G.K., Wedell, N., Hurst, G.D.D., et al., 2008. You can't keep a good parasite down: evolution of a male-killer suppressor uncovers cytoplasmic incompatibility. Evolution 62 (5), 1258–1263. https://doi.org/10.1111/j.1558-5646.2008.00353.x.
Hunter, M.S., Perlman, S.J., Kelly, S.E., 2003. A bacterial symbiont in the Bacteroidetes induces cytoplasmic incompatibility in the parasitoid wasp *Encarsia pergandiella*. Proc. R Soc. B Biol. Sci. 270 (1529), 2185–2190.
Hurst, G.D.D., Majerus, M.E.N., 1993. Why do maternally inherited microorganisms kill males? Heredity 71 (1), 81.
Hurst, G.D.D., Majerus, M.E.N., Walker, L.E., 1992. Cytoplasmic male killing elements in *Adalia bipunctata* (Linnaeus) (Coleoptera: Coccinellidae). Heredity 69 (1), 84–91. https://doi.org/10.1038/hdy.1992.97.
Hurst, G.D.D., Hammarton, T.C., Obrycki, J.J., Majerus, T.M.O., Walker, L.E., Bertrand, D., et al., 1996. Male-killing bacterium in a fifth ladybird beetle, *Coleomegilla maculata* (Coleoptera: Coccinellidae). Heredity 77 (2), 177–185. https://doi.org/10.1038/hdy.1996.122.
Hurst, G.D.D., Bandi, C., Sacchi, L., Cochrane, A.G., Bertrand, D., Karaca, I., et al., 1999a. *Adonia variegata* (Coleoptera: Coccinellidae) bears maternally inherited *Flavobacteria* that kill males only. Parasitology 118 (2), 125–134.
Hurst, G.D.D., Schulenburg, J.H.G.V., Majerus, T.M.O., Bertrand, D., Zakharov, I.A., Baungaard, J., et al., 1999b. Invasion of one insect species, *Adalia bipunctata*, by two different male-killing bacteria. Insect Mol. Biol. 8 (1), 133–139. https://doi.org/10.1046/j.1365-2583.1999.810133.x.

Hurst, G.D.D., Johnson, A.P., Schulenburg, J.H.G.V., Fuyama, Y., 2000. Male-killing *Wolbachia* in *Drosophila*: a temperature-sensitive trait with a threshold bacterial density. Genetics 156 (2), 699–709.

Ironside, J.E., Smith, J.E., Hatcher, M.J., Sharpe, R.G., Rollinson, D., Dunn, A.M., 2003. Two species of feminizing microsporidian parasite coexist in populations of *Gammarus duebeni*. J. Evol. Biol. 16 (3), 467–473. https://doi.org/10.1046/j.1420-9101.2003.00539.x.

Jaenike, J., 2007. Spontaneous emergence of a new *Wolbachia* phenotype. Evolution 61 (9), 2244–2252. https://doi.org/10.1111/j.1558-5646.2007.00180.x.

Jaenike, J., Unckless, R., Cockburn, S.N., Boelio, L.M., Perlman, S.J., 2010. Adaptation via symbiosis: recent spread of a *Drosophila* defensive symbiont. Science 329 (5988), 212–215.

Kageyama, D., Traut, W., 2004. Opposite sex-specific effects of *Wolbachia* and interference with the sex determination of its host *Ostrinia scapulalis*. Proc. R Soc. B Biol. Sci. 271 (1536), 251–258. https://doi.org/10.1098/rspb.2003.2604.

Kageyama, D., Nishimura, G., Hoshizaki, S., Ishikawa, Y., 2002. Feminizing *Wolbachia* in an insect, *Ostrinia furnacalis* (Lepidoptera: Crambidae). Heredity 88 (6), 444–449. https://doi.org/10.1038/sj.hdy.6800077.

Kageyama, D., Ohno, M., Sasaki, T., Yoshido, A., Konagaya, T., Jouraku, A., et al., 2017. Feminizing *Wolbachia* endosymbiont disrupts maternal sex chromosome inheritance in a butterfly species. Evol. Lett. 1 (5), 232–244. https://doi.org/10.1002/evl3.28.

Kiuchi, T., Koga, H., Kawamoto, M., Shoji, K., Sakai, H., Arai, Y., et al., 2014. A single female-specific piRNA is the primary determiner of sex in the silkworm. Nature 509 (7502), 633.

König, K., Zundel, P., Krimmer, E., König, C., Pollmann, M., Gottlieb, Y., et al., 2019. Reproductive isolation due to prezygotic isolation and postzygotic cytoplasmic incompatibility in parasitoid wasps. Ecol. Evol. 9, 10694–10706.

Landmann, F., Orsi, G.A., Loppin, B., Sullivan, W., 2009. *Wolbachia*-mediated cytoplasmic incompatibility is associated with impaired histone deposition in the male pronucleus. PLoS Pathog. 5 (3), e1000343. https://doi.org/10.1371/journal.ppat.1000343.

Laven, H., 1967. Eradication of *Culex pipiens fatigans* through cytoplasmic incompatibility. Nature 216 (5113), 383–384.

Lawson, E.T., Mousseau, T.A., Klaper, R., Hunter, M.D., Werren, J.H., 2001. *Rickettsia* associated with male-killing in a buprestid beetle. Heredity 86 (4), 497–505.

Leclercq, S., Thézé, J., Chebbi, M.A., Giraud, I., Moumen, B., Ernenwein, L., et al., 2016. Birth of a W sex chromosome by horizontal transfer of *Wolbachia* bacterial symbiont genome. Proc. Natl. Acad. Sci. 113 (52), 15036–15041.

Legrand, J.J., Legrand-Hamelin, E., Juchault, P., 1987. Sex determination in Crustacea. Biol. Rev. 62 (4), 439–470.

LePage, D.P., Metcalf, J.A., Bordenstein, S.R., On, J., Perlmutter, J.I., Shropshire, J.D., et al., 2017. Prophage WO genes recapitulate and enhance *Wolbachia*-induced cytoplasmic incompatibility. Nature 543 (7644), 243–247.

Lindsey, A.R.I., Werren, J.H., Richards, S., Stouthamer, R., 2016. Comparative genomics of a parthenogenesis-inducing *Wolbachia* symbiont. G3 (Bethesda) 6 (7), 2113–2123.

Lindsey, A.R.I., Rice, D.W., Bordenstein, S.R., Brooks, A.W., Bordenstein, S.R., Newton, I.L.G., 2018. Evolutionary genetics of cytoplasmic incompatibility genes *cifA* and *cifB* in prophage WO of *Wolbachia*. Genome Biol. Evol. 10 (2), 434–451. https://doi.org/10.1093/gbe/evy012.

Lynn-Bell, N.L., Strand, M.R., Oliver, K.M., 2019. Bacteriophage acquisition restores protective mutualism. Microbiology 165 (9), 985–989.

Ma, W.-J., Schwander, T., 2017. Patterns and mechanisms in instances of endosymbiont-induced parthenogenesis. J. Evol. Biol. 30 (5), 868–888. https://doi.org/10.1111/jeb.13069.

Ma, W.-J., Pannebakker, B.A., van de Zande, L., Schwander, T., Wertheim, B., Beukeboom, L.W., 2015. Diploid males support a two-step mechanism of endosymbiont-induced thelytoky in a parasitoid wasp. BMC Evol. Biol. 15, 84.

Majerus, T.M.O., Majerus, M.E.N., 2010. Intergenomic arms races: detection of a nuclear rescue gene of male-killing in a ladybird. PLoS Pathog. 6 (7), e1000987. https://doi.org/10.1371/journal.ppat.1000987.

Mann, E., Stouthamer, C.M., Kelly, S.E., Dzieciol, M., Hunter, M.S., Schmitz-Esser, S., 2017. Transcriptome sequencing reveals novel candidate genes for *Cardinium hertigii*-caused cytoplasmic incompatibility and host-cell interaction. mSystems 2 (6), e00141-17.

Martin, G., Sorokine, O., Moniatte, M., Bulet, P., Hetru, C., Van Dorsselaer, A., 1999. The structure of a glycosylated protein hormone responsible for sex determination in the isopod, *Armadillidium vulgare*. Eur. J. Biochem. 262 (3), 727–736. https://doi.org/10.1046/j.1432-1327.1999.00442.x.

Martin, J., Chong, T., Ferree, P.M., 2013. Male killing *Spiroplasma* preferentially disrupts neural development in the *Drosophila melanogaster* embryo. PLoS One 8 (11), e79368. https://doi.org/10.1371/journal.pone.0079368.

Moran, N.A., McCutcheon, J.P., Nakabachi, A., 2008. Genomics and evolution of heritable bacterial symbionts. Annu. Rev. Genet. 42, 165–190.

Moreira, L.A., Iturbe-Ormaetxe, I., Jeffery, J.A., Lu, G., Pyke, A.T., Hedges, L.M., et al., 2009. A *Wolbachia* symbiont in *Aedes aegypti* limits infection with dengue, Chikungunya, and *Plasmodium*. Cell 139 (7), 1268–1278.

Nakamura, Y., Yukuhiro, F., Matsumura, M., Noda, H., 2012. Cytoplasmic incompatibility involving *Cardinium* and *Wolbachia* in the white-backed planthopper *Sogatella furcifera* (Hemiptera: Delphacidae). Appl. Entomol. Zool. 47 (3), 273–283. https://doi.org/10.1007/s13355-012-0120-z.

Narita, S., Kageyama, D., Nomura, M., Fukatsu, T., 2007. Unexpected mechanism of symbiont-induced reversal of insect sex: feminizing *Wolbachia* continuously acts on the butterfly *Eurema hecabe* during larval development. Appl. Environ. Microbiol. 73 (13), 4332–4341.

Narita, S., Kageyama, D., Hiroki, M., Sanpei, T., Hashimoto, S., Kamitoh, T., et al., 2011. *Wolbachia*-induced feminisation newly found in *Eurema hecabe*, a sibling species of *Eurema mandarina* (Lepidoptera: Pieridae). Ecol Entomol. 36 (3), 309–317. https://doi.org/10.1111/j.1365-2311.2011.01274.x.

Negri, I., Pellecchia, M., P.J M, Patetta A, Alma A., 2006. Feminizing *Wolbachia* in *Zyginidia pullula* (Insecta, Hemiptera), a leafhopper with an XX/X0 sex-determination system. Proc. R Soc. B Biol. Sci. 273 (1599), 2409–2416. https://doi.org/10.1098/rspb.2006.3592.

Nguyen, D.T., Morrow, J.L., Spooner-Hart, R.N., Riegler, M., 2017. Independent cytoplasmic incompatibility induced by *Cardinium* and *Wolbachia* maintains endosymbiont coinfections in haplodiploid thrips populations. Evolution 71 (4), 995–1008. https://doi.org/10.1111/evo.13197.

Nur, U., Werren, J.H., Eickbush, D.G., Burke, W.D., Eickbush, T.H., 1988. A "selfish" B chromosome that enhances its transmission by eliminating the paternal genome. Science 240 (4851), 512–514.

Oliver, K.M., Moran, N.A., Hunter, M.S., 2005. Variation in resistance to parasitism in aphids is due to symbionts not host genotype. Proc. Natl. Acad. Sci. 102 (36), 12795–12800.

Oliver, K.M., Degnan, P.H., Hunter, M.S., Moran, N.A., 2009. Bacteriophages encode factors required for protection in a symbiotic mutualism. Science 325 (5943), 992–994.

Pannebakker, B.A., Pijnacker, L.P., Zwaan, B.J., Beukeboom, L.W., 2004. Cytology of *Wolbachia*-induced parthenogenesis in *Leptopilina clavipes* (Hymenoptera: Figitidae). Genome 47 (2), 299–303. https://doi.org/10.1139/g03-137.

Penalva, L.O.F., Sánchez, L., 2003. RNA binding protein *sex-lethal* (*Sxl*) and control of *Drosophila* sex determination and dosage compensation. Microbiol. Mol. Biol. Rev. 67 (3), 343–359.

Penz, T., Schmitz-Esser, S., Kelly, S.E., Cass, B.N., Müller, A., Woyke, T., et al., 2012. Comparative genomics suggests an independent origin of cytoplasmic incompatibility in *Cardinium hertigii*. PLoS Genet. 8 (10), e1003012. https://doi.org/10.1371/journal.pgen.1003012.

Perlmutter, J.I., Bordenstein, S.R., Unckless, R.L., LePage, D.P., Metcalf, J.A., Hill, T., et al., 2019. The phage gene *wmk* is a candidate for male killing by a bacterial endosymbiont. PLoS Pathog. 15 (9), e1007936. https://doi.org/10.1371/journal.ppat.1007936.

Pickup, J., Ironside, J.E., 2018. Multiple origins of parasitic feminization: thelygeny and intersexuality in beach-hoppers are caused by paramyxid parasites, not microsporidia. Parasitology 145 (3), 408–415.

Pike, N., Kingcombe, R., 2009. Antibiotic treatment leads to the elimination of *Wolbachia* endosymbionts and sterility in the diplodiploid collembolan *Folsomia candida*. BMC Biol. 7 (1), 54.

Plantard, O., Rasplus, J.-Y., Mondor, G., Le Clainche, I., Solignac, M., 1998. *Wolbachia*-induced thelytoky in the rose gallwasp *Diplolepis spinosissimae* (Giraud) (Hymenoptera: Cynipidae), and its consequences on the genetic structure of its host. Proc. R Soc. B Biol. Sci. 265 (1401), 1075–1080.

Pomerantz, A.F., Hoy, M.A., Kawahara, A.Y., 2015. Molecular characterization and evolutionary insights into potential sex-determination genes in the western orchard predatory mite *Metaseiulus occidentalis* (Chelicerata: Arachnida: Acari: Phytoseiidae). J. Biomol. Struct. Dyn. 33 (6), 1239–1253. https://doi.org/10.1080/07391102.2014.941402.

Provencher, L.M., Morse, G.E., Weeks, A.R., Normark, B.B., 2005. Parthenogenesis in the *Aspidiotus nerii* complex (Hemiptera: Diaspididae): a single origin of a worldwide, polyphagous lineage associated with *Cardinium* bacteria. Ann. Entomol. Soc. Am. 98 (5), 629–635. https://doi.org/10.1603/0013-8746(2005)098[0629:PITANC]2.0.CO.

Rances, E., Yixin, H.Y., Woolfit, M., McGraw, E.A., O'Neill, S.L., 2012. The relative importance of innate immune priming in *Wolbachia*-mediated dengue interference. PLoS Pathog. 8 (2), e1002548.

Richardson, K.M., Schiffer, M., Griffin, P.C., Lee, S.F., Hoffmann, A.A., 2016. Tropical *Drosophila pandora* carry *Wolbachia* infections causing cytoplasmic incompatibility or male killing. Evolution 70 (8), 1791–1802. https://doi.org/10.1111/evo.12981.

Rocha, E.P.C., 2008. The organization of the bacterial genome. Annu. Rev. Genet. 42 (1), 211–233. https://doi.org/10.1146/annurev.genet.42.110807.091653.

Rodgers-Gray, T.P., Smith, J.E., Ashcroft, A.E., Isaac, R.E., Dunn, A.M., 2004. Mechanisms of parasite-induced sex reversal in *Gammarus duebeni*. Int. J. Parasitol. 34 (6), 747–753.

Russell, J.E., Stouthamer, R., 2011. The genetics and evolution of obligate reproductive parasitism in *Trichogramma pretiosum* infected with parthenogenesis-inducing *Wolbachia*. Heredity 106 (1), 58.

Santos-Garcia, D., Rollat-Farnier, P.-A., Beitia, F., Zchori-Fein, E., Vavre, F., Mouton, L., et al., 2014. The genome of *Cardinium* cBtQ1 provides insights into genome reduction, symbiont motility, and its settlement in *Bemisia tabaci*. Genome Biol. Evol. 6 (4), 1013–1030.

Sasaki, T., Kubo, T., Ishikawa, H., 2002. Interspecific transfer of *Wolbachia* between two lepidopteran insects expressing cytoplasmic incompatibility: a *Wolbachia* variant naturally infecting *Cadra cautella* causes male killing in *Ephestia kuehniella*. Genetics 162 (3), 1313–1319.

Sasaki, T., Massaki, N., Kubo, T., 2005. *Wolbachia* variant that induces two distinct reproductive phenotypes in different hosts. Heredity 95 (5), 389–393. https://doi.org/10.1038/sj.hdy.6800737.

Schulenburg, J.H.G.V., Habig, M., Sloggett, J.J., Webberley, K.M., Bertrand, D., Hurst, G.D.D., et al., 2001. Incidence of male-killing Rickettsia (α-Proteobacteria) in the ten-spot ladybird beetle Adalia decempunctata L. (Coleoptera: Coccinellidae). Appl. Environ. Microbiol. 67 (1), 270–277.

Shropshire, J.D., On, J., Layton, E.M., Zhou, H., Bordenstein, S.R., 2018. One prophage WO gene rescues cytoplasmic incompatibility in *Drosophila melanogaster*. Proc. Natl. Acad. Sci. 115 (19), 4987–4991.

Shropshire, J.D., Leigh, B., Bordenstein, S.R., Duplouy, A., Riegler, M., Brownlie, J.C., et al., 2019. Models and nomenclature for cytoplasmic incompatibility: caution over premature conclusions - a response to Beckmann et al. Trends Genet. 35 (6), 397–399. https://doi.org/10.1016/j.tig.2019.03.004.

Sinkins, S.P., Walker, T., Lynd, A.R., Steven, A.R., Makepeace, B.L., Godfray, H.C.J., et al., 2005. *Wolbachia* variability and host effects on crossing type in *Culex* mosquitoes. Nature 436 (7048), 257.

Stille, B., Dävring, L., 1980. Meiosis and reproductive strategy in the parthenogenetic gall wasp *Diplolepis rosae* (L.) (Hymenoptera, Cynipidae). Hereditas 92 (2), 353–362. https://doi.org/10.1111/j.1601-5223.1980.tb01720.x.

Stouthamer, R., Kazmer, D.J., 1994. Cytogenetics of microbe-associated parthenogenesis and its consequences for gene flow in *Trichogramma* wasps. Heredity 73 (3), 317–327. https://doi.org/10.1038/hdy.1994.139.

Stouthamer, R., Werren, J.H., 1993. Microbes associated with parthenogenesis in wasps of the genus Trichogramma. J. Invertebr. Pathol 61, 6–9.

Stouthamer, R., Luck, R.F., Hamilton, W.D., 1990. Antibiotics cause parthenogenetic *Trichogramma* (Hymenoptera/Trichogrammatidae) to revert to sex. Proc. Natl. Acad. Sci. 87 (7), 2424–2427.

Stouthamer, R., Breeuwer, J.A.J., Luck, R.F., Werren, J.H., 1993. Molecular identification of microorganisms associated with parthenogenesis. Nature 361 (6407), 66–68. https://doi.org/10.1038/361066a0.

Sugimoto, T.N., Ishikawa, Y., 2012. A male-killing *Wolbachia* carries a feminizing factor and is associated with degradation of the sex-determining system of its host. Biol. Lett. 8 (3), 412–415.

Takano, S., Tuda, M., Takasu, K., Furuya, N., Imamura, Y., Kim, S., et al., 2017. Unique clade of alphaproteobacterial endosymbionts induces complete cytoplasmic incompatibility in the coconut beetle. Proc. Natl. Acad. Sci. 114 (23), 6110–6115.

Terry, R.S., Smith, J.E., Dunn, A.M., 1998. Impact of a novel, feminising microsporidium on its crustacean host. J. Eukaryot. Microbiol. 45 (5), 497–501. https://doi.org/10.1111/j.1550-7408.1998.tb05106.x.

Thomas, S., Verma, J., Woolfit, M., O'Neill, S.L., 2018. *Wolbachia*-mediated virus blocking in mosquito cells is dependent on XRN1-mediated viral RNA degradation and influenced by viral replication rate. PLoS Pathog. 14 (3), e1006879. https://doi.org/10.1371/journal.ppat.1006879.

Timmermans, M.J.T.N., Ellers, J., 2009. *Wolbachia* endosymbiont is essential for egg hatching in a parthenogenetic arthropod. Evol. Ecol. 23 (6), 931.

Tram, U., Sullivan, W., 2000. Reciprocal inheritance of centrosomes in the parthenogenetic hymenopteran *Nasonia vitripennis*. Curr. Biol. 10 (22), 1413–1419.

Tram, U., Ferree, P.M., Sullivan, W., 2003. Identification of *Wolbachia*–host interacting factors through cytological analysis. Microbes Infect. 5 (11), 999–1011.

Tsuchiyama-Omura, S., Sakaguchi, B., Koga, K., Poulson, D.F., 1988. Morphological features of embryogenesis in *Drosophila melanogaster* infected with a male-killing *Spiroplasma*: developmental biology. Zoolog. Sci. 5 (2), 375–383.

Tulgetske, G.M., 2010. Investigations into the Mechanisms of *Wolbachia* Induced Parthenogenesis and Sex Determination in the Parasitoid Wasp. Trichogramma, UC Riverside.

Van Meer, M.M.M., Witteveldt, J., Stouthamer, R., 1999. Phylogeny of the arthropod endosymbiont *Wolbachia* based on the *wsp* gene. Insect Mol. Biol. 8 (3), 399–408. https://doi.org/10.1046/j.1365-2583.1999.83129.x.

Van Melderen, L., Saavedra De Bast, M., 2009. Bacterial toxin–antitoxin systems: more than selfish entities? PLoS Genet. 5 (3), e1000437. https://doi.org/10.1371/journal.pgen.1000437.

Van Wilgenburg, E., Driessen, G., Beukeboom, L.W., 2006. Single locus complementary sex determination in Hymenoptera: an "unintelligent" design? Front. Zool. 3 (1), 1.

Veneti, Z., Bentley, J.K., Koana, T., Braig, H.R., Hurst, G.D.D., 2005. A functional dosage compensation complex required for male killing in *Drosophila*. Science 307 (5714), 1461–1463.

Verhulst, E.C., van de Zande, L., 2015. Double nexus–*Doublesex* is the connecting element in sex determination. Brief. Funct. Genomics 14 (6), 396–406.

Verhulst, E.C., van de Zande, L., Beukeboom, L.W., 2010. Insect sex determination: it all evolves around *transformer*. Curr. Opin. Genet. Dev. 20 (4), 376–383.

Walker, T., Johnson, P.H., Moreira, L.A., Iturbe-Ormaetxe, I., Frentiu, F.D., McMeniman, C.J., et al., 2011. The *w*Mel *Wolbachia* strain blocks dengue and invades caged *Aedes aegypti* populations. Nature 476 (7361), 450–453.

Weeks, A.R., Breeuwer, J.A.J., 2001. *Wolbachia*-induced parthenogenesis in a genus of phytophagous mites. Proc. R Soc. B Biol. Sci. 268 (1482), 2245–2251.

Weeks, A.R., Marec, F., Breeuwer, J.A., 2001. A mite species that consists entirely of haploid females. Science 292 (5526), 2479–2482.

Weinert, L.A., Araujo-Jnr, E.V., Ahmed, M.Z., Welch, J.J., 2015. The incidence of bacterial endosymbionts in terrestrial arthropods. Proc. R Soc. B Biol. Sci. 282 (1807), 20150249.

Werren, J.H., 1997. Biology of *Wolbachia*. Annu. Rev. Entomol. 42 (1), 587–609. https://doi.org/10.1146/annurev.ento.42.1.587.

Werren, J.H., Hurst, G.D., Zhang, W., Breeuwer, J.A., Stouthamer, R., Majerus, M.E., 1994. Rickettsial relative associated with male killing in the ladybird beetle (*Adalia bipunctata*). J. Bacteriol. 176 (2), 388–394.

Werren, J.H., Baldo, L., Clark, M.E., 2008. *Wolbachia*: master manipulators of invertebrate biology. Nat. Rev. Microbiol. 6 (10), 741–751. https://doi.org/10.1038/nrmicro1969.

Wilkes, T.E., Darby, A.C., Choi, J.-H., Colbourne, J.K., Werren, J.H., Hurst, G.D.D., 2010. The draft genome sequence of *Arsenophonus nasoniae*, son-killer bacterium of *Nasonia vitripennis*, reveals genes associated with virulence and symbiosis. Insect Mol. Biol. 19 (s1), 59–73. https://doi.org/10.1111/j.1365-2583.2009.00963.x.

Wu, K., Hoy, M.A., 2012. *Cardinium* is associated with reproductive incompatibility in the predatory mite *Metaseiulus occidentalis* (Acari: Phytoseiidae). J. Invertebr. Pathol. 110 (3), 359–365.

Yamada, R., Iturbe-Ormaetxe, I., Brownlie, J.C., O'Neill, S.L., 2011. Functional test of the influence of *Wolbachia* genes on cytoplasmic incompatibility expression in *Drosophila melanogaster*. Insect Mol. Biol. 20 (1), 75–85. https://doi.org/10.1111/j.1365-2583.2010.01042.x.

Zchori-Fein, E., Faktor, O., Zeidan, M., Gottlieb, Y., Czosnek, H., Rosen, D., 1995. Parthenogenesis-inducing microorganisms in *Aphytis* (Hymenoptera: Aphelinidae). Insect Mol. Biol. 4 (3), 173–178. https://doi.org/10.1111/j.1365-2583.1995.tb00023.x.

Zchori-Fein, E., Gottlieb, Y., Kelly, S.E., Brown, J.K., Wilson, J.M., Karr, T.L., et al., 2001. A newly discovered bacterium associated with parthenogenesis and a change in host selection behavior in parasitoid wasps. Proc. Natl. Acad. Sci. 98 (22), 12555–12560. http://www.pnas.org/content/98/22/12555.abstract.

Zchori-Fein, E., Perlman, S.J., Kelly, S.E., Katzir, N., Hunter, M.S., 2004. Characterization of a 'Bacteroidetes' symbiont in Encarsia wasps (Hymenoptera: Aphelinidae): Proposal of 'Candidatus Cardinium hertigii'. Int. J. Syst. Evol. Microbiol. 54, 961–968.

Zeng, Z., Fu, Y., Guo, D., Wu, Y., Ajayi, O.E., Wu, Q., 2018. Bacterial endosymbiont *Cardinium* cSfur genome sequence provides insights for understanding the symbiotic relationship in *Sogatella furcifera* host. BMC Genomics 19 (1), 688. https://doi.org/10.1186/s12864-018-5078-y.

Zhishan, W.U., Hopper, K.R., Ode, P.J., Fuester, R.W., Jia-hua, C., Heimpel, G.E., 2003. Complementary sex determination in hymenopteran parasitoids and its implications for biological control. Insect Sci. 10 (2), 81–93.

Zug, R., Hammerstein, P., 2012. Still a host of hosts for *Wolbachia*: analysis of recent data suggests that 40% of terrestrial arthropod species are infected. PLoS One 7 (6), e38544.

FSC
www.fsc.org
MIX
Papier | Fördert
gute Waldnutzung
FSC® C083411

Kolibri 360 GmbH
Ferdinand-Jühlke-Straße 7
99095 Erfurt, Deutschland
produktsicherheit@kolibri360.de